Aspects of the Kinetics and Dynamics of Surface Reactions

(La Jolla Institute-1979)

AIP Conference Proceedings
Series Editor: Hugh C. Wolfe
Number 61

Aspects of the Kinetics and Dynamics of Surface Reactions

(La Jolla Institute—1979)

Edited by:
Uzi Landman
Georgia Institute of Technology

American Institute of Physics
New York 1980

L.C. Catalog Card No. 80-68004
ISBN 0–88318–160–6
DOE CONF- 7908108

PREFACE

Surface science is a frontier area of intellectual scientific endeavor of coupled fundamental and practical interests. The title Surface science is intentionally used here to emphasize the interdisciplinary character of this field in which chemical and physical phenomena intertwine. In addition the diversity, and in many cases the unpredictable nature of surface processes dictate the need for a close contact between scientists involved in experimental investigations and those preoccupied with theoretical studies. In an attempt to provide a forum for such a dialogue, Walter Kohn, Harry Suhl and myself organized a workshop on "The Physics of Surfaces: Aspects of the Kinetics and Dynamics of Surface Reaction," supported by the La Jolla Institute and held at the University of California, San Diego, La Jolla, August 1-4, 1979. This volume of essays is the proceedings of the above workshop.

Surface science studies encompass both static and dynamic aspects of surface processes. The advent and proliferation of surface spectroscopies, methods of analysis, and the development of theoretical techniques have led to a much improved knowledge of surface structure on the atomic scale and a significant progress in the fundamental understanding of surface excitations (electronic and vibronic), interaction processes (such as chemisorption and physisorption) and scattering events. In planning the workshop we felt that it would be both timely and fruitful to concentrate on the kinetics and dynamics of surface reactions, in an attempt to provide a framework for the study of catalytic surface reactions. The role of a surface scientist studying catalytic reactions is like that of a matchmaker, who must first know well the properties of the partners, what excites them and how they would respond and interact under various conditions, in order to achive a successful, meaningful and lasting match. The selection of topics of discussion at the workshop was guided by the above and proceedings provide a spectrum of experimental and theoretical approaches directed at a basic understanding of surface reactions.

In the course of planning it became apparent that while significant advances have been made in the theoretical formulation and understanding of surfaces (solid surfaces in particular) the kinetics and dynamics of surface reactions present challenging new, though long recognized (Montroll), areas of theoretical and experimental studies. The basic problem is that of the evaluation of expressions for reaction rate constants starting from a microscopic description of the system. At the level of rate constants the fine detailed starting microscopic description (e.g., microscopic equations of motion) should contract to expressions given in terms of a small number of characteristic accessible, physical parameters and/or functions of these parameters. In performing such a reduction stochastics models and techniques play a major role (Montroll, Shlesinger and Landman). It is often convenient to discuss reactions in terms of elementary reaction steps (Madix, Schmidt), such as adsorption (Lang, Varma) diffusion (Gomer, Shlesinger and Landman) and desorption (Landman). The kinetic scheme describing the reaction is then represented by a set of simultaneous rate equations first order in time and in general non-linear (Suhl) for the reactants concentrations. In this context theoretical problems of fundamental interest include (by no means an exhaustive list): the evaluation of couplings between reactants and the substrate surface; calculation of potential surfaces for the reactants in the presence of a surface; adiabatic versus non-adiabatic potential surfaces and couplings; diffusion (and multi-state diffusion) on surfaces; correlation functions including inter-adsorbate (direct and/or substrate mediated) interactions and current (field emission or electrical) fluctuations; calculation of reactive cross sections and product energy distributions; inter-active two-dimensional lattice gas models; the solution of non-linear kinetic schemes; trajectory analysis of impinging reactants and energy deposition and redistribution probabilities; the effect of crystalline features (such as steps) and defects or impurities on the rates of elementary reaction processes; photocatalytic effects; the magnetic phase dependence of reaction rates. It is worthwhile noting the present relatively advanced state of under-standing of gas phase homogeneous reaction using statistical methods (such as RRKM) and classical trajectory analysis (Hase). Building upon the experience and methodologies developed for gas phase reactions the study of surface reactions and interaction processes (Harrison), particularly unimolecular reactions such as desorption, dissociation and association to an adsorbed species, seems feasible (Hase). Furthermore, the employment of theories and methods developed originally in the context of molecular pro-cesses, in the study of surface interaction processes could prove useful (Clinton). At this junction we should emphasize that recent quantitative data, and retrospective reconsideration of older observations, provides compelling evidence about the significant roles played by catalyst defects, irregularities, and additives in controlling the path of catalytic reactions. In addition, catalyst deactivation, poisonion, is inherent to many of the catalytic processes of interest. The customary course of development of theoretical models proceeds by first considering an idealized situation introducing defects and inhomogeneities in a latter stage. Due to the intrinsic role played by these factors in governing the kinetics of certain catalytic reaction, it may be advisable to in-corporate them at the initial stages of the formulation. In add-ition the interplay between the local and extended character of catalytic behavior (active sites, local defect modes versus collective modes, band structure effects) should be emphasized.

Modern surface-sensitive probes (such as LEED, XPS, UPS, AES, RHEED, EXAFS, X-ray fluorescence, IR, EELS, TDS, and volumetric methods) play an important role in the characterization of catalytic surfaces, in investigations of their properties and the binding characteristics of adsorbates. These methods can also be used to identify and characterize chemical and physical changes and kinetic steps in the course of a reactive interaction with the surface (Brundle, Mehta _et al_., Plummer _et al_., Stern, Williams _et al_.) Such measurements can be made dynamically or by "freezing" the reaction at various stages of its evolution. Apart from the above, new methods specifically designed for the study of surface reaction kinetics and dynamics have been developed. Among these, molecular beams methods (Madix, Schmidt), (Steady state and in particular Molecular Beam Relaxation Spectroscopy) play an important role. In addition, laser-induced fluorescence has been used to monitor the kinetics of a catalytic reaction (Talley and Lin). In order to enable a coherent and complete understanding of catalytic reactions it is important that the elementary reaction steps and their underlying microscopic mechanisms be studied in detail. The study of diffusion on surfaces is facilitated by the measurement of the time auto-correlation function of field emission current fluctuations (Gomer) and via Field Ion Microscopy techniques (Shlesinger and Landman). Finally, electrochemical reactions (Retzloff _et al_.) and the measurement of the adsorption interference in binary mixtures flowing through a porous solid (Madey and Photinos) are discussed.

Through the ingeneous use of certain surface spectroscopies and newly developed techniques much needed quantitative information is being provided concerning the rates of elementary kinetic processes, modifications of the electronic and vibronic spectrum and on atomic arrangements and rearrangement caused by reaction processes at surfaces. The interpretation and analysis of these experiments open exciting avenues for theoretical studies. Drawing upon the now established procedures of gas phase reactive molecular beam techniques, we expect that the next generation of experimental surface reaction investigations would aim at reactant and product state selective methods. Such measurements would provide data which would aid the formulation and critical testing of theoretical models of surface catalytic reactions.

In closing I would like to take this opportunity to express, for the organizers, our thanks to the speakers and participants at the workshop, and for all of us our gratitude to the La Jolla Institute for its generous support and assistance. The workshop and these proceedings were financed by independent research funds of the La Jolla Institute.

Uzi Landman
School of Physics
Georgia Institute of Technology

VIII.

When applied physicists are confronted by complex technological problems, they first of all attempt to reduce them to a number of simpler problems. These are then idealized until a solution becomes possible, hopefully without too much loss of realism. This approach has been successfully applied to several aspects of the physics of bulk solids. For instance, idealization to a uniform, perfect crystal structure is a good zeroeth-order approximation and also furnishes a powerful starting point for study of both the static and dynamic aspects of a solid.

Interface physics, especially when it is also interphase physics, has not yet been comparably successful, for experimental as well as theoretical reasons. To achieve conditions for a gas-solid interface experiment that are chemically as ideal as those in the interior of a solid requires great effort: the needed technology is simply not routinely available. Theoretically, the obvious non-uniformity normal to the surface, accompanied in general by further non-uniformities within the plane of the surface itself, is a major complication, even in the absence of adsorbates. Furthermore, the non-uniformities within the plane of the interface probably play a major role in chemisorption and surface reactions and, therefore, cannot be shrugged off as inessential. Finally, the kinetics of adsorbates in the interplay with the dynamics of the underlying solid represent a major experimental and theoretical challenge of a new kind.

As outlined in Dr. Landman's introduction, this conference was convened to focus on some of the essential ingredients needed for a theory of heterogeneous catalysis. Almost all the contributions, whether concerned with quite specific reactions, or aimed at general methodology have as their underlying theme the understanding of catalysis in general. The establishment of quantitatively useful universal principles of catalysis is currently out of our reach; such universality may not even exist. The only way to find out is through patient work of the kind reported here. Work that could set the stage for an inspired breakthrough.

Harry Suhl
Department of Physics
University of California
San Diego

TABLE OF CONTENTS

SOME HISTORICAL REMARKS ON THE CATALYTIC PROCESS AND ON STOCHASTIC MODELS OF CHEMICAL KINETICS

ELLIOTT MONTROLL

INSTITUTE FOR FUNDAMENTAL STUDIES, DEPARTMENT OF PHYSICS AND ASTRONOMY
UNIVERSITY OF ROCHESTER, ROCHESTER, NEW YORK, 14627.

I. INTRODUCTION

Practically all biological processes are monitored by chemical reactions catalyzed by enzymes. Thus, the catalytic process is a necessity for the existance of the biosphere. The chemical industry, and hence our modern life style is also based upon catalysis. The plan of this paper is to first present several historical vignettes to display the critical role the existence and absence of appropriate catalysts have had on certain important world events. Then a brief history of the evolution of stochastic models of chemical kinetics will be presented. This will be followed in Section III by some examples of stochastic processes basic to the mechanism of surface catalysis. A rather detailed report on stochastic models in chemical kinetics is given in references 1 and 2.

The presentation here is rather qualitative. More quantitative expositions of the stochastic style are given in the papers of Shlesinger[3] and Landman[4] in this volume.

II. SOME HISTORICAL REMARKS ON THE WORLD OF CATALYSIS

1. Fritz Haber to the Rescue. As we approach the year 2000, futurology becomes increasingly fashionable. Every year more books and articles predict the nature of the world of the year 2000. Typical doomsday prophets warn us that the supply of various important natural resources might be exhausted by the year 2000 and certainly by the year 2100. They are not the first to make such prophecies.

William Crookes, the inventor of the Crookes tube, was the dean of futurologists in the 1890's. In 1892, shortly after the recognition of radio waves by Hertz he gave a remarkable lecture on the requirements for wireless telegraphy across space, and the manner that it might evolve.

In his more pessimistic presidental lecture (1898) to the British Association for the Advancement of Science, he stated that the "... world would be faced by starvation because of failure of wheat supplies to keep abreast of population unless science found new sources of fertilizer... Chilean Guana supply of nitrates would be exhausted in about twenty years." Chilean nitrate was a restorable resource required for fertilizer in peace and gunpowder in war. In Crookes' mind the demand for the resource for both purposes increased more rapidly than the birds restorative competance.

Barbara Tuchman, in the "Guns of August" noted that in 1914 "... Germany had not planned on the need to hold out for long and upon entering the war had a stockpile (of guana) for making gun powder for six months and no more..." As the Battle of the Marne ended so

ISSN:0094-243X/80/610001-37$1.50

did the optimistic expectations of the German general staff.

It was even hinted to the Kaiser that perhaps some negotiations be started with the enemy. As these rumors drifted from military to civilian offices in Berlin they eventually reached the laboratory of Fritz Haber. He immediately came to the rescue, reviewing to the responsible authorities the potentiality of his work with O.C. Birkeland on the fixation of nitrogen through the reaction

$$3H_2 + N_2 \underset{iron}{\rightarrow} 2NH_3.$$

Within six months his magic iron catalyst was supplying ammonia for the preparation of nitrates at such a rate that the motivating crisis was forgotten, battles continued to be fought, men continued to die. Ammonia could be converted to nitrates through several different paths, one being the Ostwald process

$$4NH_3 + 5O_2 \underset{Pt\ gauze}{\rightarrow} 6H_2O + 4NO$$

$$4NO + 4KOH = 3O_2 \rightarrow 4KNO_3 + 2H_2O \quad .$$

In this process the platinum gauze is first heated to 700°C by an electric current, with the ensuing exothermic reaction then keeping the temperature at the required level.

The concept of a catalytic process was exploited, but of course not discovered by Haber. A hundred years earlier in the period 1810-15, Humphrey Davy observed that Pt hastens the combination of many gases. He also found that alcohol is easily converted to acetic acid by finely divided Pt and that the reaction

$$S + 2HNO_3 \rightarrow H_2SO_4 + 2NO$$

proceeds well with that catalyst. Davy's former "apprentice" Michael Faraday recombined H_2 and O_2 using finely divided Pt (1834)

$$2H_2 + O_2 \underset{Pt}{\rightarrow} 2H_2O \quad .$$

From these beginnings Berzelius attempted to systematize the subject.

Haber's success invited many imitators. In the U.S. the largest nitrogen fixation plants operated as the Muscle Shoals Project. The role of the catalyst was in many places vigorously investigated, to the degree that Paul Emmett[6] observed that "... the catalytic synthesis of ammonia developed in the period 1908-12 in Germany has probably been studied more intensely during the past 50 years, than any other catalytic reaction..." The iron compounds with the nitrogen according to

$$2Fe + N_2 \rightarrow 2FeN \; .$$

The importance of one of the main processes to be described in this paper, the random walk of atoms and molecules on surfaces is evident in Emmett's review[6] of the Haber process. On adsorption on the appropriate metal surface, H_2 and N_2 both dissociate with the nitrogen atoms becoming tightly bound as indicated. The mobile hydrogens execute a "random walk"

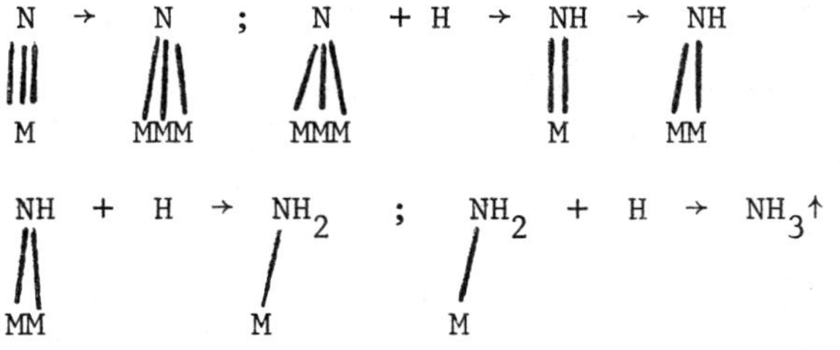

on the metal surface, meeting and combining with a fixed nitrogen atom so that upon these attachments an NH_3 is formed which then becomes released from the surface. The random walk of atoms or molecules on a surface is an important phase of the catalytic process.

2. Fischer-Tropsch in World War II. The Germans attacked Poland and France in 1939 fully recognizing that their agression would soon isolate them from the oil fields of the world. In preparation for that event, they arranged to fuel their war machine through the conversion of a locally available substitute, coal. Without the Fischer-Tropsch process (or some equivalent) for the synthesis of gasoline from coal they could not possibly have embarked on a large scale war. The first step in the synthesis is the formation of CO by the water gas reaction

$$C + H_2O \rightarrow CO + H_2$$

followed by the catalytic conversion of CO and H_2 to hydrocarbon

$$nCO + (2n+1)\ H_2 \xrightarrow[\text{cobalt}]{} C_nH_{2n+2} + nH_2O \quad .$$

With a *cobalt catalyst* the process is run at 250°C under a pressure of one to ten atmospheres. The total 1945 German war production of gasoline was at the level of about a one day consumption in the U.S. in 1974.

Today the South Africans are attempting to achieve independence from OPEC by the application of Fisher- Tropsch to native coal. While this alternative also exists for the U.S., the champions of the cause must be prepared to pay a price. We would have to increase our annual coal production from about 0.6×10^9 tons to 3×10^9 tons, a factor of five. The cost of plant construction could be of the order of 500 billion dollars. The production of the new coal cars required to transport the coal would itself be a major undertaking. Subsurface miners who in their career would have had the good fortune to escape major accidents would still have a problem, since, if they smoke, they have a several orders of magnitude greater probability of dying from penumonia in the age range 55-60 than the average smoker. Hence, the number of added deaths from increased use of coal for a decade could exceed that resulting from a nuclear power plant explosion at a plant sited in the typical manner at some distance from un urban center. Detailed coal production and accident statistics[7] are given in Table 1.

	Annual production	Annual deaths	Annual injuries
1970	612×10^6 short ton	260	11,552
1960	433×10^6 short ton	325	11,902
1950	560×10^6 short ton	643	37,264

Table 1. Coal Statistics 1950-1970.

Since coal is one of the convenient concentrated forms of carbon, it is likely that its role will increase again in the future to some degree and that the interest in finding better catalysts for conversion of coal and coal products to various chemicals will also grow. The reaction $CO + 3H_2 \xrightarrow{Ni} CH_4 + H_2O$ will continue to be much investigated.

3. Rasputin and the Blood Coagulation Cascade. The existence of appropriate catalysts permitted the Germans to fight devastating wars that have changed modern history. The absence of an appropriate catalyst was a factor in the decline of the Russian monarchy. A fateful date in Russian History was 14 August 1904, the birthday of Czarevitch Alexis, the male heir of Nicholas II. The excitement of his birth was dulled upon the discovery that the child was a hemophiliac. The common heredity tree characteristic[8] of a hemophiliac is given in Figure 1.

Through an unusual chain of events the unwashed, unshaven Eastern mystic[9] Rasputin, with penetrating eyes appeared in the royal household. According to B. Pares[10], a specialist on the fall of the Russian monarchy, "... it was the boy's illness that brought Rasputin to the palace. What was the nature of his influence in the family circle?... it was that he could undoubtedly bring relief to the boy." The nature of the relief was conjectured by the biologist J.B.S. Haldane[11] who noted "... it is also possible that by hypnosis or a similar method that he was able to produce a contraction of the small arteries... (and) was able to relax the victim."

The Philadelphia dentist, Oscar Lucas of Jefferson Hospital, motivated by the conjectured Rasputin style, extracted teeth of hemophiliacs under hypnosis[9,12]. In the period 1962-64 he established the remarkable record of 150 such extractions without recourse to the blood transfusions traditional for such cases.

After the Russian defeat in its war with Japan in 1905, after the tragedy of Bloody Sunday (January 22, 1905), after the general strikes and the abortive Revolution later in the year, some thoughtful advisors convinced the Czar that basic policy changes had to be made to restore some confidence in the monarchy. Peter Stolypin, appointed Prime Minister in November 1906 developed a pattern that seemed to make the autocracy work in a reasonable manner. An elected parliment, the Duma, brought some representation from the people. A set of land

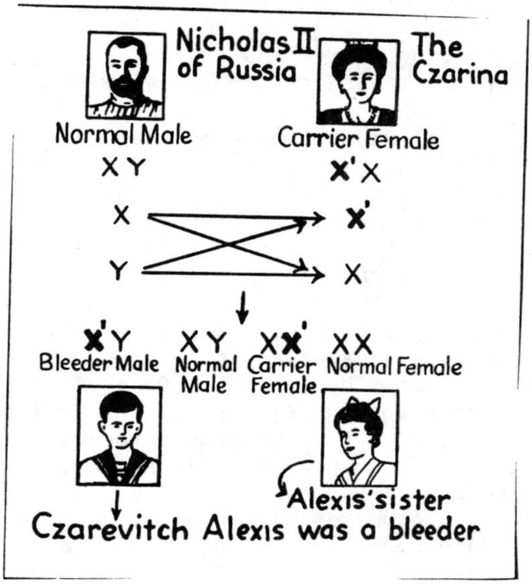

Figure 1. Illustration to show the genetic explanation of the transmission of hemophilia. (X' represents the chromosome carrying hemophilia.) From reference 8.

reforms instigated by Stolypin made land available to the peasants to a degree that by 1914 nearly 9,000,000 peasant families were tilling their own soil. The extreme right members of the aristocracy became hysterical over the possibility that their power and wealth would decline in the new order. On the extreme left[13], even "for Lenin and his dwindling herd of exiles the Stolypin era was a time of fading hope ..." for a future revolution. This progressive tone declined with the assasination of Stolypin by the double agent of police and revolutionaries, Demitri Bogrow.

"Just as Stolypin was simultaneously a symbol of residual vitality in the wasting aristocracy, the main instrument of its future recovery, Rasputin was both an ominous symptom of its decay and the ultimate agent of its collapse."[13] The Czarina was convinced that Rasputin had the power to stop her son's bleeding attacks and thus to perserve his life whenever it was threatened. "Any trace of doubt that may have lingered in her mind vanished in 1912 when the Czarevitch, who was near death from uncontrolled internal hemmorages rallied after Rasputin sent a telegram promising that the boy would be well... only saints could perform miracles, obviously therefore Rasputin was a saint."[13] The Czar was inclined to agree.

While Czar Nicholas was a weak man, the Czarina was a strong woman obsessed with the idea that all Russia's power should be in control of the Royal family and be passed on to her delicate son. Men like Stolypin were a menace. "Rasputin and the Czarina eventually created a hugh political organization to implement their will... operating on a basis of patronage and favors..." Rasputin's approval or recommendation became linked to almost every important appointment and dismissal.

An example of one of Rasputin's (and the Czars favorites) was War Minister general S khomlinov, a former dashing young cavalry officer of the Turkish War of 1877. He was infuriated with any discussion of modern warfare[5]. "... look at me for instance. I have not read a military manual in the last twenty-five years. In 1913 he dismissed five instructors in the staff college for preaching the vicious heresy of fire tactics ... Sukhomlinov had won and kept himself in favor by being at once obsequious and entertaining, by funny stories and acts of buffoonery, avoidance of serious and unpleasant matters, and careful cultivation of "The Friend... Rasputin." "... with invincible belief in the bayonets supremacy over the bullet he made no effort to build up factories for increased production of shells, rifles, and cartridges... Sukhomlinov did not even use up the funds the government appropriated for munitions. Russia began World War I with 850 shells per gun compared with a reserve of 2000-3000 used by western nations[5]." Futhermore many soldiers did not have guns. As battles progressed the gunless who might have survived longer than some with guns patiently waited for their comrades demise for a rifle.

As World War I continued, the role of the Czarina-Rasputin team became progressively deeper and more disasterous as can be learned from references 9, 10, 13 and 14. Now what is the nature of the disease of Czarevitch Alexis that led to this ridiculous situation?

Hemophilia is a consequence of the absence of factor VIII in the blood coagulation enzyme cascade exhibited in Figure 2. This remarkable cascade type of controlled response to tissue damage was deduced independently by MacFarland[15] and Davie and Ratnoff.[16]

To better understand its function consider the following reaction chain

$$\begin{array}{llll} & E_1 & & \\ S_1 \rightarrow & E_2 & + & P_2 \\ & \downarrow & & \\ S_2 \rightarrow & E_3 & + & P_3 \\ & \downarrow & & \\ S_3 \rightarrow & E_4 & + & P_4, \text{ etc.} \end{array} \tag{1}$$

The substrate S_1 with the aid of the catalyst E_1 forms another catalytic enzyme E_2; S_2 catalyzed by E_2 forms still another enzyme E_3, etc. Each stage of the cascade functions as an amplifier. Hence, neglecting the possibility of reversible reactions the cascade dynamics is characterized by the set of equations

$$\begin{aligned} dE_2/dt &= k_1E_1 \\ dE_3/dt &= k_2E_2 \quad \text{etc.} \end{aligned} \tag{2}$$

Surface contact
↓
(1) XII → XIIa
↓
(2) XI → XIa
↓
(3) IX → IXa
↓
(4) VIII → VIIIa
↓
(5) X → Xa
↓
(6) V → Va?
↓
(7) II → IIa (Thrombin)
↓
(8) I → Ia (Fibrin)

Figure 2. Sketch of the Blood Coagulation Cascade

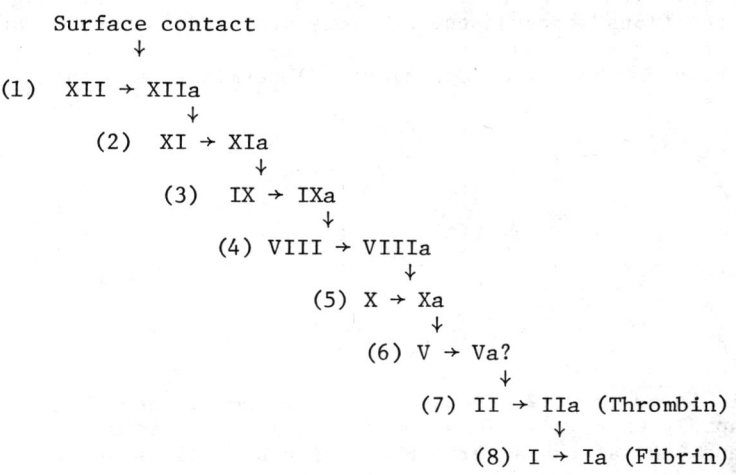

upon postulating the substrate concentrations S_1, S_2, ... to be so large that the reaction rate is determined entirely by the amount of enzyme available. If a factor such as VIII in Figure 2 is missing, the chain of reactions stops at the previous step since no catalyst is available for its continuation.

The solution of the set of equations (2) terminating with E_{n+1} is:

$$E_2(t) = k_1 E_1 t$$
$$E_3(t) = k_1 k_2 E_1 t^2/2 \qquad (3a)$$
$$E_{n+1}(t)/E_1 = (tk\ )^n/n!$$

with

$$k = (k_1 k_2 \ldots k_n)^{1/n} \ .$$

Hence, if n is large, say 12, then E_{n+1} is very small when tk<land grows very rapidly when tk<1. Thus for a certain time essentially nothing happens, and at t=k a veritable avalanche follows as is characteristic of an all or none control system. Since blood plasma must perform many functions some of the minor components participate in several cascades. Thus factor XII also appears in several other control cascades including the complement chain, the catalytic chain responsible for immunological response.[17]

The set of equations (1) is somewhat oversimplified, since under some conditions, reactions become reversible allowing for a more delicate control[18]. A possible reason for so many stages in the coagulation cascade may be to make it more of a fail-saf process so that blood does not coagulate upon reception of false signals in the circulatory system[19]. Normally the components of the flow velocity, the pressure, oxygen concentration, charge on platelets and fibrinogen molecules, etc., suffer fluctuations from their mean values. Since these fluctuations may be in the same direction as those induced by a wound the cascade may start spontaneously when it should not. By requiring the order of 12 steps in the cascade to be completed before the coagulation becomes severe, opportunities exist for termination or reversal of the cascade. The last stage in the cascade is the polymenization of fibrinogen into a fibrin network polymer which catches and holds platelets together[20]. The normal platelet charge seems to be neutralized so that they stick together in the fibrin net, plugging the wound.

It has been shown experimentally that the first few stages in the cascade may be bypassed by adding an appropriate enzyme E_j to the coagulation system. This is a weapon effectively used in nature for the incapacitation of prey. Russell's viper venom is essentially factor Xa in the coagulation cascade[21,19]. Thus, certain snake bites on mammels catalyze the last stages of the catalytic process, killing the victim by promoting internal clotting.

Catalytic enzymes are generally protein molecules. The catalytic function depends upon the conformation (shape) of the enzyme so that any action which changes its geometry may also change its catalytic activity. The basic protein structure is a linear polymer formed by combining many amino acids

```
      H O    H H    O    H    H    O    H       H    O
      | |    | |    |    |    |    |    |       |    |
  NH2-C-C-..-N-C  - C  - N  - C  - C  - N  - ...C  - C  - O-H        (4)
      |        |              |                 |
      R1       Rj             Rj+1              Rn
```

The R_j's each being one of twenty possible groups, coming associated with one of twenty amino acids. The geometrical form and hence the catalytic action depends on the nature of the R group. The determination of the sequences of particular proteins is a major research industry today. The first sequence analysis was done in 1951 by F. Sanger on bovine insulin (molecular weight-11,466). The analysis is done by breaking many molecules into fragments, doing an end analysis and a general (but not sequenced analysis of each frament). By appropriate coding algorithms the full molecule can be reconstructed.

The mathematical systematization of fragment analysis was first discussed by Bernhart, Bradley and Duda[22] in 1962. Computer codes for its application to protein sequencing were developed by Bernhard, Bradley and Shapiro in reference 13, M.J. Dayhoff and R.V. Eck independently programmed alternative schemes. By now both the chemical and computer methods of sequencing have become very sophisticated.

Through the improvement of techniques in the chemical analysis of fragments and in the data processing of the resulting analysis, the amino acid sequencing of proteins is becoming widely practiced. Year by year more sequences from more proteins of more organisms are being recorded. These have been collected and tabulated annually for a number of years by M.J. Dayhoff[24] and her collaborators and published as an "Atlas of Protein Sequences." A typical fragment of an atlas page is exhibited in Figure 3.

As these tables appear, we are exposed to a growing cryptogram from which, with ingenuity, many secrets of life might eventually be extracted. An organism is characterized by the variety of chemical reactions which govern its activity, and by the structural face it presents to the world. The reactions reflect the enzymes which catalyze them and the texture of its external structure is that of its fibrous proteins. In approaching the big book of sequences as a search for understanding through the desipherment of a cryptogram we are deploying a scientific style which would have excited one of its masters, Isaac Newton.

Traditionally Newton is considered to be the father of the canonical scientific method of making a mathematical model to abstract the most important features of a system of interest and then providing an elegant mathematical analysis of the model which beautifully and accurately explain the behavior of the system. Newton's brillian model of the dynamics of the planets and his three laws of motion firmly established this style of science.

Newton himself came to consider his more voluminous and essentially unpublished alchemical and theological investigations to be deeper and to be concerned with more important questions. There his

FIBRINOPEPTIDE A

	1	2	3	4	5	6	7	8	9	10	11	12	13	14	15	16
Human	ALA	ASP	SER	GLY	GLU	GLY	ASP	PHE	LEU	ALA	GLU	GLY	GLY	GLY	VAL	ARG
Gibbon	ALA	ASP	THR	GLY	GLU	(GLY.	GLU)	PHE	(LEU.	ALA.	GLU.	GLY.	GLY.	GLY.	VAL)	ARG
Rhesus Monkey	ALA	ASP	THR	GLY	GLU	GLY	ASP	PHE	LEU	ALA	GLU	GLY	GLY	GLY	VAL	ARG
Drill	(ALA.	ASP.	THR.	GLY.	ASP.	GLY.	ASP.	PHE)	ILE	(THR.	GLU.	GLY.	GLY.	GLY)	VAL	ARG
Rabbit	VAL	ASP	PRO	GLY	GLU	SER	THR	PHE	ILE	ASP	GLU	GLY	ALA	THR	GLY	ARG

Figure 3. Fragments of Amino Acid Sequences in Fibrinopeptide for Several Species.

Data taken from page D-95 of the 1972 edition of reference 24.

role was that of the cryptographer. It is well summarized by Maynard Keynes[25] in his essay on the 400th birthday of Newton celebrated in 1942 "...he looked on the whole universe and all that is in it as a riddle which could be read by applying pure thought to certain evidence, certain mystic clues which God has laid about the world to allow a sort of philisophical treasure hunt to the esoteric brotherhood. He believed that these clues were to be found partly in the heavens and in the constitution of the elements, but also partly in certain papers and traditions handed down by the brothers in an unbroken chain back to the original cryptic revelations in the Bible. *He regarded the universe as a cryptogram set by the creator.*"

The desipherment of certain passages in the book of sequences has already produced bits of an evolutionary tree. As Eck and Dayhoff assembled data from the literature for early editions of the Atlas, they noticed a tremendous similarity in the sequence of homologous proteins from diverse species. For example, in the case of cytochrome C (a protein involved in oxygen transport...), a protein containing 100-110 amino acid residues (the precise number depending on the species), thirty-five locations contain residues which are identical in 19 species examined. There were thirty-three positions in which either of two possible residues were found, etc. The molecular position which showed the greatest variation was number 96. From the 19 species 8 different residues were there found. The longest sequence common to all species, and probably the crucial sequence in the catalytic role of cytochrome C, extended from position 76 to 86. One might be tempted to conjecture that that sequence contains the functional unit of the molecule.

An examination of the sequences of a given protein which correspond to various species shows that some species have exactly the same sequence. From this it is concluded that they had a "common ancestor" with that sequence. A number of now extinct common ancestors can then be listed. Although the sets of "common ancestor" differ from each other in amino acid arrangement, they can be ordered so that the most similar ones are placed near each other. Through this ordering an evolutionary tree of organisms, real and extinct, can be constructed so that those that have similar sequence are close to each other on the tree. These qualitative ideas have been made more precise in the work of Eck and Dayhoff and evolutionary trees have been constructed.

III. STOCHASTIC MODELS OF CHEMICAL KINETICS

1. Smoluchowski's Model of Coagulation[26]. M. Smoluchowski was the first to introduce the concept of diffusion controlled reactions and stochastic models of kinetics in the spirit to be exploited in this report. He was concerned with the coagulation of colloids, an extension of his interest in Brownian motion. This type of process later attracted the attention of meterologists and aerodynamicists in the 1940's and 1950's through their investigations of cloud and fog formation in the atmosphere and in wind tunnels.[27,28,29]

Coagulation is affected by the sticking of pairs of suspended colloidal particles upon collision. As the process continues a colloidal suspension of many small particles is transformed into one of a small number of large coagulated units. The reaction rate is diffusion controlled. Upon collision, two suspended particles are postulated to stick, forming a new particle of the combined mass of the pair. With certain assumptions between particle size and diffusion constant it is not difficult to derive a rate equation for the number of particles of mass u, x(u):

$$\frac{\partial x(u,t)}{\partial t} = k\{ \int_0^u x(u-v,t)x(v,t)dv - 2x(u,t)N(t)\} \qquad (5)$$

with N(t) being the total number of colloidal particles (or droplets in a condensation model) at time t:

$$N(t) = \int_0^\infty x(u,t)dt \qquad (6)$$

As is usual in stochastic models involving transitions of some kind, the positive term on the right in Equation (5) represents the rate of formation of particles of mass u from smaller ones, and the negative term is the loss rate of particles of mass u through combination with others to form larger particles. The quantity k is the rate constant with others to form larger particles. The quantity k is the rate constant for the process.

An initial distribution used in cloud formation studies is the exponential one (with N_0 being a constant)

$$x(u,0) = \lambda(0)N_0 \exp -\lambda(0)u \qquad u>0 \qquad (7)$$

It is easily verified by substitution into (5) that the exponential form is preserved as a function of time with

$$x(u,t) = \lambda(t)N(t)\exp -u\lambda(t) \qquad u\ 0 \qquad (8a)$$

$$\lambda(t) = \lambda(0)/(1+ktN_0) \qquad (8b)$$

$$N(t) = N_0/(1+ktN_0) \qquad (8c)$$

Becker and Doering[30] have investigated somewhat more complicated models of the condensation process. An elegant more modern analysis, done in the spirit of the present review, has been published by W.J. Shugurd and H. Reiss[31].

Another contribution of Smoluchowski[32,33] that has proven to be important in stochastic models of chemical kinetics is his equation for the concentration distribution c(r,t) of Brownian motion particles in a force field derived from a potential U(r):

$$\partial c(r,t)/\partial t = D \operatorname{div} \{\operatorname{grad} c(r,t) + \beta c(r,t) \operatorname{grad} U\}, \ \beta = k/kT. \qquad (9)$$

2. Peter Debye Enters the Scene[34]. The most fashionable theory of chemical kinetics of the late 1930's and on into the 1940's was Henry Eyring's theory of absolute reaction rates. Various important

formulae in the theory involved Planck's constant h. This irritated Peter Debye who, in his inimitable manner would shake his head, wave his arms, and bite harder on his cigar, emphasizing that most interesting reactions occur in the liquid phase in solutions. Then the rate determining step would be the process of the reactants executing their Brownian motion in solution and finally finding each other to participate in the fast molecular rearrangement phase of the reaction. These reactions are then diffusion controlled. Had he lived another dozen years, Debye would have been pleased to see that his prejudices are equally applicable to certain surface catalyzed gas reactions with the diffusion control being a two dimensional one on a surface.

In the Debye model one considers two reacting molecular species, say "1" and "2" suspended in a fluid. They are postulated to be spheres of radii R_1 and R_2 that react when they are in their mutual sphere of influence of radius $R = R_1 + R_2$. To calculate the reaction rate we consider a large number of spheres of influence around the particles of kind 1 into which points representing particles of kind 2 diffuse. If we start with an initially uniform distribution of points, the statistics of the rate points reach various spheres of influence is the same as that of points streaming into a single sphere of influence multiplied by the number of spheres of influence. The streaming rate per sphere is precisely the diffusion current in particles per unit time into the sphere.

The current per unit solid angle is determined by the law of diffusion of particles subject to a potential U(r), if such a potential exists between particles of kinds "1" and "2". With $\beta = 1/kT$

$$i = Dn \cdot (\text{grad } c + c\beta \text{ grad } U), \tag{10}$$

D being the relative effective diffusion constant $(D_1 + D_2)$ of the two species, c is the concentration of species 2 and n is a unit vector normal to the sphere of influence. The Smoluchowski equation (9) in our spherically symmetrical situation has the alternative form

$$\partial\alpha/\partial t = Dr^{-2}\, e^{\beta U}\, \partial[r^2 e^{-\beta U}\, \partial\alpha/\partial r]/\partial r \tag{11a}$$

$$\text{with } \alpha \equiv c \exp \beta U. \tag{11b}$$

Except for very short times in the diffusion process one can establish a current i from the steady state solution of (11) determined by setting $\partial\alpha/\partial t = 0$. The appropriate boundary conditions are

$$c = w_2 \text{ at } r = \infty \text{ (the specified uniform initial distribution of particles of kind 2)} \tag{12a}$$

$$c = 0 \text{ at } r = R \tag{12b}$$

The latter condition is a statement that particles react and thus change their form upon entering the reaction zone so that the concentration of species 2 remains identically zero at the sphere boundary. The current at the reaction zone boundary may also be expressed in terms of α. Integrating (10) over the full surface of the sphere, the total current into the sphere becomes

$$I = 4\pi R^2 D \left(\frac{\partial \varepsilon}{\partial r} + \beta c \frac{\partial U}{\partial r}\right)_{r=R}$$

$$= \left[4\pi R^2 D\ e^{-\beta U(R)} (\partial\alpha/\partial r)\right]_{r=R} \tag{13}$$

The reaction rate is this current multiplied by the sphere concentration

$$\text{reaction rate} = n_1 I.$$

Since the steady state solution of (11) under the boundary conditions (12) is

$$\alpha(r) = w_2 \ \{1-[s(r)/s(r)]\} \tag{14a}$$

$$\text{with} \quad s(r) = \int_r^\infty r^{-2} dr \exp \beta u(r) \tag{14b}$$

it is found that

$$\text{reaction rate} = 4\pi w_1 w_2 (D_1 + D_2)/s(R) \quad . \tag{15}$$

In the special case $U(r) = 0$:

$$\text{reaction rate} = 4\pi w_1 w_2 (D_1 + D_2)(R_1 + R_2) \quad . \tag{16}$$

There are some situations (for example in the process of quenching of fluorescence) in which a significant number of collisions occur before the steady state diffusion current is achieved. One must then solve the full time dependent equation (12). This case has been discussed by LaMer and Umberger,[35] Montroll,[36] and Collins and Kimbell.[37] It was through discussions with Umberger and LaMer in 1945 that the author first became attracted to the subject of diffusion controlled reactions and to stochastic models of chemical kinetics. Important new results have been derived by Noolandi[38].

3. Kramers Model[39]. The diffusion regualted phase of the chemical reaction process naturally lends itself to stochastic modeling. H.A. Kramers in 1940, in a somewhat more subtle manner demonstrated the possibility of constructing a stochastic model for the *completion phase* of a reaction *after the reactants reach the reaction zone.*

In the Kramer's theory, it is assumed that a complex of atoms has been formed in the diffusion phase and that the completion of the subject reaction is expressible in terms of a single variable x, achieving a critical value. This might, for example, represent the relative displacement of a given atom in the complex exceeding a certain magnitude. The reaction would correspond to the transition of the value of x from the value x_A associated the potential minimum A in Figure 4 to a value x_B associated with a new minimum B as indicated in the figure by passage over a barrier C which corresponds to $x = x_C$.

Let us assume that the height of the potentive barrier $Q \gg kT$ and that the well at B is of great depth. Under these conditions, ideas of equilibrium statistical mechanics are applicable; that is, the relative probability that a complex has a value between x and x+dx is is unchanged by a chemical reaction. Kramers further postulated that the variable x of interest is influenced by not only the potential U(x) of the figure, but also by a number of other variables in a manner that can be characterized by a random force. The statistics of the dynamics of x are then expressed by the one dimensional Smoluchowski equation

$$\partial w/\partial t = D\ \partial/\partial x[\partial w/\partial x \ +\ \beta w \partial U/\partial x] \tag{17}$$

The "diffusion" constant D is identified with

$$D = \frac{1}{2}\, n\, \langle x^2 \rangle \tag{18}$$

where n is the number of displacements of x per unit time and $\langle x^2 \rangle$ is the average of the square of these displacements.

The current past the point C in the direction of B in Figure 4 is the 1D analogue of (13)

$$i = -D\left\{\frac{\partial w}{\partial x} + \beta w \frac{\partial U}{\partial x}\right\}_c = -De^{-\beta U(x_c)} \frac{\partial}{\partial x}[\alpha(x)]_{x=x_c} \tag{19a}$$

$$\alpha(x,t) \equiv w(x,t)\ \exp \beta U(x) \tag{19b}$$

$$\partial\alpha/\partial t = De^{\beta U(x)}\ \partial[e^{-\beta U(x)}(\partial\alpha/\partial x)]/\partial x \tag{19c}$$

If the barrier Q is sufficiently high, the current and the reaction rate are determined by the steady current associated with $\partial\alpha/\partial t = 0$. Then

$$[\alpha(x)-\alpha(x_A)]/[\alpha(x_B)-\alpha(x_A)] = \int_{x_A}^{x} e^{\beta U(x)} dx / \int_{x_A}^{x_B} e^{\beta U(x)} dx \tag{20}$$

and

$$i = D(w_A - w_B\ e^{-U(x_B)}) / \int_{x_A}^{x_B} e^{\beta U(x)} dx \tag{21}$$

where $w_A \equiv w)x_A)$ etc. The main contribution of the integrand to the integral above comes values of x in the neighborhood x_c. In this range

$$U(x) = Q - \frac{1}{2} w_C^2 (x-x_c)^2 + \ldots \quad w_c = -\left.\frac{\partial^2 U}{\partial x^2}\right|_{x=x_c} \tag{22}$$

The limits in (21) can be extended to $(-\infty,\infty)$. Then

$$\int_{x_A}^{x_B} e^{\beta U(x)} dx \sim e^{\beta Q} \int_{-\infty}^{\infty} dx\ \exp[\ -\frac{1}{2}\beta w_c^2 (x-x_c)^2] = \frac{e^{\beta Q}}{w_c} \left(\frac{2\pi}{\ }\right)^{\frac{1}{2}}$$

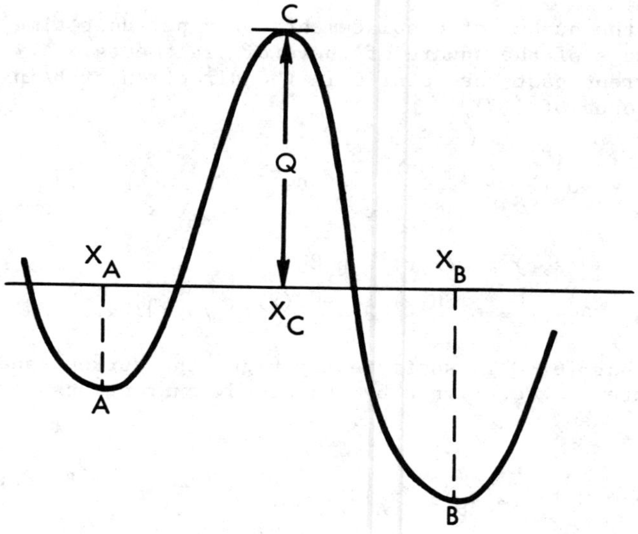

Figure 4. Potential Barrier

A deep well at B allows us to neglect the term in (21) proportional to $\exp[-\beta U(x_B)$. Then the current density across the barrier C from A to B is

$$i = Dw_A w_c e^{-Q\beta}(\beta/2\pi)^{\frac{1}{2}} \tag{23}$$

The probability that a given complex goes from well A to well B per unit time is

$$P = i/v_A \tag{24}$$

where v_A is the total number of complexes in well A. It is related to w_A by

$$v_A = w_A \int_{well} e^{-\beta U(x)} dx \tag{25}$$

In the neighborhood of A, U(x) can be approximated by the harmonic potential

$$U(x) = \frac{1}{2}w_A(x-x_A)^2 \quad \text{where } w_A = -\partial^2U/\partial x^2]_{x=x_A} \tag{26}$$

Then after letting $(x-x_A)=y$ and extending the range of integration to $(-\infty,\infty)$

$$v_A \sim w_A \int_{-\infty}^{\infty} e^{-\frac{1}{2}\beta w_A y^2} dy = \frac{w_A}{w_c}\left(\frac{2\pi}{B}\right)^{\frac{1}{2}}$$

so that

$$P = D\frac{w_A w_c}{2\pi kT} e^{-Q/kT} \tag{27}$$

Kramers and Chandrasekhar[40] have given a detailed analysis of the range of validity of this equation as well as an extension of these results to cases for which Equation (27) is no longer appropriate.

4. Ladder Climbing Model. A quantum mechanical "ladder climbing model" for the dissociation of polyatomic molecules was introduced by Montroll and Shuler[41] in 1958. The basic idea is that such a molecule has a number of quantum mechanical vibrational energy states that might be displayed as a set of rungs on a ladder. Through collisions with other molecules in the embedding medium, a given molecule in a given vibrational state may suffer a transition to high or lower levels (or rungs on the ladder). Through many collisions, the molecule of interest will negotiate a random walk up and down the ladder. It is postulated in the model that, when the molecule reaches a prespecified highly excited state, it will dissociate. Then the mathematical description of the model will be a certain random walk (or master) equation from which one might calculate the first passage time distribution to the dissociating level from the initial occupation of state distribution. The differential equation is obtained

from studying excitative processes in polyatomic molecules.

The experiments that motivated theoretical research on excitation processes in polyatomic molecules in the 1950's were studies of propagation of shocks in gases. To a certain degree such experiments evolved from work on explosions during World War II; on the other hand such shocks were expected in the Mach cone generated by flight in the supersonic regime. As experimental techniques developed they were used to investigate the properties of the molecules in the shocked gases.

A typical experiment is conducted as indicated in Figure 5.

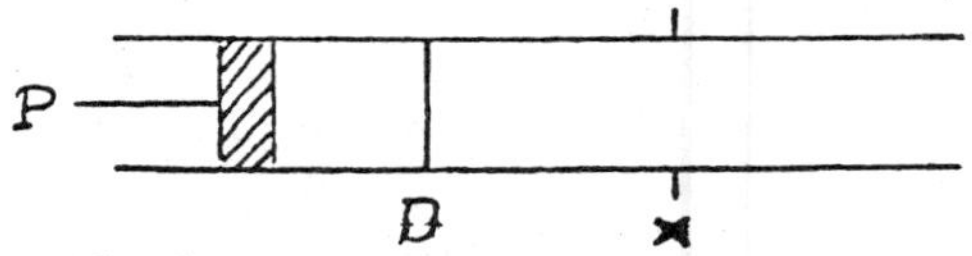

Figure 5. Shock generated by piston P breaks diaphram D. Pressure is applied by a piston P to a gas to the left of a diaphram D resulting in the sudden breaking of the diaphram. The pressure of the gas to the right of the diaphram is measured at one or several points such as the indicated x, as a function of time. In various experiments the gas might be diatomic or a monatomic gas mixed with a small amount of diatomic so that many more collisions existed between diatomic and monatomic particles than between a pair of diatomics.

Figure 6 represents schematically the usual variation of the pressure at a probe point x as a function of time

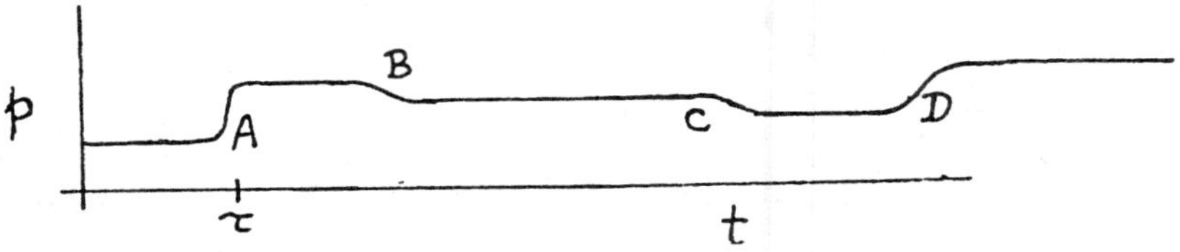

Figure 6. Variation of pressure with time at a probe point x in a shock tube. The sharp variations in the p vs. t curve have the following significance; at time τ associated with A the shock wave generated by the breaking of the diaphram reaches x. The pressure is raised as the translational degrees of freedom of the gas become excited as x. After, on the average of 10-100 collisions rotational degrees of freedom become excited by taking some energy away from the translational motion with the ensuing decline in pressure. After several thousand more collisions the vibrational degrees of freedom also become excited with a further decline in pressure at C. Finally as collisions further excite the vibrational modes, there may be sufficient excitation for some of the molecules to dissociate, thus raising the pressure since more particles will exist to contribute to the pressure.

A simple stochastic model may be constructed for the translational vibrational equilibration by considering diatomic molecules to be described as harmonic oscillators with vibrational energy levels

$$E_n = (n + \frac{1}{2})h\omega \ . \tag{28}$$

The equilibrium distribution of the energy of an assembly of such oscillators in contact with a heat bath at temperature T (the temperature of the translational degrees of freedom) has the Planck oscillator form

$$P_n = (1-e^{-\theta})e^{-n\theta} \quad , \quad \theta \equiv h\omega/kT \tag{29}$$

The dynamical state of the oscillator is characterized by a variable x, the deviation of the displacement from equilibrium. In the small displacement regime the coupling of the oscillator with the translational degrees of freedom is postulated to be linear in the displacement.

A commonly employed model for a stochastic processes is the master equation for the probability P(j,t) that a system is in a state j at time t

$$\frac{\partial P(j,t)}{\partial t} = \sum_{j'} [W(j,j')P(j',t)-W(j',j)P(j,t)] \tag{30}$$

The object W(j,j′) represents the transition rate from state j′ to state j. The summation extends overall possible states of the system. The positive term on the right hand side of (30) is the sum of the rates all possible transitions from other states to j and the negative tern is the rate of transitions from state j to any other states.

If the master equation (30) is to describe the manner that a physical in contact with a heat bath at temperature T relaxes from a nonequilibrium distribution to an equilibrium one, then as t→P(j,t) should approach the cannonical ensemble distribution

$$P(j,\infty) = [\exp-\beta E_j]/Z \ , \quad \beta=1/kT \tag{31a}$$

$$Z = \sum_j \exp-\beta E_j . \tag{31b}$$

As sufficient condition for this to be achieved is that

$$W(j,j')\exp-\beta E_j = W(j',j)\exp-\beta E_j \text{ for all } j,\ j'. \tag{32}$$

This is so called detailed balance condition can be checked by setting the left hand side of (30) to zero and noting that the right hand side also vanishes when (31) and (32) are employed.

When the j's of Equation (30) represent various quantum states, it is sometimes referred to as the "Pauli equation." The transition probabilities W(j,j′) are obtained by the Fermi "golden rule." Let the j's represent the states of an isolated system and suppose that some perturbation f is introduced to couple the various states. Then the W(j,j) is proportional to the square of the matrix element (j f j). Hence, if the unperturbed system is a harmonic oscillator and the coupling term of the oscillator with a heat bath is proportional to the displacement x of the oscillator from its mean value this matrix element would be the integral

$$(j|x|j') \sim \int_{-\infty}^{\infty} e^{-\frac{1}{2}x^2} H_j(x)\, x\, e^{-\frac{1}{2}x^2} H_{j'}(x)dx \tag{33}$$

where the product of the Hermite polynomial H_j and the Gaussian term is the well known wave function of a harmonic oscillator. Since (33) vanishes unless $j=j'\pm1$ the only non vanishing W(j,j') are W(j,j±1). Using the known values of the integral (33) and the detailed balance condition one finds the master equation

$$\frac{dP(j,t)}{dt} = \kappa\{je^{-\theta}P(j-1,t)-[j+(j+1)e^{-\theta}]P(j,t)+(j+1)P(j+1,t)\} \tag{34}$$

first derived by Landau and Teller[41] but not solved by them.

Considerable literature[40,42] has been developed for finding mean first passage time required for a system described by a master equation such as (30) to suffer a transition from our initial state $j=j_0$ to a final one $j=N$.

5. Random Walks on Lattices[43,44]. A stochastic model whose two dimensional version seems to have special promise and already some application to catalysis is the random walk on lattice model, now frequently referred to as the CTRW (continuous time random walk). Consider a simple hypercubic space lattice of d dimensions with N^d lattice points (sometimes with $N_1N_2...N_d$ lattice points). A typical lattice point is represented by $s\equiv(s_1,s_2,...,s_d)$ where each s_j ranges through the values 1, 2, ..., N(or in the asymmetric form s_j ranges through 1, 2, ..., N_j). In the most primative model one employs periodic boundary conditions such that

$$(s_1+n_1N,\ s_2+n_2N,\ ...,\ s_d+n_dN) \equiv (s_1,{}_2s\ ,...,\ s_d) \tag{35}$$

With d=1, the lattice becomes a circle of N points; d=2 a torus of N^2 points, etc.

We characterize random walks on space lattices by an alternation of steps and pauses. The probability of a given step being a displacement s is defined to be p(s), which is normalized so that

$$\sum_s p(s) = 1. \tag{36a}$$

The pausing time distribution is given by $\psi(t)$, also normalized so that

$$\int_0^{\infty} \psi(t)dt = 1. \tag{36b}$$

Important properties of the random walk can generally be determined from the functions

$P_n(s)$ = probability that a walker arrives at s after n-th step

$P(s,t)$ = probability density that a walker is at s at time t.

The function $P_n(s)$ satisfies the recurrence formula

$$P_{n+1}(s) = \sum_{s'} p(s-s')P_n(s') \tag{37}$$

If we postulate our walker to be initially at the origin so that

$$P_0(s) = \delta_{0,s} \tag{38}$$

then the generating function

$$G(s,z) = \sum_{n=0}^{\infty} P_n(s)z^n \tag{39a}$$

satisfies the "Green's function" equation

$$G(s,z) - z\Sigma p(s-s')G(s',z) = \delta_{z,0} \quad . \tag{39b}$$

It is easy to show, because of the translational invariance of our system, that

$$G(s,z) = \frac{1}{N^d} \sum_{r_1=1}^{N} \cdot\cdot \sum_{r_j=1}^{N} \frac{\exp -2\pi i r\cdot s/N}{1-z\lambda(2\pi r/N)} \tag{40a}$$

where

$$r\cdot s = r_1 s_1 + r_2 s_2 + \ldots r_d s_d \quad \text{and} \tag{40b}$$

$$\lambda(2\pi r/N) = \sum_{s} p(s)e^{2\pi i s\cdot r/N} \quad . \tag{40c}$$

The function (k) is called the structure function of the random walk. By expanding the denominator of (40a) and comparing coefficients of z^n in (40a) and (39a) we find that

$$P_n(s) = \frac{1}{N^s} \sum_{\{r_j\}} [\lambda(2\pi r/N)]^n \exp(-2\pi i r\cdot s/N) \tag{41}$$

In the limit $N\to\infty$

$$G(s,z) = \frac{1}{(2\pi)^d} \int_{-\pi}^{\pi} \cdot\cdot \int \frac{\exp -ik\cdot s}{1-2\lambda(k)} \tag{42a}$$

$$P_n(s) = \frac{1}{(2\pi)^d} \int_{-\pi}^{\pi} \cdots \int [\lambda(k)]^n \exp{-ik\cdot s} \tag{42b}$$

As an example of the determination of $\lambda(k)$ consider a random walk on a two dimensional lattice such that steps to any of the four points nearest neighbor to a walkers instaneous position are equally likely. T e four possible displacements are then $s=(0,\pm1)$ or $s=(\pm1,0)$. In each of these cases $p(s)=1/4$. Hence

$$\lambda(k) = \frac{1}{4}(e^{ik_1} + e^{-ik_1} + e^{ik_2} + e^{-ik_2}) \tag{43a}$$

$$= \frac{1}{2}(\cos k_1 + \cos k_2)$$

On a d=3 simple cubic lattice

$$\lambda(k) = \frac{1}{3}(\cos k_1 + \cos k_2 + \cos k_3). \tag{43b}$$

The functions (42a) have been seriously investigated for the $\lambda(k)$ corresponding to the cubic and other lattices. They are called lattice Green's functions.

In order to find the relation between $P(s,t)$ and $P_n(s)$ we must find the probability density function $\psi_n(t)$ for the n-th step to have been taken at time t. Then

$$\psi_1(t) \equiv \psi(t) \tag{44a}$$

$$\psi_n(t) = \int_0^t \psi(t-\tau)\psi_{n-1}(\tau)d\tau \qquad n=2,3,\ldots \tag{44b}$$

The Laplace transform of $\psi_n(t)$ is then

$$\tilde{\psi}_n(u) \equiv \int_0^\infty e^{-ut}\, \psi_n(t)dt = [\tilde{\psi}(u)]^n \quad . \tag{45}$$

It has been shown by Montroll and Weiss[45] that the Laplace transform of $P(s,t)$ is related to $\tilde{\psi}(u)$ and $G(s,z)$ through

$$\tilde{P}(s,u) = u^{-1}[1-\tilde{\psi}(u)\ G\ s,\tilde{\psi}(u)] \tag{46a}$$

so that

$$P(s,t) = \frac{1}{2\pi i}\int_{c-i\infty}^{c+i\infty} u^{-1}e^{ut}\,[1-\tilde{\psi}(u)]\,G[s,\tilde{\psi}(u)]\,du. \tag{46b}$$

Hence from $\psi(t)$ and $p(s)$, $P(s,t)$ becomes known.

We have seen in previous sections that models in chemical kinetics are often expressed in terms of first passage times and currents. It can thus be expected that first passage time distributions and mean first passage times would be important in applications of the lattice walk model. By following a scheme proposed by Feller[48] for the investigation of recurrent events, we proceed to derive formulae for the relevant quantities. We introduce a slight generalization for our notation by letting s now represent available states for a system (which in the special case of lattice walks are lattice points). The we define:

$P_n(s,s_0)$ = probability of system being in state s after nth transition, starting in state s_0 (47a)

$F_n(s,s_0)$ = probability of system being in state s for the first time after nth transition, starting initially at s_0. (47b)

Two generating functions are basic to the process under consideration

$$G(s,s_0;z) = \delta_{s,s_0} + \sum_{n=1}^{\infty} z^n P_n(s,s_0) \tag{48a}$$

$$F(s,s_0;z) = \sum_{n=1}^{\infty} z^n F_n(s,s_0). \tag{48b}$$

The following relations can be shown to exist between the P's and F's

$$F(s_0,s_0;z) = 1-[G(s_0;s_0;z]^{-1} \quad \text{and} \tag{49a}$$

$$F(s,s_0;z) = G(s,s_0;z)/G(s,s;z) \quad \text{if } s=s_0. \tag{49b}$$

Since these relations between first passage time generating functions and usual transition probability generating functions are valid for any set of states $\{s\}$, they are independent of dimensionality of the system of interest.

The relation between $\tilde{F}(s,u)$, the Laplace transform of the first passage time distribution $f(s,s_0;t)$ in the CTRW and $F(s,s_0;z)$ is obtained by noting that

$$f(s,s_0;t) = \sum_{n=0}^{\infty} F_n(s,s_0)\psi_n(t). \tag{50}$$

since the right hand side of this equation sums all the independent ways of the accomplishment of the transition $s_0 \to s$ in time t. By taking Laplace transforms and applying (45) we have

$$\tilde{F}(s,s_0;u) = \Sigma\, F_n(s,s_0)[\,\tilde{\psi}(u)]^n \equiv F[s,s^0;\tilde{\psi}(u)] \tag{51}$$

as required

We now collect several important first passage time results concerning random walks on d-dimensional simple hypercubic lattices.

First consider a lattice with $N \times N \times .. N^d$ lattice points and periodic boundary conditions (35). Each step is postulated to be to a nearest neighbor point with passage to each such point being equally likely with probability 1/2d. Simple steps to more distant points are prohibited.

An ergodic theorem exists for these walks. The mean number of steps required for a walker to return to his starting point is[46]

$$\langle n(0)\rangle = N^d \tag{52}$$

The number to reach for the first time a point displaced by a non-vanishing vector s from the starting point is

$$\langle n(s)\rangle = s(N-s) \quad \text{for } d=1 \text{ and } s=1,2,\ldots,N-1 \tag{53a}$$

$$\langle n(s)\rangle \sim N^2(s/\pi)\log|s| \quad \text{for } d=2 \tag{53b}$$

$$\langle n(s)\rangle \sim N^3\{P_3(0,1) - \frac{3}{2\pi|s|} + 0(|s|^{-2})\} \quad \text{for } d=3 \tag{53c}$$

with

$$P_3(0,1) = 1.51638\ldots \tag{53d}$$

In a lattice walk of n steps the walker visits new points for the first time as n increases. In terms of dimensionality, d, the number of distinct points visited in n steps, $s_d(n)$ is[46,47,49,50], for n large

$$s_1(n) \sim (8n/\pi)^{\frac{1}{2}} \tag{54a}$$

$$S_2(n) \sim \pi n/\log n \text{ (square lattice} \tag{54b}$$

$$s_3(n) \sim 0.6295n \quad \text{(simple cubic lattice)} \tag{54c}$$

Notice that since

$$ds_1/dn \sim (2/n\pi)^{\frac{1}{2}} \tag{55a}$$

$$ds_2/dn \sim (\pi/\log n) \quad [1-(\log n)^{-1}] \tag{55b}$$

$$ds_3/dn \sim 0.6295, \tag{55c}$$

the rate of a walker encountering new points diminishes rapidly in one dimension, remains constant in three and varies very slowly in two dimensions.

Our lattice random walk may also be investigated by considering the generalized master equation[51]

$$dP(s,t)/dt = \int_0^t \phi(t-\tau)\ \{-P(s,\tau) + \sum_{s'} p(s-s')P(s',\tau)\}d\tau \tag{56}$$

where the Laplace transform $\tilde{\phi}(u)$ of $\phi(t)$ is related to that of our pausing time distribution, $\tilde{\phi}(u)$ by

$$\tilde{\phi}(u) = \tilde{\psi}(u)[1-\tilde{\psi}(u)] \tag{57}$$

If we take the Laplace transform of (56) and apply the convolution theorem, we find that

$$P(s,t) = (1/2\pi i) \int_{c-i\infty}^{c+i\infty} e^{ut} du[u+\tilde{\phi}(u)]^{-1} G[s,\tilde{\phi}/(u+\tilde{\phi})] \tag{58}$$

with G being defined as the lattice Green's function satisfying (42a). Then (46b) is equivalent to (58) when $\tilde{\phi}(u)$ and $\tilde{\phi}(u)$ are related to (57).

If $\psi(t) = \lambda e^{-\lambda t}$, then $\phi(t) = 2\lambda\delta(t)$ and (56) becomes a standard master equation

$$dP(s,t)/dt= -\lambda P(s,t) + \lambda \sum_{s'} p(s,s')P(s',t). \tag{59}$$

If (with $\lambda^2 > 4a$)

$$\psi(t) = 2a(\lambda^2-4a)^{-\frac{1}{2}} e^{-\frac{1}{2}\lambda t} \sinh[\tfrac{1}{2}t(\lambda^2-4a)^{\frac{1}{2}}] , \tag{60a}$$

which is the equivalent of the difference between two exponentials,

$$\phi(t) \quad ae^{-\lambda t} . \tag{60b}$$

When (56) is differentiated with respect to t,

$$a^{-1} P_{tt}(s,t) + (\lambda/a)P_t(s,t) = a[-P(s,t) + \sum_{s'} p(s-s')P(s',t)] \tag{61}$$

If steps are taken to nearest-neighbor points only and if the lattice spacings are made very small, then by proceeding to the continuum limit this equation takes the form of the telegraphers equation of Helmholz, Kelvin, and Heaviside (with $P_t \equiv \partial P/\partial t$, etc.)

$$a^{-1}P_{tt} + (\lambda/a)P_t = DP_{xx} + \beta P_x$$

the D and β being appropriately defined. It is known that at early times an initial pulse propagates as a wave, while at later times it propagates as a diffusion packet. This phenomenon was observed in the early days of telegraphy. Signal diffusion reduced the data rate in long cables such as the early Atlantic cable.

6. Adam, Delbrück, and the Importance of Dimensionality on Biological Diffusions Processes. Adam and Delbrück[54] considered "...the problem of bringing an object (a diffusible molecule) from a large or distant area of production, P, to a small target area, Q, where it is to be used" in a biological process. Three classes of examples, each with its own length scale were examined:

(i) Short distances of the order of microns in individual cells
(ii) distances of the order of meters in hormone transport
(iii) very large distances of the order of kilometers in interaction of insects, say in the form of sex attractants.

It was observed that ordinary molecular diffusion was far too slow for the required functioning of involved organisms. Processes could be hastened at all of the above levels through appropriate convective processes. In case (i) protoplasmic streaming, (ii) blood circulation, (iii) wind currents hasten the process. Each of these forms of convection still retains some random component and the effective form of diffusion is a turbulent diffusion with its own diffustion constant being much greater than the molecular diffusion constant of an undisturbed system.

It is the second form of expediting the required transport that is especially interesting to us here since it is closely related to features of surface catalysis. An organism that responds to the 3D diffusable molecule must have an appropriate receptor. Once the molecule lands on the receptor it must be further processed, being transported to the trigger point to start the next part of a response process. In case (iii) the molecule must be transported to a nerve end to transmit the message to the brain of the organism to alert it to the existance of the source of the diffusable molecule. Generally the receptor is a 2D surface and the received molecule is free to diffuse on the surface. Two extreme situations are:

(a) very tight binding to the surface-undesirable since it would take a very long time to finally diffuse to the trigger site if it did not land there originally.
(b) very weak binding to the surface-undesirable since a high probability would exist for escape from the surface before diffusion would take it to the trigger site.

For a given surface area some intermediate "happy medium" degree of binding would be optimum to allow for sufficient mobility to find a trigger site before it escaped from the surface.

It has been emphasized by Adam and Delbrück in this intermediate case, surface diffusion is to be much preferred over 3D volume diffusion. This is evident from a consideration of Equation (53) which indi-ates that the order of $N^2 \log N$ steps are required to reach a special site in the center of a 2d array of N^2 points compared with the order of N^3 steps to reach the center of a 3D array of N^3 points. The advantage of reducing from 2D to 1D is small since the number of steps required to reach the center of a 1D array of N points is (with $s=1/2N$ in (53a)) of order N^2. The extra log factor in 2D is not significant and a planer array is a better target than a linear one.

The case examined in detail in reference (54) is the silkworm moth Bombyx mori. The female exudes a sex attractant bornbykal (a 16-carbon straight chain alcohol with two double bonds in a particular configuration). These are perceived over long distances by the male "... by means of very numerous sense organs, sensilla, each a hair 100μ long and 2μ in diameter. These hairs are arranged in a basket array on the surface of the main stem and the branches of the moth's antennae. Each hair is innervated by the terminal process of one or two sensory nerve cells, which run up the length of the hair; ... shielded from direct contact with outside air... . Direct contact occurs only at minute pores, 150Å in diameter, distributed along the surface of the sensillum and spaced about 4500Å apart from each other. The pores constitute only about 1/1000 of the surface of the sensillun. There are 3200 pores on each hair and about 10,000 sensilla per antenna."

While the bornbykal seems to execute a random walk on a 2d surface to a sensitive pore site the surface is not smooth but is an array of steep hills and deep valleys. Thus a system has evolved in nature that is analogous to that used industrially by supporting a metal catalyst on a highly irregular matrix such as silica gel or alumina. This support technique invented by Wilder Bancroft and others more than fifty years ago allows one to pack a large surface area into a small space.

The last stage of the natural receptor system of the silkworm is effected by a 1d random walk since signal conduction in nerves is controlled by ion transport through the membrane cover of a nerve cell. A certain dispersion is associated with the propagating signal. Hence the complete process of chemical stimulation contains elements of 1, 2, and 3 dimensional random walks.

IV. STOCHASTIC PROCESSES IN CATALYSIS

In this section, various models discussed in the last will be considered in the context of random processes on surfaces that seem to be relevant in catalysis. A more systemmatic approach to this topic is given in the lectures of Shlesinger and Landman.[3,4]

The science of thermodynamics provides a set of criteria for the possible occurance of particular chemical reactions upon the mixing of the required molecular species; but it gives no indication of the time required for the completion of the reaction to a specified degree. Reaction rates are strongly influenced by appropriate catalysts. The role of a catalyst, we have already indicated (and in this section catalyst will refer to surface catalyst) is to provide a medium for participating molecules to confront each other or an active site in a state favorable for completion of the reaction. Two molecules speedily whizzing by each other in the gaseous phase have little chance for attachment; however, adsorbed on a surface their more restricted motion affords a greater opportunity for their finding of each other in an appropriate state.

Three classes of processes are central to the mechanism of surface catalyst (a) the adsorption of the relevant molecules on the surface (b) the motion of the molecules on the surface until they find each other or in the case of dissociation reactions, an active site which promotes the reaction (c) the detailed dynamics of the collision process which results in molecular attachment or, in the case of dissociation reactions the dynamics of the interaction of the molecule with the active site which excites the molecule to its state of dissociation.

The first process we consider is photosynthesis. While it is not normally discussed as a catalytic process its stochastic aspects have many of the features of the more traditional examples. The schematic chemical equation for photosynthesis is

$$m\hbar\omega + nH_2O + nCO_2 \rightarrow nO_2 + (CH_2O)_n \tag{63}$$

where the $\hbar\omega$ represents a photon absorbed by a photosynthetic unit and the complex $(CH_2O)_n$ represents a sugar molecule (with n being 5 for a pentose sugar and 6 for a hexose). The total number of photons required for the formation of a single oxygen molecule is generally about 8-10. References (55), (56), and (57) are excellent reviews of models of photosynthesis.

The story of photosynthesis starts with experiments by Joseph Priestley (1771), the discoverer of oxygen. After observing the death struggles of mice in bell jars with and without burning candles, he made similar experimental arrangements for mint plants and discovered that they not only survived in the air poisoned by the burning candles but indeed purified it so that a mouse could live in it. Incidentally as a "practical" application of his researches on gases Priestly obtained patents on carbonated beverages, which he later leased with the Swiss entrepreneur Schwepps who used them to found his carbonated beverage empire.

The Dutch physician Jan Ingenhausz (1773) who repeated and elaborated on the Priestly experiments was the first to notice that green plants could only purify air in the presence of sunlight. He did his experiments during a short visit to England to vaccinate the royal family. George III of England seems to have been sufficiently satisfied with the results to send Ingenhauz to Vienna to vaccinate Cousin Maria Theresa and her family. He was then retained as court physician at Schoenbrunn at a salary of 5000 gol gulden per year and was given sufficient leisure to continue his scientific researches[58]. In one of these he anticipated Brown in the observation of Brownian motion by observing that when finally powdered charcoal floated on an alcohol surface it executed a highly erratic random motion. By good fortune later scientists neglected his work, otherwise they would have been confronted with the verbal complexity of Ingenhauszen motion.

Through a further elecidation of experiments of the Priestly type the Swiss chemist Jean Senebier (1782) identified the role of various gases in photosynthesis and proposed a basic chemical equation analogous to (63) for the process. A plant energy storage term was added later by Julius Mayer, the discoverer of the first law of thermodynamics. An interesting controversy existed for many years concerning the question of whether the O_2 came from the water or the CO_2 molecules. The more commonly accepted equations

$$CO_2 + h\omega \rightarrow C + O_2$$

$$C + H_2O \rightarrow CH_2O$$

would choose CO_2 to be the source. Bredig's (1914) conjecture

$$2H_2O + h\omega \rightarrow O_2 + 4H$$

$$4H + CO_2 \rightarrow (CH_2O) + H_2O$$

however was verified by Rubin and Kamin (1941) in one of the first applications of tracer techniques to biochemistry. Through studies with enriched O^{18} they proved the water to be the source of the O_2. J.C.H. Smith (1943) showed that all the CO_2 went into the sugar; the process in the sunflower plant taking, for example, between one and two hours.

The detailed elaboration of the mechanism of photosynthesis might then be considered as the problem of the green box

$$\begin{matrix} H_2O \\ CO_2 \end{matrix} + \text{photons} \rightarrow \begin{matrix} \text{Green} \\ \text{box} \end{matrix} \rightarrow \begin{matrix} O_2 \\ \text{Sugar} \end{matrix} .$$

As with the resolution of numerous biological puzzles, rapid progress follows the employment of appropriate organisms. In this case the chosen are the pigmented bacteria chlorella and the green algae pryenoidosa.

For our purpose it is sufficient to model a photosynthetic unit to be a periodic two dimensional periodic array of chlorophyll molecules, each molecule containing a porph-rin group with a long closed free electron path of alternate double and single bonds (see Figure 7). The electron distribution function associated with the free electron channel of a molecule in the ground state is essentially a uniform one with no net current. Upon the absorption of a photon by a chlorophyll molecule, a directed electron current is established along the closed channel. This excited state is named an exciton. The energy transfer from one such molecule to another was first discussed by Forster[59].

Since the closed path current is analogous to that in a radio antenna, one might call molecules such as chlorophyll, antenna molecules. A lattice of these is effectively a periodic array of matched and (judging from experimental data on the difference of fluorescene of a photosynthetic unit compared with a chlorophyl solution composed of the same number of chlorophyll molecules), indeed highly directional antennas. As an antenna molecule radiates (in single quanta of excitation) the transmitted signal is picked up by one of the neighboring molecules exciting it into an exciton state. The quanta associated with the originall- absorbed photon effectively executes a random walk through the lattice as the exciton is transferred from one chlorophyl atom to another. In addition to the lattice of chlorophyll molecules, a photosynthetic unit contains a number of trapping centers which annihilate excitons and employ the associated energy to activate enzyme molecules to complete the chemical reaction (1).

If photons are treated conceptually as a molecular species, each of the three processes characteristic of surface catalysis (as listed at the beginning of this section are evident in photosynthesis. The absorption of photons by the photosynthetic unit corresponds to (a), the random walk of the excitons to (b), and the last phase of trapping and formation of sugar and oxygen to (c). This model was proposed by

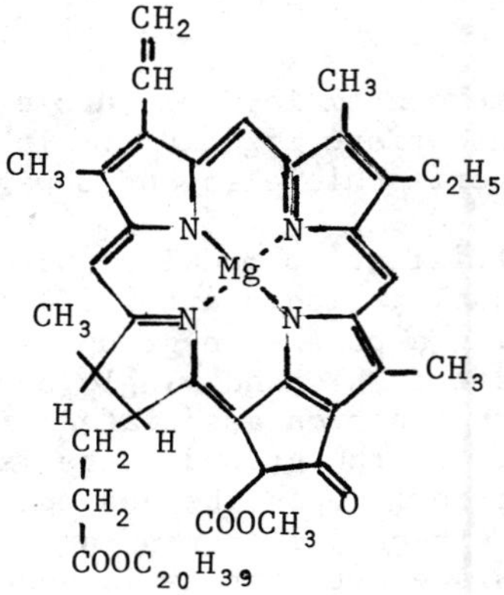

Figure 7. Structure of Chlorophyll a Molecule

Duysens[55] and part (b) investigated in references 59-61.

Phase (b) is modeled as a simple random walk process. Figure 8 represents a two dimensional lattice with a set of periodically spaced traps characterized by the x's. In the model a photon falls with equal likelihood on any of the non-trapping points. It is required to find the mean first passage time (or better passage time distribution) for the photon excitation (the exciton) to migrate to one of the traps. The number of steps $<n>$ required to reach a trap for the first time depends upon N, the ratio of the total number of lattice points to the number of traps, and the dimensionality of the lattice (and to a weaker extent the structure of the lattice. It is found that

a) On linear array of points

$$<n> = N(N+1)/6 \tag{64a}$$

b) On square lattice

$$<n> \sim (N/\pi)\log N + 0.195056N - 0.1170 + 0(N^{-1}) \tag{64b}$$

c) On single cubic lattice

$$<n> \sim 1.5164N + 0(N^{\frac{1}{2}}) \tag{64c}$$

Results have also been found for other lattice in reference (44).

Experimentally exciton trapping and migration and monitored by observing the fluorescent yield of the optically excited systems. A certain fraction of the photons coverted to excitons are re-emitted through fluorescence before they can be trapped. Such observations indicate that the receptor configuration of a photosynthetic unit is essentially two dimensional. Since, in a 1d system an exciton would have to make $0(N^2)$ steps before trapping, it would suffer a much longer migration time before trapping than one in a 2d system which would require only 0(N log N) steps. Hence the fluorescent yield would be greater in a 1d than in a similar 2d system (for the same N). With each step by an exciton, a small probability of fluorescence exists. The appearance of an appropriate number of excitons at a trap drives the catalytic enzyme in the required manner to process the conversion of CO_2 and H_20 into 0_2 and sugar. Photon energy transfer via exciton migration to trips has also been investigated in several organic solids. Experimental results have led to a controversy as to whether this type of energy transfer is by hopping (as discussed above) or through wave propagation, evidence for both types of propagation being available. V. M. Kenkre[62] has recommended a resolution of the conflict through the employment of the telegraphers equation (62). An experiment that follows exciton propagation in the early time regime would imply the existance of a wave while one that detects effects associated with the long time regime would indicate a diffusion energy transport process.

Now let us consider some qualitative aspects of surface catalysis of dissociation reactions. In the first step in the process molecules become absorbed on a surface. Then by surface diffusion they reach active sites where energy from the surface excites the molecule to a degree that dissociation occurs. The following conflict arises in the development

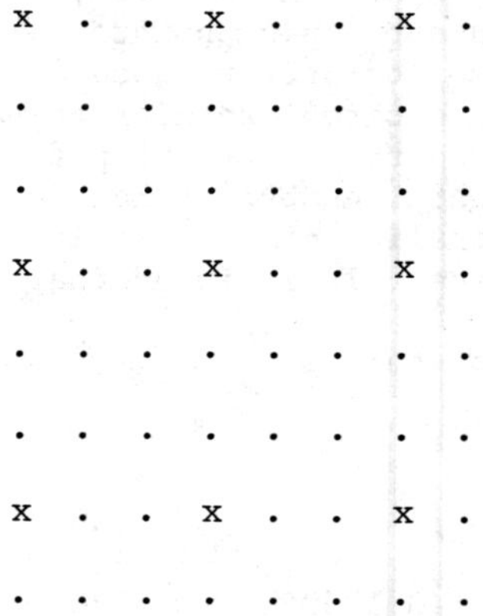

Fig. 8. A lattice of periodically spaced traps.

of a strategy to investigate these events. An effective catalyst is generally characterized by a rough ill defined surface while good basic science is usually done on regular, well defined clear reproducible surfaces. We refer here only to work done in this style. In the spirit common to the investigators we trust that the understanding of the well defined will improve that of the complicated real world technology.

Field emission and field ion microscopies, developed by E. W. Müller[63,64] have proven to be effective instruments for the investigation of clean well defined (usually tungsten) surfaces. With these techniques, R. Gomer[65] and his associates have made numerous studies of diffusion on surfaces. This work is discussed by Gomer[66] in this volume. Studies have also been made by T. T. Tsong et al[67].

G. Ehrlich[68,69] and his collaborators have followed the motion of dimers in channels on tungsten surfaces. An especially interesting feature of this random walk process is the manner that the bond length of the dimer stretches and contracts during the random walk. The temperature variation of the diffusion constant for the process seems to be best explained from a consideration of a random walk with internal state charges[70,71]. This topic is discussed in this volume in some detail by Landman and Shlesinger[3].

While in our works the diffusion constant (or diffusion constant matrix in the case of systems with internal states) is taken to be determined from experiment, H. Suhl[72,73] and collaborators have attempted to determine it in some special cases from first principles of solid state theory. The basic idea is that an atom or molecule on a metal surface suffers an interaction with crystal-phonons and with the electron plasma of the metal. At a given moment the surface absorbed species might be considered to be trapped in a potential well. Then, the random forces from the local electromagnetic field in the metal excite the motion of the absorbed species to the degree that it climbs out of the well and falls into a neighboring one. Suhl and his collaborators use the Kramers theory described above to discuss this process, calculating the parameters required in Kramers model from solid state many body theory.

A different form of diffusion theory enters into the practical problems of catalytic beds and slurries. This concerns diffusion through porous media and is reviewed in reference[74].

Considerable progress toward the understanding of the role of surface defects in catalysis has been made by G. A. Somerjai[75]. By cleaving, polishing and heat treating platinum and other surfaces he has been able to prepare various well characterized steps (one atomic layer each) on the surface. Hydrocarbon catalysis has been investigated as a function of the detailed nature of the steps on Pt.

From the kinetics of N_2 dissociation at step edges on tungsten observed by Besoke and Wagner[76,77], C. H. Wu and the author have attempted to determine the diffusion constant of N_2 on tungsten. This work will be published elsewhere.

A detailed theory of the influence of surface defects on bond breaking of attached molecules has not yet been made. In some cases localized lattice modes of vibration at defect sites may couple strongly with certain vibrational modes of the molecule. Our ladder climbing kinetics described in section (III.4) might even be more appropriate for this type surface excitation of molecules than it is for gas kinetics. The vibrational-rotational coupling sometimes important in gas kinetics probably has little significance in surface excitation kinetics. In the original Montroll-Shuler model, bond breaking was postulated to occur when in the climb through the ladder of energy states a preassigned excited state was reached for the first time. This idea has been extended by Freed and Heller[78] and Kenkre and Seshadri[79] to allow for bond breaking with decreasing probability at successive lower energy levels.

The preparation of this report was partially supported by U.S. DOE Contract No. EG-77-5-05-5489 with Georgia Institute of Technology.

REFERENCES

1. E. W. Montroll, Energetics in Metallurgical Phenomena (Ed. W. Mueller, Gordon and Breach, NY, 1967) Vol. 3, 139.
2. D. A. McQuarrie, J. Appl. Prob. 4, 413 (1967).
3. U. Landman and M. Shlessinger (this volume).
4. U. Landman (this volume).
5. Barbara Tuchman, Guns of August (MacMillan, NY, 1962).
6. P. H. Emmett, The Physical Basis for Heterogeneous Catalysis (Ed. E. Drauglis and R. I. Jaffee, Plenum Press, NY, 1974), p. 3.
7. Historical Statistics of the United States (U.S. Dept. of Commerce, Bureau of the Census, 1976), Vol. 1, p. 606.
8. B. Jaffe, Outposts of Science (Simon and Schuster, NY, 1935), p. 27.
9. R. K. Massie, Nicholas and Alexandra (Athaneum Publishers, NY, 1967).
10. B. Pares, The Fall of the Russian Monarchy (Vintage Books, NY, 1961).
11. J. B. S. Haldane, Heredity and Politics (Norton, NY, 1938).
12. O. Lucas, A. Finkelman and L. M. Tocantins, J. of Oral Surgery, Anesthesia and Hospital Dental Service, 20, 489 (1962).
13. E. Taylor, The Fall of the Dynasties (Doubleday, Garden City, NY, 1963) (see especially Chapter 9).
14. Barbara Tuchman, The Proud Tower (MacMillan, New York, 1962).
15. R. G. MacFarlane, Nature 202, 498 (1964).
16. W. E. Davie and O. D. Ratnoff, Science 145, 1310 (1964).
17. P. L. Mollison, Proc. Roy. Soc. London, Ser. B 173, 377 (1969).
18. H. C. Hemker and P. W. Hemker, Proc. Roy. Soc. London, Ser. B, 173, 411 (1969).
19. E. W. Montroll, Advances in Chem. Phys. 26, 145 (1974).
20. Shirley A. Johnson, Blood Clotting Enzymology (W. H. Seegers, Editor, Academic Press, NY, 1967), p. 379.
21. R. Briggs, Proc. Roy. Soc. London, Ser. B 173, 277 (1969).
22. S. A. Bernhard, D. F. Bradley and W. L. Duda, IBM J. Res. Develop. 7, 246 (1963).
23. D. F. Bradley, C. R. Merril and M. B. Shapiro, Biopolymers 2, 415 (1964).
24. M. J. Dayhoff and R. V. Eck, Atlas of Protein Sequences and Structures, National Biochemistry Research Foundation, Silver Spring, MD, 1966 and succeeding years.
25. J. M. Keynes, Newton Tercentenary Celebrations, Roy. Soc. p. 27 (Cambridge 1942).
26. M. Smoluchowski, Z. Physik Chem. Vol. 92, 129 (1918).
27. W. A. Schumann, Quar. J. Roy. Meteor. Soc. 66, 195 (1940).
28. M. Tsuji, Memoires of the Faculty of Sci., Kyusen Univ. Ser. B 1, 74 (1953).
29. F. Gilmore, The Dynamics of Condensation and Vaporization (Office of Naval Research Rept., Cal. Inst. of Tech., Pasadena, CA, 1951).
30. R. Becker and W. Doering, Ann. Physik 24, 719 (1935).
31. W. J. Shugard and H. Reiss, J. Chem. Phys. 65, 2827 (1976).
32. M. Smoluchowski, Ann. Phys. 48, 1103 (1915).
33. M. Smoluchowski, J. Physik. Chem. 92, 129 (1917).
34. P. Debye, Trans. Electro. Chem. Soc. 82, 265 (1948).
35. J. Q. Umberger and V. K. La Mer, J. An. Chem. Soc. 67, 1099 (1945).
36. E. W. Montroll, J. Chem. Phys. 14, 202 (1946).
37. F. C. Collins and G. E. Kimbell, J. Coll. Sci. 4, 425 (1949).
38. K. M. Hong and J. Noolandi, J. Chem. Phys. 68, 5163, 5172 (1978).
39. H. A. Kramers, Physica 7, 284 (1940).

40. S. Chandrasekhar, Rev. Mod. Phys. 15, 1 (1943).
41. E. W. Montroll and K. E. Shuler, Advances in Chem. Phys. 1, 361 (1958).
42. L. Landau and E. Teller, Physik Z. Sowjetunion 10, 34 (1936).
43. I. Openheim, K. Shuler and G. Weiss, Stochastic Processes in Chemical Kinetics: The Master Equation (MIT Press, 1977).
44. E. W. Montroll, Proc. of Sympos. in App. Math. of Am. Math. Soc. 16, 193 (1964).
45. E. W. Montroll and G. H. Weiss, J. Math. Phys. 6, 167 (1965).
46. G. S. Joyce, Phil. Trans. Roy. Soc. London 232, 583 (1972).
47. T. Morita and T. Horiguchi, J. Phys. A5, 67 (1972).
48. W. Feller, An Introduction to Probability, Theory and Applications (Wiley, NY, 1951).
49. A. Dvoretzsky and P. Erdos, Proc. 2nd Berkeley Symp. on Math. Stat. and Prob. (Univ. of Cal. Press, 1951), p. 33.
50. G. H. Vineyard, J. Math. Phys. 4, 1191 (1963).
51. V. M. Kenkre, E. W. Montroll and M. F. Shlesinger, J. Stat. Phys. 9, 43 (1973).
52. G. Adam and M. Delbrück, Structural Chemistry and Molecular Biology (Ed. A. Rich and N. Davidson, Freeman, 1968), p. 198.
53. J. Barber (Editor) Topics in Photosynthesis (Elsevier, 1977).
54. E. I. Rabinowitch and Govingee, Photosynthesis (Wiley, 1969).
55. L. N. M. Duysens, Progr. Biophys. 14, 1 (1964).
56. P. W. van der Pas, Dictionary of Scientific Biography (Ed. C. C. Gillispie, Scribners, 1973), VII, p. 11.
57. Th. Förster, Ann. Phys. (Leipz.) 2, 55 (1948).
58. D. L. Dexter, J. Chem. Phys. 21, 836 (1953).
59. R. M. Pearlstein, Brookhaven Nat. Lab. Symp. 19, 19 (1967).
60. R. J. Knox, J. Theoret. Biol. 21, 244 (1968).
61. E. W. Montroll, J. Math. Phys. 10, 753 (1969).
62. V. M. Kenkre, Stat. Mechanics and Stat. Methods in Theor. and Applications (Ed. U. Landman, Plenum, 1977), p. 441.
63. E. W. Müller and T. T. Tsong, Field Ion Microscopy, Amer. Elsevier (NY, 1969).
64. R. Gomer, Field Emission and Field Ionization (Harvard Univ. Press, 1961).
65. R. Gomer, Solid State Physics 30 (Ed. H. Ehrenreich, F. Seitz and D. Turnbull, Academic Press, 1975), p. 93.
66. R. Gomer, this volume.
67. T. T. Tsong, P. Cowan and G. Kellog, Thin Solid Films, 25, 97 (1975).
68. G. Ehrlich, CRC Crit. Rev. Solid State Phys. 4, 205 (1974).
69. D. A. Reed and G. Ehrlich, J. Chem. Phys. 64, 4616 (1975).
70. U. Landman, E. W. Montroll and M. F. Shlesinger, Phys. Rev. Letters, 38, 285 (1977).
71. U. Landman and M. F. Shlesinger, Phys. Rev. B16, 3389 (1977); B19, 6207, 6220 (1979).
72. E. G. d'Agliano, W. L. Schaich, P. Kuman and H. Suhl, Nobel Symp. 24, Medicine and Natural Sciences, Collective Properties of Physical Systems (Eds. B. Lundquist and S. Lundquist, Acad. Press, NY, 1973), p. 200.
73. E. G. d'Agliano, P. Kumer, W. Schaich and H. Suhl, Phys. Rev. B11, 2122 (1975).
74. C. N. Satterfield and T. K. Sherwood, The Role of Diffusion in Catalysis (Addison-Wesley, 1963).

75. G. A. Somorjai, The Physical Basis for Heterogeneous Catalysis (Eds. E. Drauglis and R. I. Jaffee, Plenum Press, NY, 1974), p. 395.
76. K. Besocke and H. Wagner, Surface Science 87, 457 (1979).
77. H. Wagner, in Springer Tracts of Modern Physics Vol. 85 (Eds. G. Hohler and E. A. Nickisch, Springer, Heidelberg, 1979).
78. K. F. Freed and D. F. Heller, 56, 4155 (1972).
79. V. Seshadri and V. M. Kenkre, Phys. Rev. A17, 223 (1978).

THE KINETICS OF ELEMENTARY REACTIONS ON SINGLE-CRYSTAL SURFACES

Robert J. Madix
Stanford University, Stanford, CA. 94305

ABSTRACT

The techniques used to study the kinetics of reaction on single-crystal surfaces are briefly summarized. Several elementary reactions including CO adsorption, desorption and oxidation are discussed in terms of transition state theory and detailed balance arguments. Overall, transition state theory adequately accounts for the observed pre-exponential factors in a variety of reactions. More refinements in the theory are needed.

INTRODUCTION

This conference is indeed a significant one, as even in recent years very few studies have been concerned with questions regarding reaction kinetics and dynamics on solid surfaces. This meeting brings together both experimentalists and theorists, and it could well serve as the springboard for the development of this field. With such a purpose in mind I would like to present an overview of our present state of knowledge of the kinetics of elementary reaction processes on solid surfaces. All of the results to be described were obtained on single crystal surfaces employing the methods of surface science to characterize surface composition and structure. (1-4)

At the outset it is important to realize that the surfaces on which reactions take place are two dimensional lattice networks and exhibit specific binding sites for adsorbed species. These sites may be, for example, a four-fold hollow site, a two-fold bridging site or an a-top site on a surface with a square array of atoms, as shown as sites A, B and C, respectively in Figure 1. Though differences in binding sites will not be considered specifically in this article, it is important to note that their presence forces one to treat the adsorbed phase as a two-dimensional lattice gas with its corresponding complexities (5). Example of the importance of lattice gas effects will be given below.

There are, of course, a large number of interesting subjects to be addressed within the framework of surface reactivity. The first of these is the adsorption, and, by microscopic reversibility, desorption event. The dynamics of adsorption are poorly understood, and a phenomenological description employing the so-called precursor state (6) has been developed extensively. For a description of the reverse event, desorption, transition state theory holds promise, but little is known about the partition function of the adsorbed state, and even less is known about the transition states. In one case, however, the desorption of CO, a fairly complete understanding is emerging, as will be discussed below. For more complex reaction events on surfaces the elementary steps themselves must be isolated

ISSN:0094-243X/80/610039-18$1.50

and identified. One may then study reaction events such as unimolecular decompositions, bimolecular reactions, and disproportionations.

As a framework for interpreting experimental results, transition state theory seems most appropriate at this time. Presently, the question of primary concern is whether or not the order of magnitude observed for entropies of activation agree with those "predicted" by simple transition state arguments. More refined theories must surely follow. One could also simply ask whether or not the pre-exponential factors for reactions on surfaces agree reasonably with those observed for analogous gas phase reactions, the important point being whether or not the experience and intuition developed in gas phase kinetics translates to reactions on surfaces. Nonetheless, the formalized transition state approach will be employed here. The rate constant, $\vec{k}$, is given by

$$\vec{k} = \kappa \frac{kT}{h} \frac{f^{\neq}}{\pi_i f_{i,a}^{\nu_i}} e^{-\Delta\varepsilon_o/kT} \tag{1}$$

where these symbols have their usual significance, and the subscript a denotes the adsorbed state. Johnston (7) has carefully developed this subject, and the reader should note that the partition functions included in Equ. (1) appear for each degree of freedom in the transition state not involved as a reaction coordinate. The transmission coefficient can be viewed as a correction factor to account for the events in which reaction trajectories in phase space having the correct transition state configuration do not result in product formation. The partition function for the adsorbed state must take into account the lattice gas features of the surface.

As an introduction some of the salient results of recent studies will be summarized before proceeding with the details.

1. Desorption of CO from metals shows a frequency factor appreciably greater than $10^{13} s^{-1}$. This effect is due to the high entropy of the transition state for desorption in which the CO molecule is effectively a two-dimensional gas. Lattice gas effects for adsorbed CO must be considered in the adsorbed phase for a realistic description of the desorption process.

2. Unimolecular decompositions show frequency factors from 10^{12} to $10^{16} s^{-1}$. These values can be rationalized on the basis of simple transition state theory. Values of 10^{12}— $10^{14} s^{-1}$ are observed for cyclic transition states.

3. Bimolecular reactions occur between species bound to the surface. No evidence has been presented to support "spectator stripping" reactions in which a gaseous molecule reacts directly with an adsorbed species. In the case of atomic collisions, such processes may occur (8,9).

4. Random walk of adsorbed reactants to reactive centers occurs after adsorption, particularly when one of the reactant species is relatively immobile.

EXPERIMENTAL

In addition to precluding the adsorption of unwanted species during reaction the use of low ambient pressures assures that secondary encounters of reaction products with the surface do not occur. Typically, products are observed mass spectrometrically in line-of-sight with the surface, and those products not entering the mass spectrometer are removed by pumping. Proper calibration of the experimental apparatus allows one to quantitatively determine the amounts of products formed (10,11). This is an important consideration for identifying reaction intermediates.

There are five experimental methods currently in use to study the kinetics of surface reactions. These methods are: (1) a steady state molecular beam or static (batch) reactor (12, 13), (2) pressure jump (14), (3) molecular beam relaxation spectroscopy (MBRS) (15, 16), (4) temperature programmed reaction spectroscopy (TPRS) (17), and (5) temperature jump (18). Steady state measurements are useful in measuring overall reaction probabilities, but generally they reveal little about the reaction mechanism after adsorption occurs. Pressure jump techniques are useful for studying bimolecular reactions. Reactions can be studied over a wide range of temperatures below the desorption temperature of one of the reactants. Reactant coverages between zero and unity (fraction of saturation coverage) are accessible, and reaction times between one and 100 seconds can easily be measured. To reach reaction times between 10^{-5} and 10^{-1}s molecular beam modulation is employed to provide periodic pressure variation. The temperature range over which such studies are conducted can be relatively narrow for accurate studies. Reactant concentrations as low as 10^{-4} molecules/cm^2 can be studied, so this method allows one to study the "ideal gas" limit of reaction behavior. MBRS and pressure jump, when combined, provide a range of surface reaction times from 10^{-5} to 10^{2}s.

The kinetic methods employing temperature transients require prior adsorption of the reactants. This is normally done well below the desorption temperature of the reactant, though this need not be the case if stable reaction intermediates are formed upon adsorption at higher temperatures (17). A temperature jump then produces an isothermal reaction to form products. This technique is quite useful for studying reactions which change in order as the surface concentration decreases (18). For complex surface reactions heating the surface linearly with time (TPRS) easily separates reaction channels differing as little as 1 kcal/gmole in activation energy. As the temperature is ramped, reactions activate approximately when $T = 16E_{act}$(kcal/gmole). Elementary surface reactions can be identified, and several methods can be used to extract accurate rate constants for elementary reaction steps. This powerful method is described in detail elsewhere (17,19).

RESULTS AND DISCUSSION

The selected results compiled in Table I illustrate several trends of the kinetics of surface reactivity. Since the covalent solids silicon and germanium are well-represented by localized dangling orbitals at the surface, their reactions with small molecules are of great interest. Atomic oxygen reacts readily, showing a reaction probability of about 0.5 (20). By generating the atomic beam with a radio-frequency discharge, the relative reactivity of atomic and molecular oxygen was readily measured. Molecular oxygen showed a much lower reaction probability with a very low activation energy (24). This difference was rationalized on the basis of a reaction trajectory which demanded strong overlap of both oxygen atoms in the molecule with surface dangling bonds in order to produce dissociation. The reaction is very exothermic, and the potential energy surface must exhibit deep wells for the proper O-Ge alignment. The increased reactivity of ozone supports this hypothesis, since the overlap of the terminal oxygens with the adjacent germanium surface atoms is stronger, and the O-O bond strength is also weaker. These differences should ease the geometric restrictions for reaction and broaden the attractive portion of the chemical well, since the same overlap of the O-Ge produces relatively greater weakening of the O-O bond in ozone. In NOCl the weaker bond with the chlorine atom was selectively broken during collision with heated silicone, illustrating the dominance of the effectively longer range interaction of the Cl atom with the surface. Interestingly, NO_2 was completely unreactive (23). Desorption of GeO from the surface subsequent to reaction showed a very high frequency factor of $10^{16} s^{-1}$. This value was explained in terms of loosened rotations and vibrations in the transition state (25). Similarly high values have been recently observed for CO desorption from metals (18, 37, 38).

The adsorption and desorption of molecular CO can be understood rather well by transition state theory. The adsorption of CO occurs with near unity probability on metals, so that the surface dividing reactants and products may be taken as a plane parallel to the solid surface. The transition state is then a two-dimensional gas, and the reaction coordinate is the distance of the molecule from the solid surface. Then

Table I Pre-exponential Factors for Some Elementary Reactions[o]

Elementary Step	Surface	Transition State	$\nu^{\dagger}$	E_{act} kcal/gmole	Ref.
Covalent Surfaces					
$O_{(g)} + Y_{(s)} \rightarrow O_{(a)}$ (Y = Si, Ge)	Ge(111) Si(111)	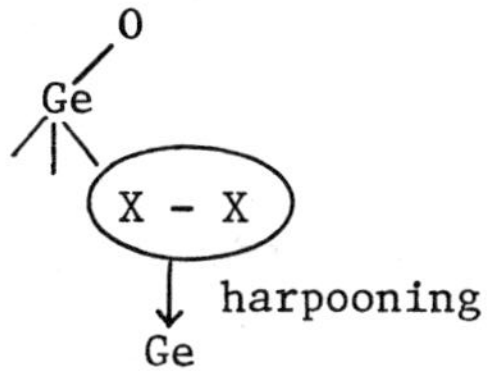	0.5	∿0	(20)
$X_2(Cl_2, Br_2, I_2) + Y_{(s)} \rightarrow 2X_{(a)}$	Ge(111) Si(111)		0.3	∿0	(21) (22)
$O_{3(g)} + Ge_{(s)} \rightarrow O_{(a)} + O_{2(g)}$	Ge(111)	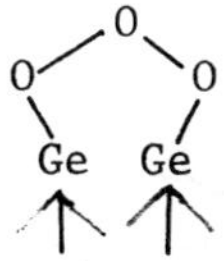	0.3	∿0	
$NO\ Cl_{(g)} + Si_{(s)} \rightarrow NO_{(g)} + Cl_{(a)}$	Si(111)	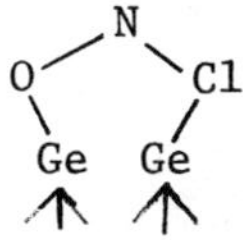	0.3	∿0	(23)
$O_{2(g)} + Y_{(s)} \rightarrow 2O_{(a)}$	Ge(111)	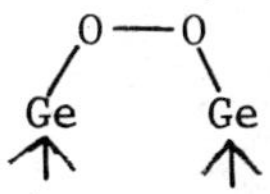	0.02	<1	(24)

o These values are selected from the literature to represent trends.

* The transition state is pictured primarily to show the molecular coordinates involved in the reaction. No precise geometry is intended.

† All values are s^{-1} unless otherwise noted. Values on this page are reaction probabilities.

Table I Pre-exponential Factors for Some Elementary Reactions (cont.)

Elementary Step	Surface	Transition State	ν	E_{act} kcal/gmole	Ref.
Covalent Surfaces					
$GeO_{(a)} \rightarrow GeO_{(g)}$	Ge(111)	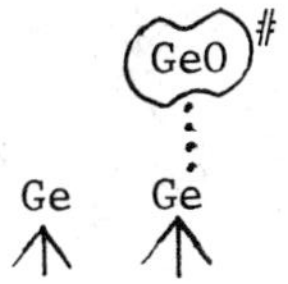	10^{16}	55	(24)
$O_{(g)} + Ge_{(s)} \rightarrow GeO_{(a)}$	Ge(111)	--	--	--	(25)
$Ge_{(s)} + 2\ Cl_{(a)} \rightarrow Ge\ Cl_{2(g)}$	Ge(111)	Surface diffusion	10^{7}	25	(26)
$Si_{(s)} + 2\ Cl_{(a)} \rightarrow Si\ Cl_{2(g)}$	Si(111)	Surface diffusion	10^{8}	40	(26)
$C(graphite) + O_{(a)} \rightarrow CO$	Basal plane	Surface diffusion	10^{7}	30	(27)
$C(graphite) + O_{(a)} \rightarrow CO$	Prism plane	Bulk diffusion and solution	--	--	(28)

# Denotes a 2D-gas

Table I Pre-exponential Factors for Some Elementary Reactions (cont.)

Elementary Step	Surface	Transition State	ν	E_{act} kcal/gmole	Ref.
Metals					
$H_2 + M \rightarrow 2\,H_{(a)}$	Ni(111)	M–H–H–M	0.04^a	∿0	(29)
	Ni(100		0.06^a	∿0	(30)
	Pd(111)		--^{a}	∿0	(31)
	Fe(100)		$10^{-1\,a}$	∿0	(32)
	Cu(110)		$10^{-1\,a}$	3-5	(33)
	Ni(110)(2×1)C		$\sim 10^{-5\,a}$	--	(34)
$CO_{(g)} + M \rightarrow CO_{(a)}$	(approximately unity for all metals)				(35)
$2H_{(a)} \rightarrow H_{2(g)}$	Ni(100)	M–H–H–M	0.16 cm^2s^{-1}		(30)
	Ni(111)		0.4 cm^2s^{-1}		(32)
	Pd(111)		--		(31)
	Pt(110)		0.4 cm^2s^{-1}		(36)
	Fe(100)		10^{-1} cm^2s^{-1}		(32)
	Cu(110)		10^{-6} cm^2s^{-1}		(45)
$CO_{(a)} \rightarrow CO_{(g)}$	Ni(110)	(C ≡ O)$^{\#}$ over M	10^{15-16}		(37,38)
	Ru(0001)		10^{16-18}		(18)

a: these values are reaction probabilities.

Table I Pre-exponential Factors for Some Elementary Reactions (cont.)

Elementary Step	Surface	Transition State	ν	E_{act} kcal/gmole	Ref.
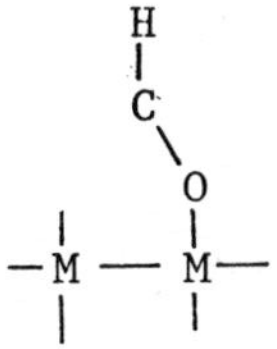 → $CO_{2(g)}$ + $H_{(a)}$	Cu(110)	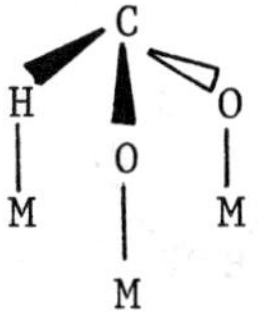	10^{14}	32	(39)
	Ag(110		10^{15}	30	(40)
	Ni(100)p(2×2)C		10^{15}	28	(41)
	Fe(100)		10^{14}	31	(42)
	W(100)(5×1)C		10^{14}	36	(43)
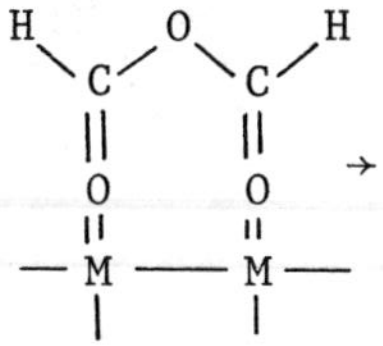 → $2H_{(a)}$ + $CO_{(a)}$ + $CO_{2(g)}$	Ni(110)	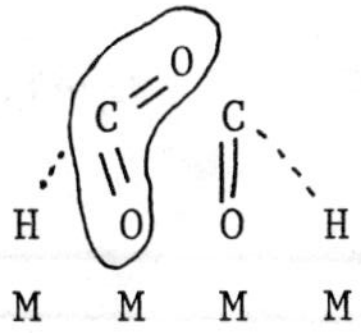	10^{16}	25.5	(44)
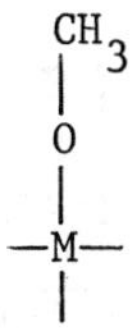 → $H_2CO_{(g)}$ + $H_{(a)}$	Cu(110)	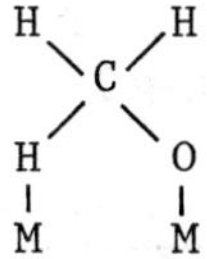	$5x10^{12}$	22.1	(45)

Table I Pre-exponential Factors for Some Elementary Reactions (cont.)

Elementary Step		Surface	Transition State	ν	E_{act} kcal/gmole	Ref.
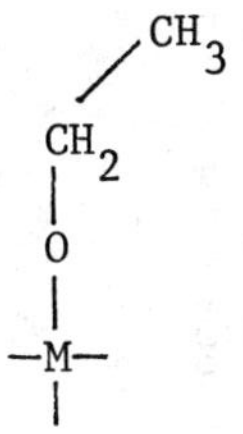	$\rightarrow CH_3CHO_{(g)} + H_{(a)}$	Cu(110)	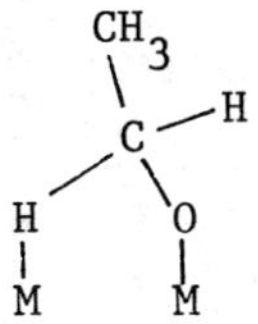	$5x10^{13}$	20.4	(46)
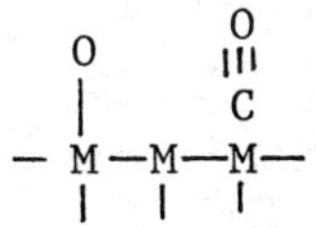	$\rightarrow CO_{2(g)}$	Ag(110)	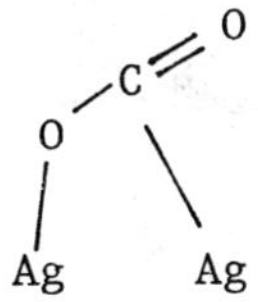	$10^{-3}\ cm^2s$	6	(47)
Xe M M	$Xe_{(a)} \rightarrow Xe_{(g)}$	W(100)	---	10^{16}		(48)

$$\vec{k}_{ads} = \frac{kT}{h} \frac{f_{tr}^2 \, f_{rot}^2 \, f_{vib}}{f_{tr}^3 \, f_{rot}^2 \, f_{vib}} e^{-\varepsilon_o/kT} \quad . \tag{2}$$

Since adsorption is unactivated

$$k_{ads} = \frac{P}{\sqrt{2\pi m \, kT}} \quad , \tag{3}$$

the kinetic theory collision rate. That is, the adsorption probability is one. From this choice of transition state the desorption frequency factor can be calculated. In the limit of zero coverage by CO

$$f_a = F_{vib} N_s \tag{4}$$

where F_{vib} is the total vibrational partition function for the adsorbed species, and N_s is the adsorption site density (5, 49, 50). The rate constant for desorption then becomes

$$\vec{k}_{des} = \frac{kT}{h} \frac{f_{tr}^2 \, f_{rot}^2 \, f_{vib}}{F_{vib} \, N_s} e^{-\varepsilon_o/kT} \tag{5}$$

For unactivated adsorption $\varepsilon_o = \Delta E_{ads}$, and values of ν_{des} as high as 10^{18} s^{-1} are possible. Weak vibrational modes due to frustrated translations and rotations appear to reduce the actual values to $10^{14} - 10^{16}$ s^{-1}.

The values of the reaction probabilities for H_2 on metals (listed as ν in Table 1) indicate that the two dimensional gas transition surface does not adequately describe the reaction event. Choosing this transition surface yields a value of the transmission coefficient of 10^{-1} for most metals (51). This lower value of κ may be due to the higher dimensionality of the potential surface which requires H-H bond breakage. Strangely low ν_{ads} values have been observed for copper (33), Ni(110)(2×1)C(34) and W(100)(5×1)C(52). These very low values are apparently due to stringent geometric requirements of the transition state (53), forcing the reactant to climb the repulsive barrier at the surface in order to react.

The pre-exponential factors for the more complex reactions listed in Table I show excellent agreement with transition state analysis. Cyclic transition states show ν values of $O(10^{12}-10^{14})$.

These values are as expected from reactions of small organic molecules in the gas phase (54). More refined analysis of differences in ν must await more precise measurements of the kinetic parameters. It is interesting to note that large entropy effects due to lattice degeneracies in the adsorbed state do not manifest themselves. This result is presumably due to the fact that the transition states are not appreciably more mobile then the adsorbed state, and the lattice-associated entropies cancel.

One of the simplest and most widely studied bimolecular reactions is the oxidation of carbon monoxide (55). Using MBRS, Engel and Ertl (56) have shown that the reaction takes place between oxygen atoms and carbon monoxide molecules bound to the surface. This reaction was recently studied using the pressure jump method on Ag(110) with similar results (48). Oxygen was preadsorbed on the silver surface, and the crystal then suddenly rotated into a beam of CO. The rate of evolution of CO_2 as a function of time is shown in Fig. (2) at two different temperatures. Remarkably, the rate increased with decreasing temperature. The rate of oxygen consumption decreased with time according to bimolecular kinetics, as shown clearly in Fig. 3. The slope of the straight lines in Fig. 3 gives the apparent rate constant, which involves competitive desorption and reaction of CO, and

$$\vec{k}_{app} = -\frac{S \cdot F\, k_r}{k_d} = \nu_{app}\, e^{-\varepsilon_{app}/RT}$$

where S is the adsorption probability of CO, F is the CO flux, and k_r and k_d are the rate constants for reaction and desorption, respectively. The values of ν_{app} and ε_{app} were $2\times 10^{-3}\,s^{-1}$ and -4.8 kJ mole^{-1}. Extraction of the values of ν_r and E_r rests upon the evaluation of S, F, ν_d and E_d. The flux was calibrated, and S was taken to be of order unity (this is a very reasonable assumption, as discussed above). Since CO adsorption is generally unactivated, E_d can be obtained from adsorption isotherms, if they exist. Actually only isotherms for Ag(111) have been published (57), and they were used to evaluate both $E_{ads} = E_d$ and f_a, the partition function of the adsorbed species. The value of ν_d was calculated from

$$\nu_d = \frac{kT}{h} \frac{f_{tr}^2\, f_{rot}^2\, f_{vib}}{f_a} \tag{6}$$

in accordance with the previous discussion. In this fashion $E_r = 23$ kJ mol^{-1} and $\nu_r = 0(10^{-3}s^{-1})$. This value of ν_r is what is expected for random walk of the adsorbed CO to the oxygen atoms for reaction. It is extremely unlikely that physically adsorbed CO was involved in the reaction, since a negative activation energy was observed. If physically adsorbed CO were reacting with adsorbed oxygen, E_r must be less than about 10 kJ mol^{-1}, which seems impossible for a CO molecule with unperturbed bonds reacting with an oxygen atom bound to the surface by more than 200 kJ mol^{-1}. It must be concluded that reaction occurs via two bound reactants.

Similar results were obtained for the CO oxidation on a stepped platinum surface Pt[9(111)(100)] (58). A steady beam of molecular oxygen of 10^{16} molecules cm^{-2}s^{-1} was used to linearize the reaction. A modulated beam of CO (10^{13-14} molecules cm^{-2}s^{-1}) reacted with the surface oxygen to form CO_2, which was detected using lock-in detection methods. The in-and-out-of-phase components of the first coefficient of the carbon monoxide are shown in Fig. 4 for reactive and unreactive scattering of CO from this surface. In the absence of surface oxygen the low temperature limit of I_o is a direct measure of the adsorption probability, S. In this case S = 0.7. The in-phase component decreased sharply at the same temperature that the out-of-phase component went through as maximum. The desorption rate constant was 10^{15} exp{-36, 200(cal/gmole)RT}s^{-1}. In the presence of adsorbed oxygen all of the CO which adsorbed reacted to form CO_2, as evidenced by the absence of any out-of-phase CO component. The rate constant for CO_2 formation was measured as 1.5×10^{-7} exp {9,700 cal/gmole/RT} cm^{2}s^{-1}.

The pre-exponential factor for the reaction $CO_{(a)} + O_{(a)} \rightarrow CO_2$ can be understood in terms of transition state theory. The adsorption of CO_2 on Pt(111) is known to be activated. Even at translational energies of 10 kcal/gmole adsorption does not occur (59), and it is very likely that internal modes must be activated to promote surface reaction. The existing evidence strongly suggests that a close collision is required to dissociate CO_2 to $CO_{(a)}$ and $O_{(a)}$, placing the transition state very near the products. The transition state must then be a lattice gas without appreciable entropy beyond lattice degeneracy. If weak vibrations are included in the partition function for adsorbed CO (as calculated from CO desorption from the clean surface), the pre-exponential factor is calculated to be 10^{-6}cm^{2}s^{-1}. This agreement is reasonable.

SUMMARY

Studies of surface reactivity have yielded kinetic rate constants for a wide variety of events. Most of the measured values can be rationalized by transition state theory. Further refinements are needed in both experiment and theory in order to develop a rigorous understanding of the subject.

ACKNOWLEDGEMENTS

The author gratefully acknowledges the support of the National Science Foundation (NSF Eng 12964) during the preparation of this paper.

REFERENCES

1. Physics Today, April, 1975.
2. G. A. Somorjai and L. L. Kesmodel, Int. Rev. of Science Phys. Chem. Ser. 2, Vol. 7 (1975) 1.
3. G. Ertl and J. Küppers, Low Energy Electrons and Surface Chemistry (Springer, N.Y., 1974).
4. R. Gomer, Ed., Topics in Applied Physics 4, Springer-Verlag, New York (1975).
5. T. Hill, Introduction to Statistical Mechanics (Addison-Wesley, Menlo Park, Calif., 1960).
6. L. D. Schmidt (this volume).
7. H. S. Johnston, Gas Phase Reaction Rate Theory (Ronald Press, N.Y., 1966).
8. G. A. Melin and R. J. Madix, Trans. For. Soc. 67, 198, 2711 (1971).
9. H. Wise and B. J. Wood, Adv. in Atomic and Molecular Physics, 3, 291 (1967).
10. J. L. Falconer and R. J. Madix, J. Catalysis 48, 262 (1977).
11. E. Ko, J. B. Benziger and R. J. Madix, J. Catalysis, in press.
12. J. B. Anderson, R. P. Andres and J. B. Fenn, Molecular Beams, Ed. J. Ross (Interscience, N. Y., 1966).
13. H. P. Bonzel and R. Ku, Surf. Sci. 33, 91 (1972).
14. M. Bowker, M. A. Barteau and R. J. Madix, to be published in Surface Science.
15. D. R. Olander, Structure and Chemistry of Solid Surfaces, Ed. G. A. Somorjai (John Wiley, N.Y. 1969).
16. R.J. Madix and J. A. Schwarz, Surf. Sci. 24, 264 (1971).
17. R. J. Madix, Chemistry and Physics of Solid Interfaces, Ed. R. Vanselow, Vol. II (CRC Press, Boca Raton, Fla., 1979), p. 63.
18. H. A. Engelhardt and D. Menzel, Surf. Sci. 57, 591 (1976).
19. R. J. Madix, Proceedings of the Second European Congress on Surface Science (Cambridge, Mass., March, 1979). In press.
20. R. J. Madix and A. A. Susu, Surf. Sci. 20, 377 (1970),
21. R. J. Madix and A. A. Susu, J. Catalysis, 28, 316 (1973).
22. R. J. Madix, R. Parks, A. A. Susu and J. A. Schwarz, Surf. Sci. 24, 288 (1971).

23. D. Ying and R. J. Madix, unpublished data.
24. R. J. Madix and M. Boudart, J. Catalysis 7, 240 (1967).
25. R. J. Madix, G. A. Melin and A. A. Susu, Entropie (Nov./Dec. 1969), p. 70.
26. R. J. Madix and J. A. Schwarz, Surf. Sci. 24, 264 (1971).
27. D.R. Olander, R. H. Jones, J. A. Schwarz and W. J. Siekhaus, J. Chem. Phys. 57, 421 (1972).
28. D. R. Olander, W. Siekhaus, R. Jones and J. A. Schwarz, J. Chem. Phys. 57, 408 (1972).
29. K. Christman, O. Schober and G. Erbl, J. Chem. Phys. 60, 4719 (1974).
30. J. Lapujaulade and K. S. Neil, J. Chem. Phys. 57, 3535 (1972).
31. H. Conrad, G Ertl and E. E. Latta, Surf. Sci. 41, 435 (1974).
32. J. B. Benziger and R. J. Madix, submitted to Surface Science.
33. M. Balooch, M. Cardillo, D. R. Miller and R. E. Stickney, Surf. Sci. 46, 35 (1974).
34. J. McCarty and R. J. Madix, J. Catalysis 38, 402 (1975).
35. R. J. Madix and J. Benziger, Ann. Rev. Phys. Chem. 29, 285 (1978).
36. J. A. Fair and R. J. Madix, to be published.
37. J. L. Falconer and R. J. Madix, Surf. Sci. 48, 393 (1975).
38. C. R. Helms and R. J. Madix, Surf. Sci. 52, 677 (1975).
39. D. H. S. Ying and R. J. Madix, in press.
40. M. A. Barteau, M. Bowker and R. J. Madix, submitted to Surface Science.
41. E. I. Ko and R. J. Madix, App. Surf. Sci. 3.
42. J. B. Benziger and R. J. Madix, to be published.
43. J. B. Benziger, E. I. Ko and and R. J. Madix, J. Catalysis, in press.
44. J. L. Falconer and R. J. Madix, Surf. Sci. 46, 473 (1974).
45. J. E. Wachs and R. J. Madix, J. Catalysis 53, 208 (1978).
46. I. E. Wachs and R. J. Madix, App. Surf. Sci. 1, 303 (1978).
47. M. Bowker, M. A. Barteau and R. J. Madix, to be published in Surface Science.
48. T. E. Madey and J. Yates, J. Chem. Phys.
49. J. Fair, to be published.
50. R. J. Madix, K. Christman and R. J. Madix, Chem. Phys. Letters 62, 38 (1979).
51. J. B. Benziger, E. I. Ko and R. J. Madix, J. Catalysis 54, 414 (1978).
52. A. Gelb and M. J. Cardillo, Surf. Sci. 64, 197 (1977).
53. S. W. Benson, Thermochemical Kinetics (Wiley-Interscience, N.Y., 1976).
54. T. E. Engel and G. Ertl, Adv. in Catalysis (1979).
55. T. E. Engel and G. Ertl, Proc. 7th Ind. Congress on Solid Surfaces, Vienna, 1978.
56. G. McElhiney, H. Papp and J. Pritchard, Surf. Sci. 54, 617 (1976).
57. J. Fair and R. J. Madix, to be published.
58. M. J. Cardillo, private communication.

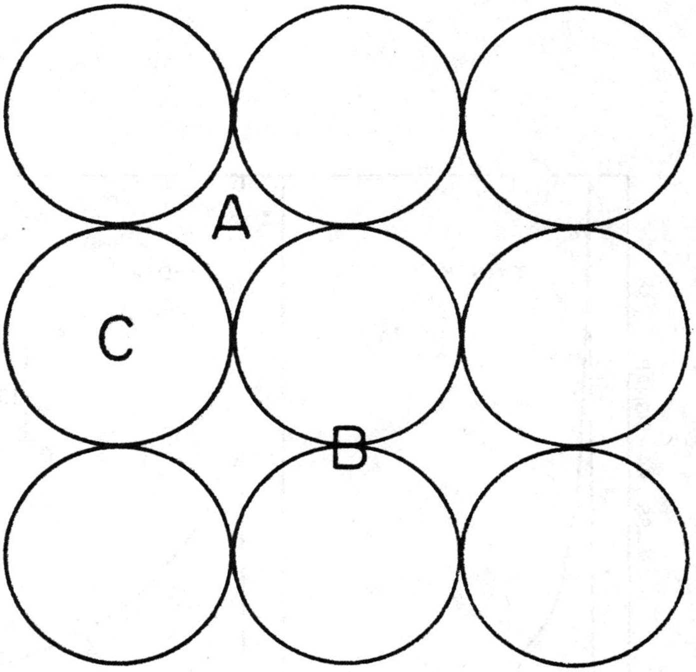

Fig. 1. Schematic of the (100) face of a face-centered cubic metal showing A. four-fold hollow, B. 2-fold bridge and C. a-top sites, respectively.

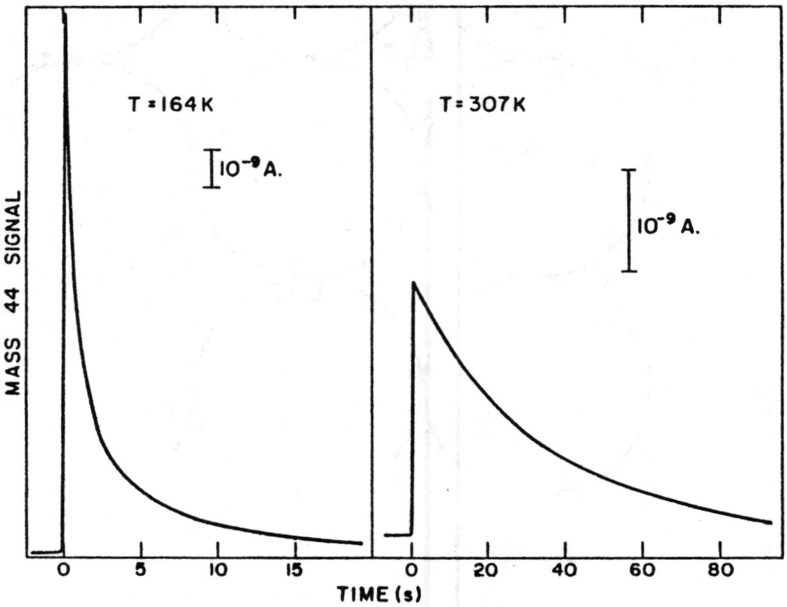

Fig. 2. Exponentially decaying rates of CO_2 oxidation at two temperatures following a pressure jump of CO into a silver (110) surface with preadsorbed oxygen. A negative activation energy was exhibited.

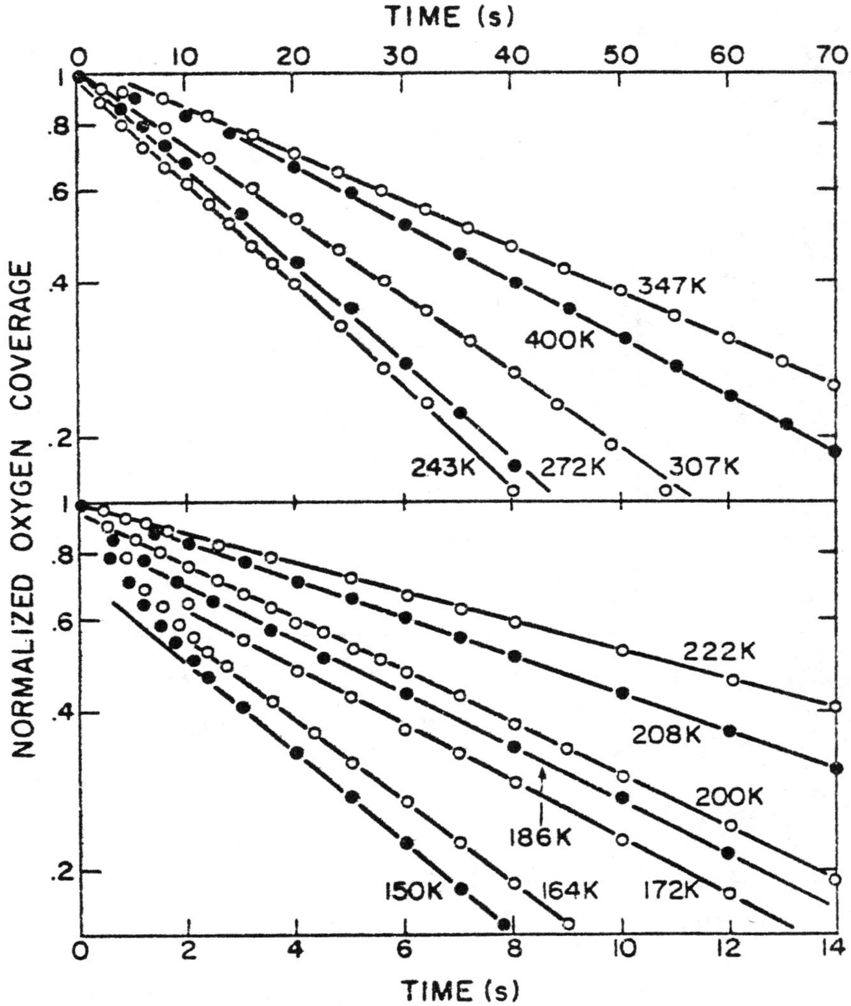

Fig. 3. Plot of the log of the rate of CO_2 production <u>vs</u> time for several temperatures.

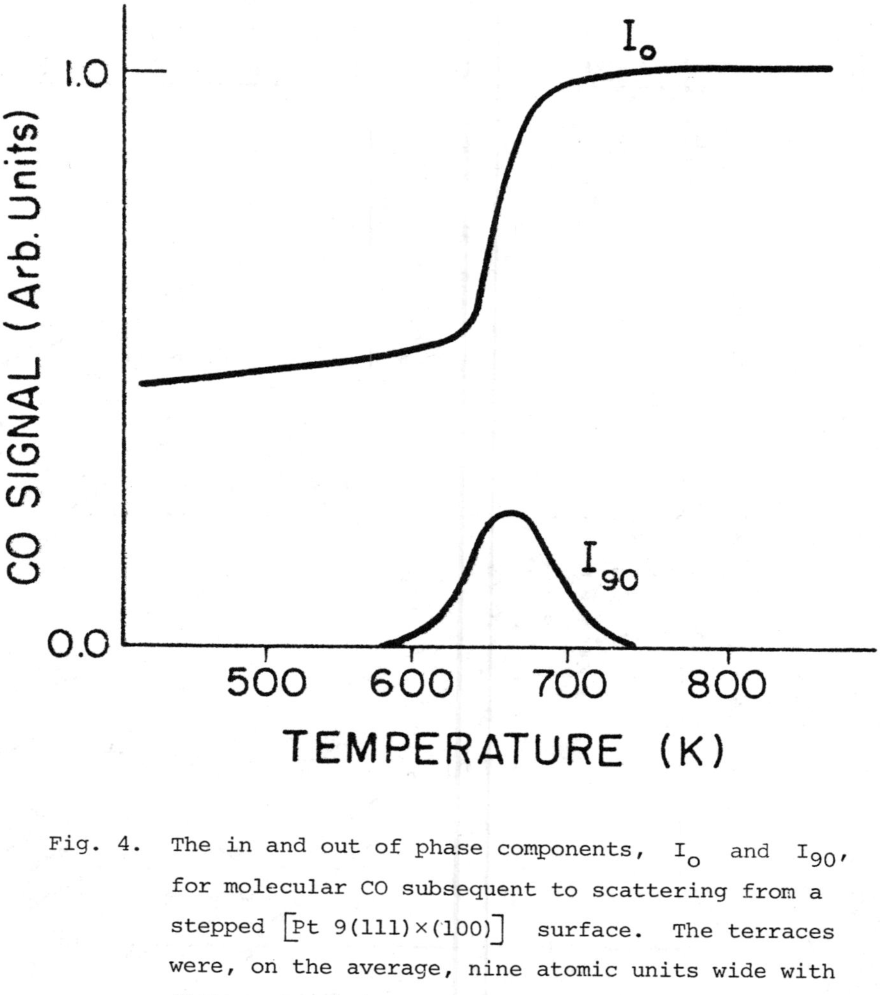

Fig. 4. The in and out of phase components, I_0 and I_{90}, for molecular CO subsequent to scattering from a stepped [Pt 9(111)×(100)] surface. The terraces were, on the average, nine atomic units wide with (111) orientation and the monatomic steps were of (100) orientation.

SOME ASPECTS OF OXYGEN ADSORPTION AND INITIAL OXIDATION OF SINGLE CRYSTAL Ni, Fe AND Ni/Fe ALLOY SURFACES

C. R. Brundle
IBM Research Laboratory, San Jose, California 95193

ABSTRACT

Some aspects of oxygen interaction with Ni(100), Fe(100) and Ni/Fe(100) surfaces are reviewed. Emphasis is on the kinetics in relationship to the mechanism of adsorption; the effects of order/disorder phenomena; the distinction between adsorption and oxidation; and in the case of the alloy the role of surface segregation. The analysis techniques concerned are Auger, XPS, UPS, LEED and X-ray fluorescence.

INTRODUCTION

The objective of this paper is to review the behavior of well-defined Ni, Fe and a Ni/Fe alloy surfaces toward oxygen in terms of oxygen uptake kinetics and the nature of the reaction products. An attempt is made to understand the kinetics in terms of the proposed adsorption mechanism; the progression from the chemisorption to the oxidation stage; and the nature of the reaction products. The techniques used by the present author are mainly X-ray photoemission (XPS) and Low Energy Electron Diffraction (LEED). The earlier work reviewed also includes Auger, Reflection High Energy Diffraction (RHEED), and X-ray fluorescence results. XPS is used to provide a surface elemental analysis, to identify the chemical nature of the products, and to provide some means of depth distribution analysis. Ultraviolet photoemission (UPS) is also used to identify chemical changes in the metal substrate. LEED is used to monitor the geometric nature of any ordered adsorbed oxygen structures or oxide products.

The main area addressed is the behavior of the sticking probability, s, as a function of oxygen coverage, θ, at temperature, T. θ is defined as unity when there is one adsorbed oxygen atom per surface metal atom. For the FCC structures of Ni(100) and Ni/Fe(100) this corresponds to 1.6×10^{15} atoms sq. cm. For the BCC structure of Fe(100) this corresponds to 1.22×10^{15} atoms sq. cm. s is followed from a θ of zero through to the "passivation" stage. Under the UHV conditions used and the oxygen pressures used ($<10^{-4}$ Torr) "passivation" corresponds to between 2-3 monolayers for the surfaces (possibly higher for Fe(100)). This does not imply that oxide growth has ceased, merely that it is now very slow ($s<10^{-6}$) such that the further thickening of the oxide layer does not occur within any reasonable time scale under the conditions used. This represents the start of the "tarnishing" stage which is usually the starting point for corrosion studies at high pressure and temperature. The discussion of s as $f(\theta,T)$ leads us into considerations of models for the chemisorption and oxidation processes and therefore also of information from XPS, UPS and LEED concerning the nature of the products as a function of θ.

The majority of the discussion centers around $Ni(100)/O_2$ since it is the best studied system. The less well-studied $Fe(100)/O_2$ system is then compared to the Ni(100) case and explanations for the differences proposed. Finally Ni(75%)/Fe(25%)(100) is discussed.

The type of surface physics instrumentation necessary for these studies is well-known and we will not review the requirements here. Relative oxygen surface coverages are obtained usually from XPS, or Auger, oxygen peak intensities. Absolute coverages are usually obtained by normalization at critical coverages (such as a well-ordered LEED structure or saturation coverage) for which other techniques have established absolute values. Elemental depth distributions can be estimated in XPS by varying the detected photoemission ejection angle, ϕ, between outgoing photoelectron and the crystal surface.[1] At low ϕ the surface sensitivity of the technique increases. Chemical identification of the reaction products are attempted using XPS and UPS chemical shifts and valence band changes. In some of the work reported here, Secondary Ion Mass Spectroscopy (SIMS) is used to distinguish OH surface species from O.[2] LEED is, of course, used to help establish adsorbate-substrate geometries.

Ni(100) OXIDATION

The primary reference for the interaction of oxygen with Ni(100) is the LEED/Auger study of Holloway and Hudson (referred to hereafter as HH).[3] Their results are summarized in Fig. 1 where the growth of the O KLL Auger signal is plotted as a function of oxygen exposure at 302K (Fig. 1a) and 147K (Fig. 1b). One can see that at 302K the initial sticking probability, s_0, is high (assumed to be unity and justified by several procedures in their paper), but it drops rapidly. Between 2L and ~25L the oxygen uptake is very slow, and then it rises again, only to finally fall off and effectively cease by ~120L. At a pressure of 10^{-6} Torr a further exposure of 10^5L does not measurably increase the signal. In fact for pressures below 10^{-5} Torr the uptake curves are dependent only on exposure and temperature, and not on pressure. The equivalent curve at 147K shows that s_0 is also unity, but it remains high much longer than at 302K and then oxygen uptake ceases sharply at the same saturation coverage as for 302K. The maximum intensity regions of the two LEED O overlayer patterns observed at 302K, p(2×2)O; c(2×2)O; plus the onset of the p(1×1)NiO are also marked in the figure (only very diffuse patterns were discernable at 147K). A schematic representation of the real-space geometries they correspond to are shown in Fig. 2. (The location of the oxygen atoms in the four-fold hollow for the c(2×2) structure at a d-spacing of 0.9Å was not available from LEED intensity measurements and calculations[4] till some time later than the Holloway and Hudson study. It is assumed that the p(2×2) structure has the same oxygen

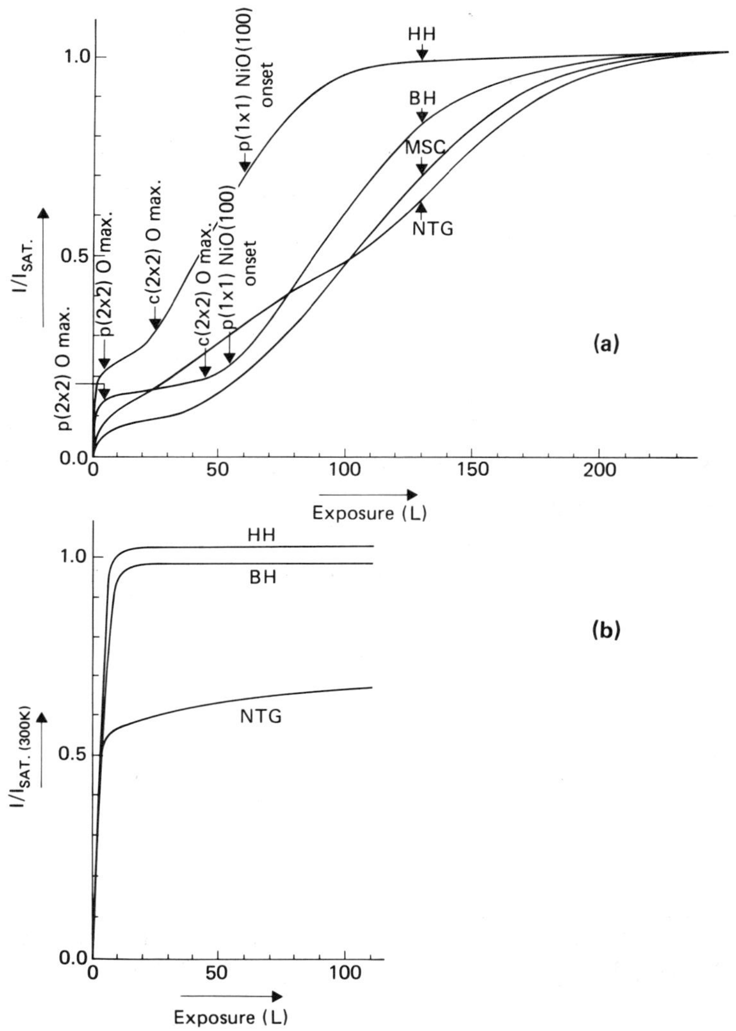

Figure 1. Oxygen uptake curves for $Ni(100)/O_2$ as a function of exposure measured by O KLL Auger intensity (HH[3]); X-ray fluorescence (MSC[5]); and XPS O1s intensity (NTG[6] and BH[7]).

(a) Ambient temperature

(b) 77K (NTG and BH) and 147K (HH)

NOTE: The normalized signal intensities, I/I_{sat}, are only directly proportional to oxygen atom concentration for X-ray fluorescence (MSC[5]). For Auger and XPS I/I_{sat} is proportional to oxygen atom concentration at low coverages, provided there is no lattice penetration (i.e., up to the plateau); for higher coverages allowance must be made for signal attenuation by inelastic electron scattering when converting to an oxygen atom concentration scale.

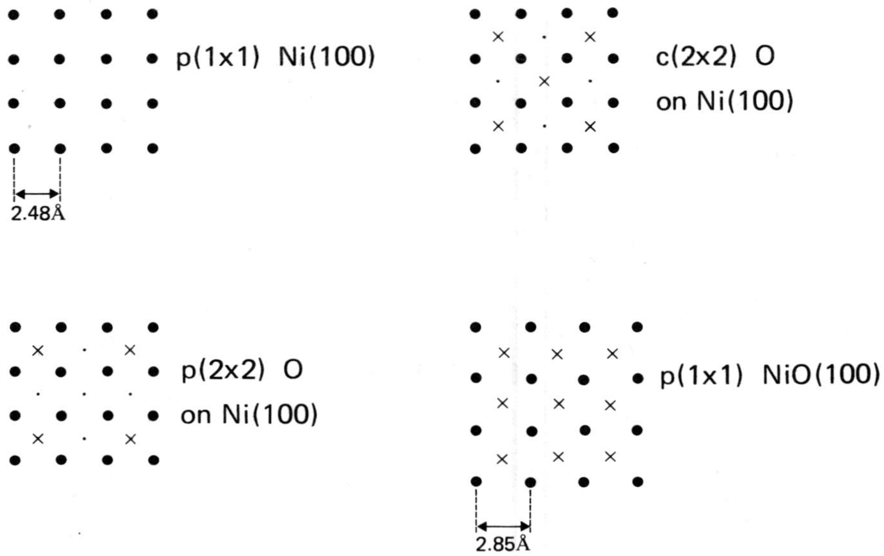

Figure 2. Schematic of Ni(100) lattice; the chemisorbed superstructures of oxygen; and the NiO(100) lattice. The location of the chemisorption site as the four-fold hollow is from LEED intensity measurements.[4]

location and d-spacing. The identification of NiO in the later stages of the reaction by LEED is straightforward since the contracted p(1×1) pattern corresponding to a ~15% expansion of the Ni-Ni spacing in the NiO(100) surface is clearly observed.) There is, however, considerable overlap of the regions in which the different structures were observed. Figure 3 converts Fig. 1 into a plot of s/s_0 against oxygen coverage, θ. Since θ is expected to be linear with Auger intensity in the adsorption regime, the θ scale up to s_{min} is obtained using the assumption that $s_0=1$ and knowing the oxygen exposures. Beyond s_{min} an increasingly large signal attenuation correction is required to convert the apparent coverage, θ', into a true coverage. HH estimated that θ_{sat} was 1.6 allowing for the attenuation corrections. Assuming that the oxide formed by them is NiO(100) this converts to 2·1 atomic layers of oxide (we will see later that this is not an entirely correct assumption).

Several important facts were established from the HH study:

1. Oxygen adsorption is dissociative throughout the whole process. (Actually the evidence offered for this is quite weak--the correlation of the maximum in the p(2×2) LEED pattern to $\theta \simeq 0.25$ oxygen atoms. The proof that the statement is correct comes from later XPS and UPS results which never show evidence for a molecular oxygen chemisorbed structure.)

2. At room temperature and higher the decrease in s was clearly associated with the chemisorption regimes prior to any oxide formation. The minimum in s occurs around 0.35 monolayer coverage, based on the assumption that $s_0=1$.

3. At room temperature and higher the subsequent increase in s was associated with the formation of NiO, since NiO was observed directly by LEED. In fact in this study s started to increase before the NiO LEED pattern was observed. Our own XPS and LEED work, however, correlates the onset of oxidation more closely with the rise in s.[2,7]

4. The coverage scale was approximately established. In particular the θ_{sat} value of ~2 layer of nucleated NiO was the first qualitatively reliable estimate for the Ni(100) surface.

5. At low enough temperature s drops very little with increasing θ until monolayer coverage is approached. The adsorption and oxidation reactions thus appear to be proceeding differently here.

HH interpreted these facts to imply a two stage mechanism for the oxygen interaction; dissociative Langmuir chemisorption (dissociation and chemisorption at the impact site; if the site is occupied the only alternative is reflection) while s decreases, followed by oxide nucleation and island growth when s rises again. Dissociative Langmuir kinetics should result in a $(1-\theta)^2$ dependence of s. For dissociative adsorption the molecule must reach a point where there are nearest neighbor empty sites (referred to as nn from now on). For random impact followed by dissociation if an nn site is struck, or reflection if it is not, plus random distribution of the product O atoms, the number of such nn sites will decrease with θ as $(1-\theta)^2$, where θ here represents the fractional coverage of total available chemisorption sites. The actual rate of decrease of s is much faster than $(1-\theta)^2$ assuming the number of available sites equals the number of surface Ni atoms and so clearly does not fit Langmuir statistics in its simplest form. We will see later that the discrepancy can be accounted for by correct consideration of order in the chemisorbed phase which in effect excludes specific adsorption sites. Clearly the low temperature results are far from Langmuir kinetics since s drops very little with increase in θ. HH made no attempt to rationalize this.

The kinetics in the oxide nucleation stage were well reproduced by a model in which two dimensional NiO islands grow across the surface from a fixed number of nucleation sites by capturing diffusing surface oxygen <u>molecules</u> and/or directly impinging oxygen at the island perimeters. The surface molecular oxygen was assumed to be physisorbed oxygen, present at a low steady state concentration. The model predicts that s should rise as the island perimeters increase, and then fall as the islands coalesce. If one assumes the surface diffusion mechanism the introduction of an empirical value for the average jump distance of the diffusing oxygen species of 2.7Å, leads to a value of ~3 kcal for the difference between heats of physisorption and desorption of molecular O_2, together with a nucleation site density of 1 in $\sim 3\times 10^5$ Ni atom sites. Since the low temperature result did not show any rise in s the model obviously can only be fit to the low temperature data in the higher coverage regime where s is falling (see Fig. 3).

Despite the considerable success of the HH work in explaining the adsorption kinetics on Ni(100) a lot of questions remained unanswered (or even unasked at this stage):

1. Since ordered chemisorption structures are formed, indicating lateral interactions, how does this order affect the kinetics?

2. Why is the low temperature data so dramatically different from the higher temperatures?

3. What is the nature of the nucleation sites for two dimensional NiO island growth and how does the nucleation process occur?

Shortly after HH, the LEED analysis of the c(2×2)O/Ni(100) structure[4] provided the location of the chemisorbed overlayer atoms. The d-spacing of 0.9Å indicates that the c(2×2)→p(1×1) NiO(100)

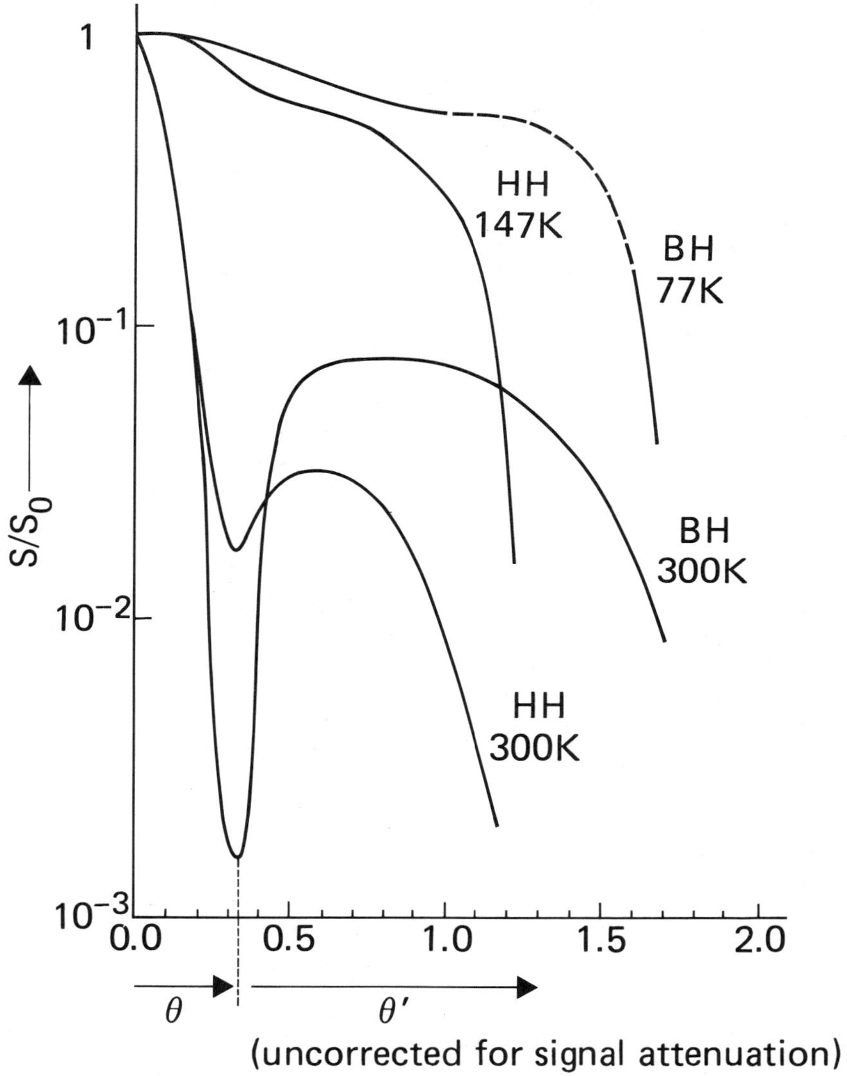

Figure 3. Conversion of data in Fig. 1 to relative sticking probability versus coverage curves. Note that the coverage scale represents true θ up to the onset of oxygen penetration but above that the apparent coverage, θ', must be corrected for attenuation to obtain the true coverage, θ.

conversion requires a vertical (relative) movement of the O atoms of 0.9Å as well as a Ni lattice expansion of ~15%.

The next important study of the system, by Mitchell, Sewell and Cohen (MSC),[5] involved a combined RHEED/X-ray emission study. X-ray emission is not a surface sensitive technique and therefore as conventionally used can give a highly accurate determination of the total oxygen content in a thick oxide film, but does not have sufficient sensitivity for monolayer studies. In the MSC studies, however, they increased the sensitivity by using grazing angle electron incidence, achieving a ~5% monolayer O detection limit. They found the same general kinetics as in the HH study, with some differences. Their plateau region was flatter and lasted to longer exposures (see Fig. 1). Though θ in the X-ray emission technique should be more reliably calibrated than in the Auger study (it does not rely on knowing s_0 and exposures accurately, but uses known thickness oxide standards), the coverages at the end of the plateau region actually agree well (0.35 monolayer in the HH study, 0.33 in the MSC study). MSC also noted that O coverage for the plateau was depressed by the presence of C contaminates in a 1:1 proportion. There is poorer agreement in the oxidation stage. The terminal value of 2.9 NiO layers given in the MSC study is more likely to be correct because there is no necessity to include an attenuation factor for the X-ray fluorescence as there was in the Auger study of HH. The onset of NiO nucleation was detected by RHEED in the MSC study somewhat earlier than in the LEED of the HH study. In addition MSC positively identified by RHEED two morphologies of NiO growing with epitaxial relationships to the Ni(100) substrate, NiO(100) and NiO(111). They also pointed out that both orientations were present in the LEED data of HH, but the NiO(111) spots had been assigned as rotated Ni(100) spots. Finally MSC performed some tests to try and establish the nature of the nucleation process that produced the NiO. They found that heat treatment at any stage after the onset of oxidation had no effect on the oxidation kinetics (i.e., it continued at the rate appropriate to the original temperature). Heating prior to the oxidation onset, however, converted the subsequent kinetics to become that appropriate to the higher temperature. Their interpretation was that proto-nucleation sites exist from the start in a number determined by temperature and that the oxygen atoms which become associated with these are in equilibrium with the chemisorbed phases, the nuclei continually dissolving and reforming until the end of the plateau stage at which point they pass a critical size and grow irreversibly. They do so because all c(2×2) sites are now filled and so diffusing and/or impinging oxygen can only supply the growing nuclei.

The MSC study, then, provides a plausible answer to the question of how the nucleation occurs but does not tell us what the nature of the sites are. In addition it provided a better calibration of the quantities of oxygen involved and showed that two NiO morphologies could be developed.

XPS studies of the Ni/O_2 system are plentiful. There are only two detailed kinetic studies on the (100) surface, however. The first one by Norton, Tapping and Goodale (NTG)[6] and the second by Brundle and Hopster (BH).[7] The studies have been complicated by the observations of two O(1s) features (a major one at ~529.5 eV and a minor one some 1.5 eV higher) the origins of which have been a source of controversy. It is now generally agreed that the ~529.5 eV feature can be characteristic of both chemisorbed overlayer oxygen atoms and of oxygen in NiO. This has been demonstrated conclusively by Hopster and Brundle by examining the LEED at the same time as the XPS as a function of exposure.[8] The origin of the high BE feature is still not agreed upon. NTG ascribe it to a defect oxide containing Ni^{3+} for room temperature oxidation using dry O_2 (following the original suggestion of Kim and Winograd for polycrystalline Ni),[9] and to some protonated oxygen species when H_2O is present along with O_2. They demonstrated, by nuclear microanalysis, that protonated species are present for saturation coverage at 295K followed by several days wet air exposure at atmospheric pressure. Such treatments produce an extremely intense high BE O1s feature. Hopster and Brundle have shown by direct correlation with the OH^- SIMS signal that at 295K the high Be O(1s) in their work is caused by protonation, and that this comes about by co-adsorption of residual H_2O with O_2 at the end of the plateau region in the s versus θ curve, and at the start of the oxidation region.[2,8] (Note that the instrument base pressure is less than 1×10^{-10} Torr during these experiments!) The protonated species can be increased in relative concentration by deliberately increasing the background H_2O pressure, and can be practically eliminated by performing the adsorption run at higher pressure (10^{-6} cf 10^{-9} Torr) which effectively reduces the relative H_2O contamination level. A high BE O1s is also observed during 77K adsorption, with much higher intensity and a slightly different BE. Under these conditions it does not represent OH species, since no OH^- is observed in SIMS.[8]

None of this discussion on whether the species is OH or Ni_2O_3 or whatever has much bearing on the s versus θ curve, however, since the total oxygen uptake and the kinetics is only slightly affected by its formation or absence. It is possible that the morphology of the oxide islands is affected somewhat (LEED), but there is insufficient data to comment further. One point that both NTG and BH agree on is that at no stage does molecular O_2 appear in the XPS (or UPS) spectrum. In the valence band region it would be expected to show molecular O_2-like levels and in the XPS the O1s should be several eV higher BE than for NiO.

The 295K at 77K uptake curves of NTG and BH are also shown in Fig. 1, for comparison with the other results reported. One can see that the general form of the curves in figure is consistent for all four studies. The one serious anomaly is apparently the low saturation limit at 77K obtained by NTG, for which we have no explanation.† Their 300K data also shows a marked additional "kink" which is not present in any of the other data. The other differences are mainly minor ones of degree of flatness in the

---

†In subsequent communications with P. Norton it has become clear that the 77K NTG curve of Fig. 1 refers to Ni(110), not (100). For Ni(100) the results are in agreement with the HH and BH studies.

plateau region (i.e., differences in s_{min}) and subsequent initial oxidation rate. Since we believe the onset of oxidation to be a nucleation phenomena we would ascribe these differences to variations in perfection of the 100 surfaces used in the four studies. That this factor has a huge effect on reaction rate can be established by simply lightly sputtering a well-annealed Ni(100) surface before the adsorption run. The plateau region then starts to be washed out as nucleation occurs earlier and faster than on the well-annealed surfaces. Similar results are obtained for unannealed evaporated Ni films.

It is also likely that the exposure scales are not exactly accurate for all the four studies, since measurement of oxygen pressures by ion gauges is known to be problematic. The variations in the value of I/I_{sat} for the plateau region does not represent a discrepancy between the measurements since I/I_{sat} is only related linearly to oxygen concentrations over the entire uptake in the X-ray fluorescence data of MSC. In the other studies, as stated earlier, signal attenuation by inelastic scattering becomes important in the later oxidation stages because of the short mean free-path lengths of the Auger and photoelectrons. This has the effect of depressing the true (i.e., unattenuated) I_{sat} signal and therefore increasing I/I_{sat} at the plateau stage. The effect should be the same in the two XPS studies, but greater for the Auger study where the KE of the O Auger electron is much less than that for the XPS O(1s) electrons. Qualitatively these are effects observed in Fig. 1a.

Of course the XPS studies can reveal more than simply the kinetics of the oxygen uptake and the proof that the chemisorbed species is dissociated oxygen. The $Ni2p_{3/2}$ signal will, in principle, reveal any significant difference in electronic structure of the surface nickel atoms as a function of oxygen uptake by exhibiting a chemical shift. In practice the fact that the $Ni2p_{3/2}$ signal is observed from several layers depth and not just the top atomic layer makes the identification of chemically shifted components at low oxygen coverages difficult. This can be overcome by working at low angles of electron ejection, ϕ, (with respect to the surface), thereby decreasing the effective escape depth. For the $Ni2p_{3/2}$ level using MgK_{α} radiation the photoelectron KE is ca. 400 eV and the inelastic mean free path length of the order of 8Å. For a ϕ of 45° 90% of the signal is generated from the top 13Å, whereas at ϕ=15° 90% is generated from the top 4Å.[1] Using a low ϕ BH[2,8] showed that at 295K there was no significant change in the electronic structure of the surface Ni atom (i.e., no chemical shift) throughout the chemisorption stage, but that Ni^{2+} became identifiable coincident with the increase in s after the plateau region. In the same study this was also about the point where an NiO LEED pattern first became discernable. From one point of view the identification of Ni^{2+} by XPS is redundant information since LEED establishes it as NiO. On the other hand it shows that the $Ni2p_{3/2}$ XPS signal will not show the characteristics of Ni^{2+} in NiO until nucleation of NiO has occurred--i.e., it is necessary to have Ni^{2+} species completely coordinated by O^{2-} species.

At low temperature no ordered structures are formed so LEED cannot be used to establish when NiO nucleation occurs. We are forced, therefore, to rely on the $Ni2p_{3/2}$ XPS results. Unfortunately we have far less data at 77K than at 300K. At the oxygen coverage equivalent to that of the ca. half-monolayer well-formed c(2×2) at 300K (ca. 2L exposure at 77K) there is no evidence for any chemically shifted $Ni2p_{3/2}$ component. At ca. 80L exposure, (ca. 1.2 monolayer of O) there is evidence for some Ni^{2+}, but much less than would have been present at 300K at this coverage. In fact, it appears that at the low temperature the surface Ni atoms undergo a chemical shift at somewhere between 0.5 and 1.2 monolayer coverage into an <u>intermediate</u> position between Ni^0 and the Ni^{2+} position of NiO.[8] This is confirmed by examining the UPS spectra[8] in addition to the Ni2p XPS core-levels. Here a clear intermediate shift in the Ni d-band density of states is observed at 77K which is not there at 300K (Fig. 4). The low temperature regime needs more detailed study, but our current interpretation of the difference in behavior from the 300K results is that NiO nucleation does not occur till $\theta=1$, and that some form of pre-oxide or new chemisorbed overlayer precedes the nucleation. This species is not identical to the c(2×2)O 300K state in its electronic structure since it causes a shift in the Ni2p XPS level and the Ni d-band peak which is not present for the c(2×2)O structure.

The exact coverage at which NiO forms at 300K and at 77K, and the exact shape of the s versus θ curves at the two temperatures have importance when considering the role of order-disorder phenomena and the role of a mobile precursor state (molecular oxygen) in the mechanism of the adsorption process. As we pointed out earlier, simple Langmuir dissociative adsorption does not fit the experimental kinetic data accurately at 300K, and is even qualitatively wrong at 77K. King and Wells[10] have considered in some detail the effects that short-range order in the chemisorbed overlayer and the presence of a precursor state have on the adsorption kinetics of N_2 on W(100). The situation there is less complex than for $Ni(100)/O_2$ since there is no equivalent nitride stage to the NiO nucleation. Up to the onset of NiO formation, however, one may draw an analogy between the two systems in considering the limits of kinetic behavior under the influence of short-range order and precursor mobility. Thus, following King and Wells for a square lattice it can be shown first that for Langmuir-like dissociative adsorption (i.e., assuming that mobility of any molecular precursor plays no role in the kinetics) there are two limits to the behavior of s as a function of θ. If there is zero short range order then

$$s = s_0(1-\theta)^2 \ . \qquad (1)$$

As mentioned earlier this behavior comes about because of the necessity of having two adjacent empty chemisorption sites, nn, for dissociative adsorption. With no short-range order the number of available nn sites is given by $(1-\theta)^2$. The opposite extreme of perfect short-range order in the chemisorbed layer (caused by a

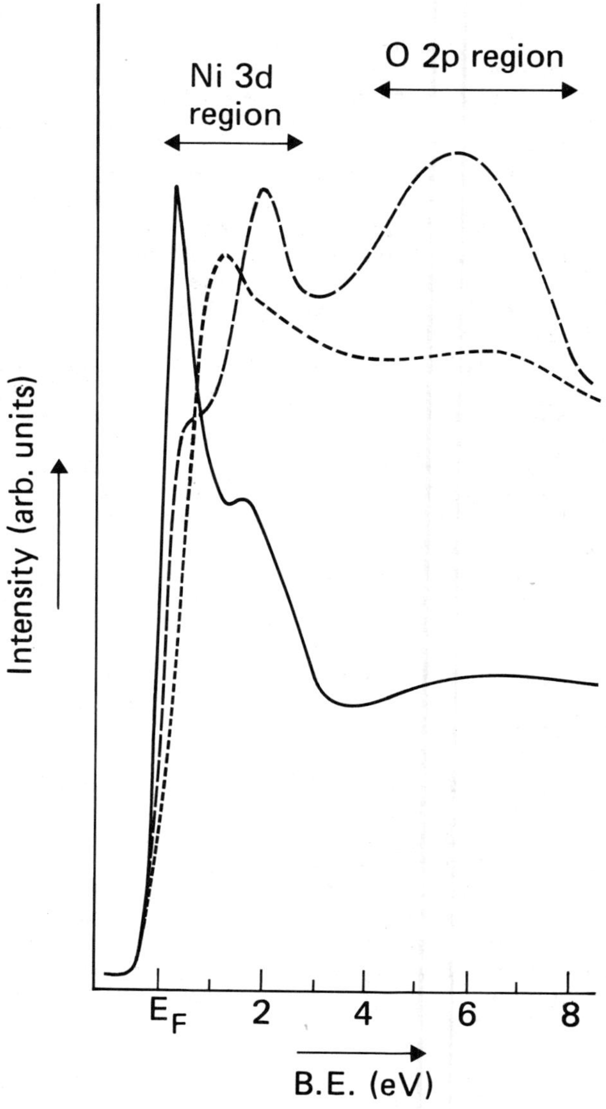

Figure 4. He II UPS data[7] for clean Ni(100) (full curve); partial oxidation at 300K (dashed curve); and partial oxidation at 77K (dotted curve).

large repulsive interaction between O atoms on nearest neighbor sites) results in the behavior

$$s = s_0(1-2\theta) \qquad (2)$$

for the c(2×2)O structure. Thus the sticking probability falls to zero at θ=0.5 with a linear dependence on θ. The paper by King and Wells should be referred to for the derivation of this equation. It is clear that for perfect order s must be zero at θ=0.5, since there are no more nn sites available at that coverage for a perfect c(2×2) structure. These limits of behavior of s/s_0 are plotted in Fig. 5, curves a and b. For intermediate degrees of order curves intermediate between these extremes are obtained.

If chemisorption occurs via a precursor molecular state where mobility is such that it can make several hops across the surface before desorption, then clearly the Langmuir behavior of s/s_0 becomes modified. It can be shown that

$$s/s_0 = [1 + K(1/\theta_{nn}-1)]^{-1} \qquad (3)$$

$$K = f'd/(fa+fd)$$

where fa = the probability for adsorption at an nn pair site
fd = the probability for desorption from an nn pair site
f'd = the probability for desorption at a filled site
θ_{nn} = the coverage of empty pair sites.

K=1 corresponds to the situation where the precursor has zero mobility since in this situation f'd=1 and fa+fd=1. Equation (3) then reduces to

$$s = \theta_{nn}\, s_0 \, ,$$

i.e., s reverts to the limits of Eqs. (1) and (2) depending on whether there is short-range order or not. K=0 corresponds to the situation where f'd=0, i.e, the precursor can make an infinite number of hops without desorbing. In this condition, where the effect of the precursor mobility is at a maximum, Eq. (3) reduces to

$$s = s_0 \text{ for positive } \theta_{nn} \, .$$

Thus for the unity value of s_0 in the Ni(100)/O case, an incident oxygen molecule will adsorb into the precursor state with a probability of unity and will diffuse over the surface until it finds an nn pair site where it can dissociatively chemisorb. The coverage at which θ_{nn} reaches zero and therefore s drops from unity to zero depends on the short-range order. For no short-range order it is very close to θ=1 (curve c, Fig. 5) and for perfect short-range order into a c(2×2) structure it is very close to θ=0.5 (curve d, Fig. 5).

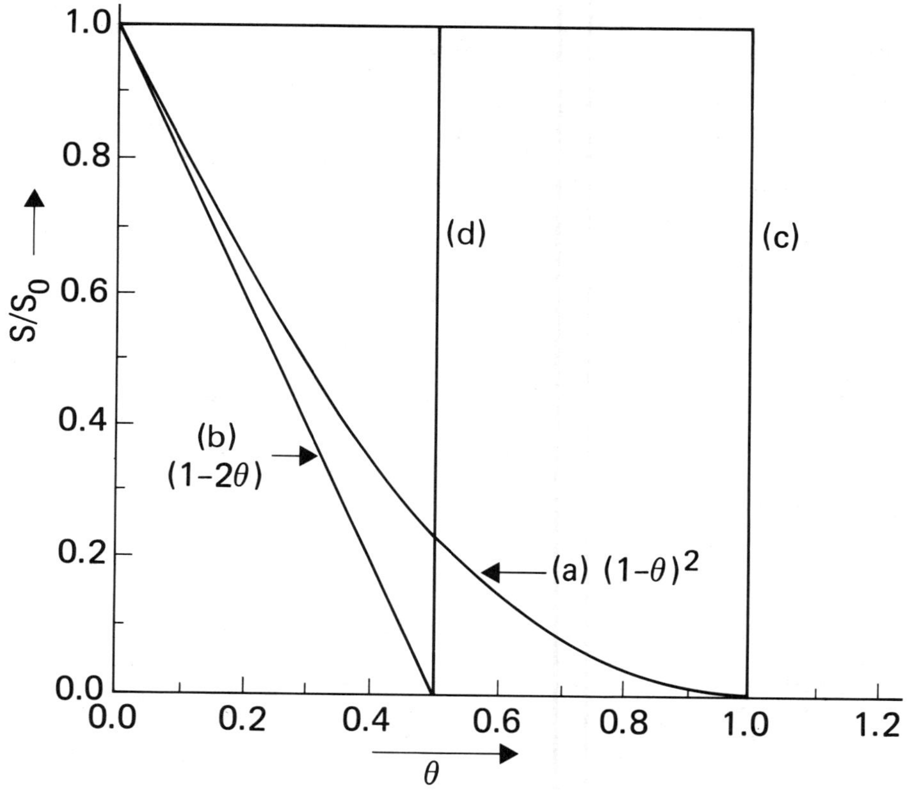

Figure 5. Model limits for the functional forms of s versus θ for the Ni(100)/O2 system. Curve (a) Langmuir kinetics for a chemisorbed product with no short-range order; (b) Langmuir kinetics for a perfect c(2×2)O chemisorbed structure; (c) infinite precursor state mobility and no short-range chemisorbed atom order; (d) infinite precursor state mobility and perfect c(2×2)O short-range order.

In considering the behavior of the $Ni(100)/O_2$ system as a function of temperature it is fairly obvious what trend we would expect if the above modeling has validity for the chemisorption part of the process (i.e., preceding any oxide nucleation). On decreasing the temperature the probability of precursor diffusion versus desorption, f_m/f_d increases. This should lead to a decrease in K. The degree of short-range order should decrease since it becomes more difficult for the strongly bound chemisorbed O atoms to make the diffusion hops necessary to form the desired c(2×2) structure. In fact we know the degree of order gets worse as the temperature is lowered since the c(2×2) pattern gets more diffuse the lower the adsorption temperature[3] and no order at all is observed at 77K.[8] We would expect, then, that the s/s_0 versus θ curve should change from something between curves b and d at a temperature high enough to give sharp c(2×2) structure, towards curve c at 77K.

The experimental data at 300K and 77K, taken from the work of BH in Figs. 1 and 3, are replotted in Fig. 6. Even though the experimental absolute coverage may be in error up to 20% and the 77K data is somewhat preliminary it is immediately obvious that the behavior at 300K and 77K lie outside the extremes of the model curves. At 300K s/s_0 falls faster than is allowed even for zero mobility of the precursor state and perfect c(2×2) ordering (cf curve b, Fig. 5). At 77K s/s_0 has only dropped to a value of 0.5 by θ=1. Nevertheless the change in the kinetics on going from 300K to 77K is as qualitatively expected; the fact that s/s_0 remains relatively unchanged during the initial adsorption stages indicates that precursor mobility is now having an appreciable effect on the kinetics. The fact that s/s_0 does not drop precipitately at θ=0.5 indicates, in agreement with the lack of a LEED pattern, that the role of short-range order has been quenched. What are the reasons for the quantitative discrepancies with the modeling? At 77K an s/s_0 which remains high even above θ=1 must indicate a rapid transfer of oxygen atoms out of the chemisorbed state into an oxidation or preoxidation state which generates fresh Ni sites at the surface (by place exchange) for further adsorption. Since the XPS and UPS actually indicate less NiO formation at 77K than at 300K for similar oxygen coverage the transition must be into the state which causes the intermediate shifts in the Ni2p and valence band spectra. We can only speculate at the moment as to what this state is. Amorphous or nonstoichiometric oxide nuclei are two possibilities. In any case it is clear that the onset of this fast oxidation or preoxidation stage distorts the kinetic models for the chemisorption stage.

At 300K the discrepancy, a much more rapid decrease in s/s_0 than suggested by the limits of the model, can be accounted for if we were to believe that the p(2×2) 1/4 monolayer structure rather than the subsequent half-monolayer c(2×2) structure were the controlling influence in the short-range order effect. If it were relatively unfavorable for dissociation to occur at nn pair sites where one of the n sites was adjacent to two filled sites, the formation of an ordered p(2×2) structure would lead to an s/s_0

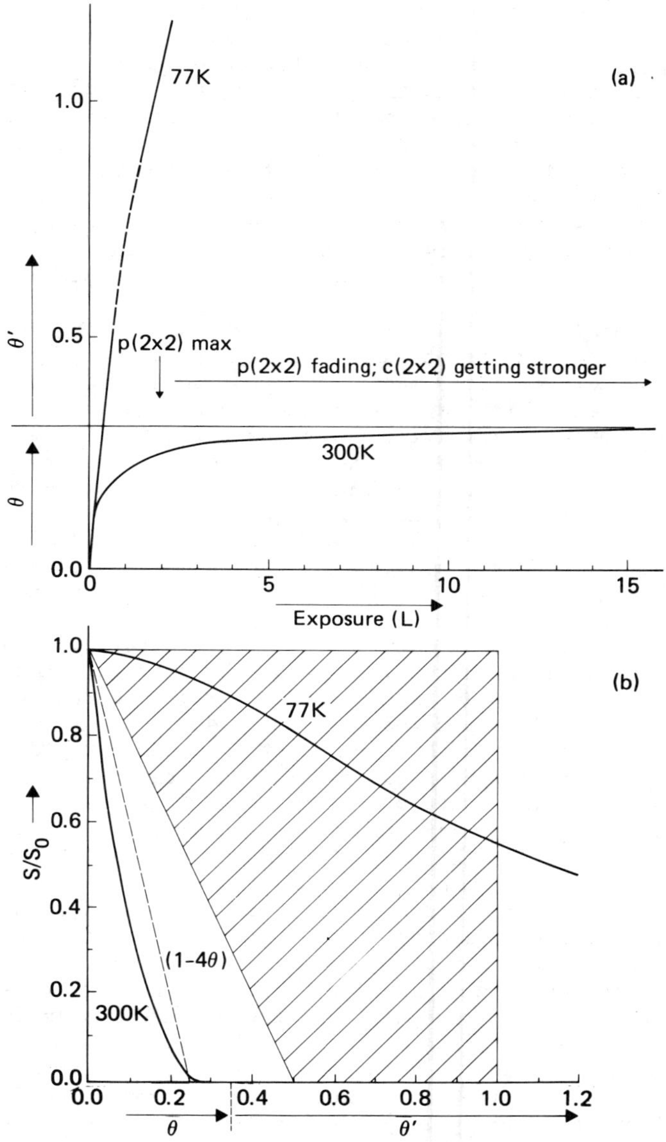

Figure 6. (a) Details of low exposure oxygen uptake curves from the XPS work of BH; (b) Conversion of (a) into an s/s_0 versus coverage plot. The expected limits from the model behavior of Fig. 5 are marked cross-hatched. The dashed curve represents the predicted behavior at 300K for perfect p(2×2) order with zero precursor mobility.

behavior which would drop towards zero at 0.25 monolayer instead of 0.5 monolayer. This is shown in Fig. 6 by the dotted curve. This is actually reasonably close to the experimental behavior, as can be seen from Fig. 6. The "tail" in the experimental curve above ~0.2 monolayer coverage would then be a consequence of the fact that there are still 50% of the total number of nn sites available for dissociative adsorption after a perfect p(2×2) lattice is formed, albeit that the probability of dissociation at these is greatly reduced.

Our overall understanding of the Ni(100)/O_2 system may be summarized as follows.

There is a fast chemisorption stage, the extent (coverage) and kinetics of which are dependent on the short-range order of the chemisorption product induced by adatom repulsion and on the precursor state mobility. It is therefore dependent on temperature. At room temperature the dominant order influence seems to be exerted by the p(2×2) structure, resulting in an s/s_0 which is <0.1 by $\theta=0.25$. The shape of the curve over most of this range indicates that the precursor state has negligible effect on the kinetics (i.e., its probability of desorption is much higher than the probability of diffusion). The c(2×2) structure forms more slowly after the rapid formation of the p(2×2) structure and apparently does not reach completion before NiO nucleation becomes appreciable (above about $\theta=0.35$ monolayer). The NiO formation is definitively recognized from both the XPS $Ni2p_{3/2}$ chemical shifts and the NiO LEED pattern. It is accompanied by an increase in s/s_0 as a function of θ attributed to oxygen atom capture at the perimeter of NiO islands 2-3 layer thick. At 77K the chemisorption stage is more prolonged because of the lack of short-range order, owing to the inability of the chemisorbed O atoms to undergo the limited diffusion necessary to produce ordered structures. Thus nn sites are still available above $\theta=0.5$. s/s_0 remains high because the effect of the precursor state is no longer negligible. The probability of diffusion to empty nn sites is now high compared to the probability of desorption. In addition an intermediate Ni/O state appears to be formed at 77K, prior to NiO formation, which produces a significant chemical shift in the $Ni2p_{3/2}$ signal and change in the Ni3d band peak maximum. This state may represent either amorphous NiO, nonstoichiometric NiO, or simply a high coverage O/Ni(100) chemisorbed layer which is electronically different from the c(2×2) or p(2×2) layers. Since s/s_0 remains high even beyond $\theta=1$ place exchange into the pre-oxide or oxide structure to produce fresh Ni surface atoms must be rapid.

The detailed nature of the nucleation process which produces oxide at 300K at 77K is not understood, though it very probably occurs at defect sites which exist from the onset of adsorption and grow into protonuclei oxide islands during the chemisorption stage.

Fe(100) OXIDATION

The adsorption of oxygen on Fe(100) is less studied than on Ni(100) and among the studies that do exist there is not complete agreement concerning the shape of the s versus θ curve; the saturation value of θ; or even the nature of the chemisorption or oxidation products formed at various values of θ and at various temperatures. I will make no attempt to thoroughly review the subject here, but just cover the points necessary to make a qualitative comparison to the Ni(100) study.

The first point of contention is whether a c(2×2)O structure is even formed, or is the first O/Fe pattern observed the p(1×1) structure. Several studies have not observed the c(2×2) structure and have ascribed that observed in earlier work to carbon impurity. At least three recent studies at 300K have observed the c(2×2)O structure prior to p(1×1)O under conditions where carbon (or sulfur) contamination could not be the culprit.[11-13] The structure was never sharp in any of these recent studies and it gives way rapidly as a function of exposure to the p(1×1). Carbon contamination will also produce a c(2×2) structure at 300K which is sharper than that of oxygen and does not transform to a (1×1). This structure may indeed have contributed to the earlier reports of c(2×2)O structure. The p(1×1) which forms at higher coverage has the same lattice spacing as the Fe(100) surface--i.e., the additional c(2×2) half-order spots merely fade and leave the same (1×1) pattern as for clean Fe(100). It may be distinguished from clean Fe, however, by the different characteristic IV curves. A theoretical analysis of the IV curves puts the O atoms at the center of the four-fold sites, 0.5Å above the plane of the Fe surface atoms.[14] The oxygen atoms are thus considered to be much deeper in the four-fold hollows in the p(1×1)O/Fe(100) than in the c(2×2)O Ni(100), though not actually in the surface plane as for the p(1×1) O/Ni(100) structure (which is, of course, an NiO(100) lattice distinguishable from Ni(100) by the difference in lattice spacing). The differences in the proposed Fe and Ni case geometries are illustrated in Fig. 7. The Fe LEED work must be regarded as much less sure than the Ni(100)/O_2 situation, however, since the c(2×2) structure was not studied (since it was not observed) and if the observed p(1×1) structure consisted not of an O overlayer but of a p(1×1) from the original Fe(100) lattice, plus FeO(1×1), the spots would be indistinguishable and any comparison of the combined intensity to a theoretical IV curve for an O superstructure would be difficult. The reason that Fe(100) at FeO(100) LEED spots are indistinguishable (as opposed to Ni(100) at NiO(100)) is simply that Fe is a bcc lattice in which the Fe atoms are more widely spaced (see Fig. 7) than in the fcc Ni(100) lattice, such that a FeO plane may be formed by dropping O atom into the four-fold sites with only a few percent expansion of the Fe(100) lattice compared to the ca. 30% expansion required in the Ni(100) lattice. I therefore consider it likely that the observed p(1×1)O structure represents oxide nucleation on Fe(100) in the same fashion as it does for Ni(100), though this point certainly needs further study. Another point which needs

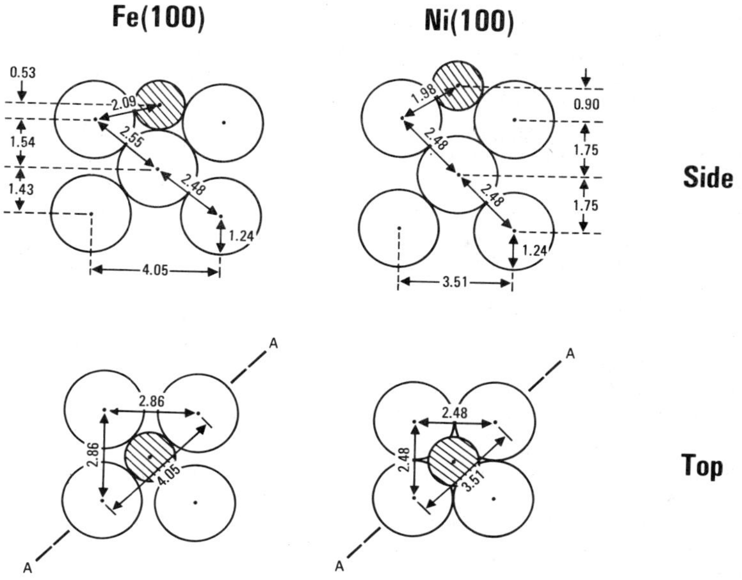

Figure 7. Comparison of the LEED determined geometries for overlayer oxygen on Fe(100)[14] and Ni(100).[4]

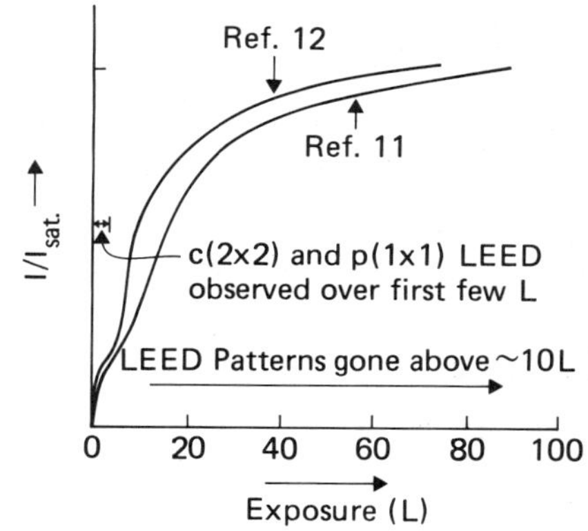

Figure 8. Oxygen uptake curves as a function of exposure for Fe(100)/O_2 at 300K measured by Auger O KLL intensities.[11,12]

mentioning is that the p(1×1) structure is very diffuse (i.e., badly ordered) at 300K and fades out at higher coverages leaving no order, unlike the oxide structures on Ni(100) which do not fade out. The LEED studies on Fe(100)/O_2 were, in fact, performd on an oxygen exposed and annealed sample to generate a well-ordered p(1×1) structure. The XPS evidence indicates that for 300K oxygen exposures leading to coverages where the p(1×1)O structure has started to fade the dominant oxide produce in Fe^{3+}.[13] At lower coverages, around those of the p(1×1) structure, a mixture of Fe^{3+} and Fe^{2+} is formed, and during the c(2×2) stage there is no evidence of any oxide present.[13] The annealing treatments at p(1×1) and higher coverages which tend to improve ordering and intensity (and also sometimes produce more complex patterns which have been attributed to other epitaxial relationships between FeO and Fe(100))[11] also increase the Fe^{2+}/Fe^{3+} ratio in XPS.

Our tentative interpretation so far then is that the O/Fe(100) system has the same stages in it as the O/Ni(100) in that a c(2×2) structure is followed by a p(1×1) oxide. However, the c(2×2) structure is weak and unstable compared to that on Ni(100) going quickly to the p(1×1) which itself does not exist over much exposure range at 300K before the epitaxial relationship to the subtrated disappears and Fe^{3+} species dominate. The reason for the instability of the c(2×2) overlayer compared to that on Ni(100) may be straightforward; it could just reflect the more open bcc Fe lattice and therefore the greater ease of nucleating an oxide structure since the Fe atoms need not separate as they must for Ni.

What are the characteristics of the s versus θ curve? Qualitatively they are the same as for Ni(100)/O_2 as can be seen from Fig. 8. It is clear that the "plateau" stage starts at lower exposures than for Ni(100) and continues for only a short duration (ca. 3L) as opposed to Ni (ca. 45L) before the rise in s. At 80K studies on Fe films show a constant s up to high coverage[15] (as for Ni), but work on Fe(100) has not yet been reported. LEED at 80K shows no ordered structures, however, in agreement with the Ni data.[13]

It seems therefore that the model for the mechanism of oxygen interaction with Fe(100) is qualitatively the same as for Ni(100), but that the oxide nucleation is faster at 300K because of the more open bcc structure.

The details of comparing the functional forms of the S versus θ curves to those predicted from the order-disorder/precursor mobility models must await better experimental data and more accurate absolute values of θ.

OXIDATION OF Ni/Fe(100)

If detailed investigations of oxygen chemisorption and the early stages of oxidation on single crystal metal surfaces are not common, the investigation of single crystal binary alloys is rare indeed. An extra dimension occurs which is not present for the individual elements; surface segregation. Even for the clean alloy surfaces it is not uncommon for the concentration of the components to be different from the bulk owing to the combined thermodynamic driving forces of strain and chemical bonding energies.[16] Under the influence of an adsorbate (oxygen in this case) the segregation effects can be greatly increased owing to the preferential oxidation of one component of the alloy over the other. The detailed behavior may be alloy composition dependent and crystal face dependent. It may also be that kinetic factors inhibit the achievement of equilibrium.

We have embarked on a series of studies on a Ni/Fe alloy of 76/24 composition. This is close to the composition of permalloy (80/20) which is used as a soft magnetic material in many practical devices and so has some practical significance. It also affords a chance to see whether the combination of the two metals produces any catalytic effects which are distinct from either of the constituents. One part of these studies is the oxidation of the fcc(100) face. The results[1,17] which are relevant for a comparison to the Ni(100) and Fe(100) cases are summarized here.

Starting with a clean well-annealed surface, the variable angle XPS measurements demonstrate that there is a slight Fe segregation at the surface (see Table I). On exposure to oxygen at 300K the segregation increases as a function of oxygen coverage. It is accompanied by preferential formation of iron oxides. At a total coverage of $\theta \sim 0.7$ a sharp c(2×2) pattern is observable; there is no evidence for any oxidation of Ni, and the metallic content in the first 4Å is now ca. 32% Fe of which approximately half is in oxide form. At $\theta \sim 1$ the metallic content in the top 4Å is 34% Fe and the majority of this (two thirds) is oxidized. The Ni has only just started to form oxide. The c(2×2)O structure is only just passing its maximum intensity, however. Continuing the oxygen adsorption process to saturation ($\theta \sim 2.5$) results in a 46/54 Fe/Ni ratio in the top 4Å compared to the original 28/72. Practically all the Fe is now in an oxidized form but the Ni is still present in both Ni^{0} and Ni^{2+} states. It is important to note that the c(2×2)O structure did not fade until at least $\theta = 1.3$, yet at that time most of the surface Fe was oxidized.

The preferential formation of Fe oxides over Ni oxide is in accordance with thermodynamic expectation since the heats of formation are higher for the iron oxides. The mechanism by which the transport of Fe to the surface occurs is not so clear, however. It is certainly not by bulk diffusion since this would be far too slow at 300K. It is conceivable that grain-boundary diffusion is sufficiently rapid to account for the segregation, but more likely some field-assisted mechanism involving the negative charge on the chemisorbed oxygen atoms is responsible. According to the Fehlner-Mott theory[18] this would cause a place-exchange to occur between cation and anion at the surface. This proposed mechanism need not be different from that in the pure metal oxide nucleation processes. It is just that for the alloy the change in relative concentration of Fe and Ni in the surface region brought about by the diffusion (which should act equally on Fe and Ni atoms) plus the thermodynamic preference of oxygen to bond to Fe and immobilize it at the surface, makes it clear that the diffusion is rapid and over several atomic layer distances. It is also clear, however, that as the oxide layer thickens the driving force for diffusion decreases and in fact the passivated stage (Θ_{sat}) does <u>not</u> correspond to the equilibrium situation. Provided Fe can reach the site of the available oxygen the thermodynamics show that the Fe will always be oxidized in preference to the Ni. Thus heating the 300K Θ_{sat} surface to 600K in vacuum produces further massive Fe segregation to the surface; the formation of Fe oxide; and the reduction of the NiO present to Ni^0 (see Table I) without any significant change in the oxygen content of the outer layer. Thus, at this temperature bulk diffusion is sufficiently fast to allow equilibrium to be reached.

Another interesting point about the Ni/Fe 300K oxidation is that correlation of the LEED behavior with the XPS identification of the products clearly shows that there is <u>lateral</u> segregation as well as surface segregation. Thus the c(2×2)O structure is still sharp at a coverage when most of the surface Fe is in the form of oxide. It must therefore originate from an overlayer structure on either clean Ni_3Fe patches or on segregated Ni patches with nucleated Fe oxide patches making up the rest of the surface.

We have made some preliminary studies of the kinetics for the Ni/Fe system (using AES this time).[18] The results are shown in Fig. 9, together with the information on the coverage regimes of the LEED patterns and the onsets of the different oxides. If this figure is compared to the uptake curves for Ni(100)/O_2 and Fe(100)/O_2, it is clear that qualitatively, at least, the Ni/Fe behavior is simply the sum of the individual Ni and Fe behavior. Thus the ordered LEED structures of chemisorbed O on Fe give way rapidly to oxide (within 10L exposure) whereas the c(2×2)O structure on Ni is stable for 50L. The reason why the behaviors of Ni and Fe in the NiFe(100) surface should be so similar to those in the individual single crystal surfaces is again not so clear. NiFe 75/25 has an fcc lattice with an almost identical lattice constant to that of fcc Ni. Fe on the other hand is bcc and in the section on the interaction of oxygen with Fe(100) I suggested that the instability of the chemisorbed O superstructure with respect to oxide nucleation could be ascribed to the more open bcc structure. Possibly it has less to do with crystal structure than with the higher heat of formation of Fe oxides (compared to NiO).

At 80K the interaction of oxygen with the NiFe surface produces a dramatically different result than at 300K.[17] There is <u>no</u> Fe segregation and <u>no</u> preferential oxidation of Fe (see Table I), even though the saturation coverage is similar to that at room temperature. The uptake curve at 80K, shown in Fig. 9, indicates why. s remains near s_0 until close to saturation coverage and then drops suddenly. The behavior at 80K is, therefore, also the sum of the individual behavior of Ni and Fe at 80K, keeping s close to unity to high coverage. For Ni(100) the reason that s could remain close the unity beyond Θ=0.5 was ascribed to the inability to establish short-range order at 80K--i.e., the stable c(2×2)O never forms. The implication for the alloy surface is that

Table I Fe Segregation Effects for Ni 76%/Fe 24% (100)/O_2

Condition	% Fe in Top ~13Å	% Fe in Top ~4Å
Clean, non-annealed, 300K	23	24
Clean, annealed, 300K	24	28
θ=0.7, 300K	25	32
θ=1, 300K	27	34
θ_{sat}=2.5, 300K	35	46
θ_{sat} 300K heated to 600K	56	85
θ_{sat} 77K=2.3	25	26

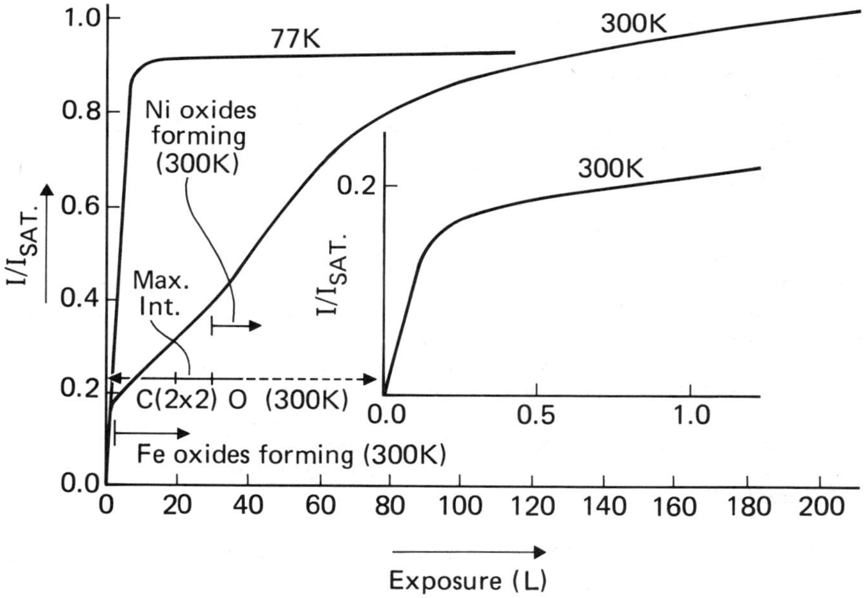

Figure 9. Oxygen uptake curves for Ni 76%/Fe 24%(100) at 300K and 77K, as determined from the O KLL Auger intensities.[17]

the resistance of the Ni component to oxide formation at 300K, which results in preferential Fe oxidation and Fe surface segregation, is due to the stability of the c(2×2)O structure and that this resistance is lost at 80K where the ordered structure is never established in favor of oxide nuclei. It therefore seems that in addition to simple oxide relative heats of formation arguments the stability of the intermediate chemisorbed structures also play an important role in determining whether there is preferential oxidation and surface segregation at the single crystal alloy surfaces.

SUMMARY

In reviewing the work on oxygen interaction with Ni, Fe and Ni/Fe I have tried to establish the following points:

1. Though there is no evidence for any significant concentration of molecular oxygen at the surfaces, the s versus θ curves during the chemisorption stages indicate that the mobility of a molecular precursor state can have a controlling influence on the kinetics. At low temperatures (80K) this is very evident from the constancy of s as a function of θ. At room temperature the effects of the precursor state on the kinetics appear to be minimal, the s versus θ curve for Ni(100)/O_2 being dominated by the short-range order established in the chemisorbed layer. It appears that the quarter-monolayer p(2×2) structure rather than the half-monolayer c(2×2) exert the controlling influence since S tends to zero at θ~0.25.

2. The chemisorption stages are followed by an oxide nucleation stage which at room temperature causes s to rise from its minimum value at the end of the chemisorption stage. The nuclei grow laterally as islands of 2-3 layer thickness by capture of oxygen at the perimeters. It is not clear whether the oxygen is captured from the diffusing precursor state or directly from the gas phase. It is also not clear exactly what process generates the nuclei though it is definitely associated with the presence of surface defects.

3. At low temperature (80K) s does not increase during oxidation since it remains high throughout. There is as yet been no firm delineation between the end of chemisorption and the onset of the oxidation stages. For Ni(100) it appears that there is an intermediate stage between chemisorption and nucleation of NiO which could represent nonstoichiometric or small grain amorphous NiO. There is a temperature dependent balance between ordered chemisorbed structures and oxide nucleation. At 80K the oxide nucleation process dominates since no ordered

chemisorbed structure develops and so s remains high even well above $\theta=1$.

4. A comparison of the behavior of Fe(100) to Ni(100) reveals that the general trends are the same, but that the chemisorbed structures are much less stable with respect to oxide formation on Fe(100). It is reasonable to suggest that this might simply be explained by the more open bcc Fe structure which requires almost no expansion of the lattice to form oxide.

5. Point 4 above is important when considering oxygen interaction on the Ni/Fe(100) surface. The preferential oxidation and consequent surface segregation of Fe at 300K is what would be predicted from the superposition of the individual Ni(100) and Fe(100) behaviors. Likewise the lack of any preferential oxidation and segregation at 80K is predicted from the individual 80K behavior. The actual distribution of products at any temperature and coverage is dependent on complex thermodynamic and kinetic factors. It is not until 600K when bulk diffusion is sufficiently rapid to allow the reaction

$$NiO + Ni_3Fe \longrightarrow FeO + 4Ni$$

to equilibrate that the thermodynamically predicted products, iron oxide and unoxidized nickel are formed. At lower temperatures the reaction is kinetically limited and though field assisted diffusion mechanisms allows preferential Fe oxidation to occur at 300K the mechanism dies out rapidly as the oxide layer thickness and so substantial amounts of NiO are also formed by θ_{sat}.

ACKNOWLEDGMENTS

It is a pleasure to acknowledge the many stimulating discussions concerning this work I have had with my IBM colleagues, D. J. Auerbach, J. Q. Broughton, F. Abraham and J. Barker and also with Professor R. J. Madix and Professor D. A. King.

REFERENCES

1. C. R. Brundle, E. Silverman and R. J. Madix, J. Vac. Sci. Tech. 16, 474 (1978).
2. H. Hopster and C. R. Brundle, J. Vac. Sci. Tech. 16, XXX (1978).
3. P. H. Holloway and J. B. Hudson, Surface Science 43, 123 (1974).
4. J. E. Demuth and T. N. Rhodin, Surface Science 45, 123 (1974).
5. D. F. Mitchell, P. B. Sewell and M. Cohen, Surface Science 61, 355 (1976).
6. P. R. Norton, R. L. Tapping and J. W. Goodale, Surface Science.
7. C. R. Brundle and H. Hopster, to be published.

8. C. R. Brundle and H. Hopster, to be published.
9. K. S. Kim and N. Winograd, Surface Science 43, 625 (1974).
10. D. A. King and M. G. Wells, Proc. Roy. Soc. A339, 245 (1974).
11. G. W. Simmons and D. J. Dwyer, Surface Sci. 48, 373 (1975).
12. C. F. Brucker and T. N. Rhodin, Surface Sci. 57, 523 (1976).
13. C. R. Brundle, IBM J. Res. 22, 235 (1978).
14. K. O. Legg, F. P. Jona, P. W. Jepsen and P. M. Marcus, J. Phys. C. 8, 4492 (1975).
15. A. M. Horgan and D. A. King, Surface Science 23, 259 (1970).
16. F. F. Abraham, N. H. Tsai and G. M. Pound, Surface Science 83, 406 (1979).
17. E. Silverman, C. R. Brundle and R. J. Madix, to be published.
18. F. P. Fehlner and N. F. Mott, Oxidation of Metals 2, 59 (1970).

PRECURSOR INTERMEDIATES IN ADSORPTION, DESORPTION AND REACTION*

L.D. Schmidt
Department of Chemical Engineering
and Materials Science
University of Minnesota
Minneapolis, MN 55455

ABSTRACT

Most processes at solid surfaces involving chemisorbed species probably proceed through weakly bound (precursor) intermediate states. In this article simple models of kinetics predicted by such states are summarized and compared with experimental results. Adsorption can easily occur with constant sticking coefficient, and desorption can be proportional to $(1-\theta)^{-1}$ if a precursor state is involved. Reaction can exhibit kinetics which are independent of the properties of chemisorbed states if reaction proceeds through a precursor.

An essential property of these intermediates is that their populations be independent of underlying chemisorbed states, i.e. that they occupy different sites. The precursor could be a physically adsorbed or weakly chemisorbed species which may be weakly bound to specific adsorption sites in contrast to the chemisorbed species which presumably occupies fixed sites. Requirements of the principal of detailed balance and implications of these intermediates for kinetics and dynamics of surface processes are discussed.

INTRODUCTION

Surface processes probably with few exceptions proceed through weakly adsorbed intermediate species. Thus there may be few "elementary" or one-step processes at surfaces. This arises, first, from the fact that surface processes occur in series with transfer from and to homogeneous phases, and second, because the adsorption state between surface and homogeneous phase is probably not the tightly bound one.

The notion of these states, introduced 25 years ago to explain coverage independent sticking coefficients, remains ephemeral because they have probably never been observed directly. Indeed, one

*This work partially supported by NSF under Grant No. DMR 75-02627

ISSN-0094-243X/80/610083-13$1.50

usually assumes that their coverages are low in deriving equations and then observes that such expressions explain rate data. These states are probably physically adsorbed or weakly chemisorbed species, and the properties of these states could possibly be populated statically using low temperatures and high pressures; however, it is worth noting that this connection can only be inferred and their properties may be quite different at higher temperatures where rates are being measured.

Precursors appear to offer reasonable interpretations of some experiments in kinetics, in dynamics such as molecular beam-surface processes, and in anomalous values of rate coefficients. However, such experiments are sometimes far from conclusive as may be interpretations and models requiring a precursor.

We consider the general process of molecule A as gas A_g, chemisorbed A_s, precursor A*, or as a reaction product B as sketched in the following kinetic scheme:

$$\begin{array}{ccccc} A_g & \underset{k_d^*}{\overset{k_a^*}{\rightleftarrows}} & A^* & \underset{k_d}{\overset{k_a}{\rightleftarrows}} & A_s \\ & & \downarrow k_R & & \\ & & B & & \end{array} \qquad (1)$$

We shall generally use a kinetic derivation, representing each step as a rate coefficient k times some function of pressures P or coverages θ^* and θ in precursor and chemisorbed states respectively. We use this model even though the surface steps, k_a and k_d, actually involve diffusion processes. An alternate approach, the successive site model, sums probabilities of various processes on each site, but expressions obtained are similar or identical to those obtained with kinetic models and the latter are somewhat simpler to interpret.

Before discussing individual rate processes it is important to discuss site and equilibrium relationships between states on a surface to distinguish between same site and independent site adsorbates. If two states A_s and A^* occupy the same sites (or if occupation of one excludes one or more neighboring sites of the other), then the coverages at equilibrium are related by the expressions

$$\theta^*/\theta = K = K_o e^{-\Delta H/RT} \; ; \; \theta^*+\theta \leq 1 \qquad (2)$$

where K is an equilibrium constant and ΔH the difference in heats of adsorption. The existence of a fixed number of sites n_o requires the above inequality in fractions of saturation θ and θ^* or equivalently that $n+n^* \leq n_o$. If different sites are involved each density can independently go to saturation, $\theta \leq 1$ and $\theta^* \leq 1$, and the equilibrium expression becomes

$$\frac{\theta^*(1-\theta)}{\theta(1-\theta^*)} = K = K_o e^{-\Delta H/RT} \; . \qquad (3)$$

The $1-\theta$ factors are required by lattice statistics to prevent either density from exceeding its saturation value.

If A* were merely a vibrational or electron excited state or a specific orientation of molecule A with a ground state coverage θ, then one would have competitive adsorption, Eq. 2, and none of the following expressions involving precursors would be valid. The states we are describing have independent populations, Eq. 3.

A qualitative picture of the chemisorbed and precursor states and the relevant one-dimensional potential surface is shown at the left. The states are drawn as between and on top of atoms, but the models, shown to the right, only require independent occupation of each state.

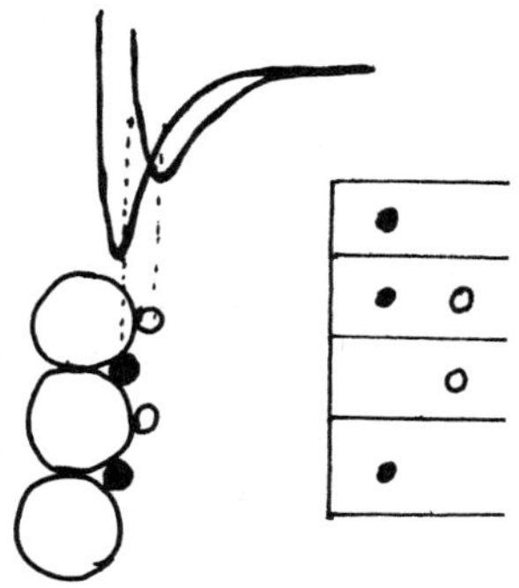

Figure 1. Model and potential curves for chemisorbed (solid circles) and precursor states (open circles).

ADSORPTION

The direct or Langmuir model of adsorption kinetics[1] assumes that the sticking coefficient S is the product of the adsorption probability on a bare surface S_o and the probability $1-\theta$ of the incident molecule striking a bare site,

$$S(\theta) = S_o(1-\theta). \tag{4}$$

The earliest adsorption rate experiments (including Langmuir's alkali metal adsorption[1] on W in the 1920's) yielded sticking coefficients which did not decrease as fast as predicted by Eq. 4. Becker[2] suggested that a precursor might be involved and Ehrlich[3] developed a kinetic model which was soon modified by a successive site model by Kisliuk.[4] With slight modifications by King,[5] Gomer,[6] and us,[7-9] these notions have been used almost exclusively to interpret $S(\theta)$.

It is assumed that under the conditions of the experiment the chemisorbed state once formed cannot desorb so that Eq. 1 becomes

$$A_g \underset{k_d^*}{\overset{k_a^*}{\rightleftharpoons}} A^* \xrightarrow{k_a} A_s. \tag{5}$$

The coverage in the A* state θ^* is assumed small so that the steady state approximation may be used.

$$\frac{d\theta}{dt} = k_a^* P - k_d^* \theta^* - k_a \theta^*(1-\theta) = 0 \tag{6}$$

or

$$\theta^* = \frac{k_a^* P}{k_d^* + k_a(1-\theta)} . \tag{7}$$

The rate of adsorption r_a is $k_a\theta^*(1-\theta)$ so that

$$S = \frac{r_a}{\text{flux}} = \frac{k_a^*\theta^*(1-\theta)}{P} = \frac{k_a}{1+\frac{k_d}{k_a^*(1-\theta)}} = \frac{S^*}{1+\frac{K}{1-\theta}} = \frac{S_o(1+K)}{1+\frac{K}{1-\theta}} \qquad (8)$$

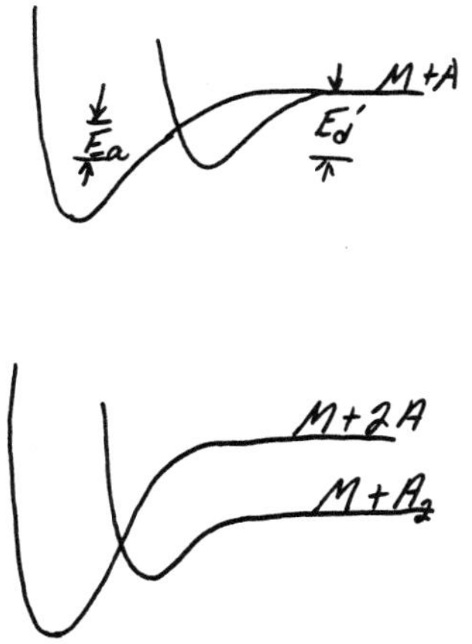

Figure 2. Potential curves for associative and dissociative adsorption.

where

$$K \equiv \frac{k_d^*}{k_a} = K_o \exp\left[\frac{(E_d^*-E_a)}{RT}\right]. \qquad (9)$$

Figure 2 shows one-dimensional potential curves for associative adsorption (Eqs. 5 and 8) and dissociative adsorption (Eqs. 10 and 11). Activation energies E_d^* and E_a implied by these models are shown in the figure.

Dissociative adsorption. For the process

$$A_{2g} \underset{k_d^*}{\overset{k_a^*}{\rightleftarrows}} A_2^* \xrightarrow{k_a} 2A_s \qquad (10)$$

the chemisorption step may have the rate $k_a\theta^*(1-\theta)^2$ if two sites are required and these are uncorrelated. This yields an expression

$$S = \frac{S_o(1+K)}{1+\frac{K}{(1-\theta)^2}} \qquad (11)$$

rather than the corresponding Langmuir model expression

$$S = S_o(1-\theta)^2. \qquad (12)$$

Successive site model. Here one sums probabilities of the precursor desorbing p_d, chemisorbing p_a, and jumping to an adjacent precursor site p_j. This leads to a closed solution of the form

$$S(\theta) = \frac{S_o}{1+\frac{K'\theta}{1-\theta}} \tag{13}$$

where

$$K' = \frac{p'_d}{p_a+p_d}$$

for the situation of Eq. 5. We note that, while this expression does not appear exactly the same as Eq. 8, it in fact has the same coverage dependence if $K=K'/(1-K')$. For dissociative adsorption the kinetic and successive site expressions are not identical however. The successive site model for dissociative adsorption, Eq. 10, yields

$$S(\theta) = \frac{S_o}{1+\frac{K'\theta}{(1-\theta)^2}} \tag{14}$$

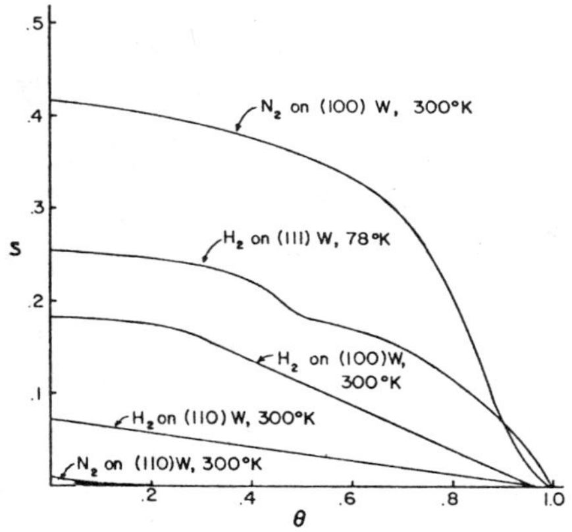

Figure 3. Typical $S(\theta)$ data for gases on W(100) at 300K except where noted.

Comparison with experiment. Measurements of $S(\theta)$ can be fit with these expressions, but two problems should be noted. First, most chemisorbed systems involve multiple binding states even on single crystal planes, and results on polycrystalline surfaces contain an average over planes and states which should not be expected to agree with any of the preceeding one-state models. Second, $S(\theta)$ data is difficult to obtain accurately because of variable temperature following heating to clean initially and because of the need to differentiate coverage versus time curves in the "uptake" method. In Figure 3 are shown representative $S(\theta)$ results for several gases on W(100), probably the most studied surface. In most cases S decreases more slowly than $1-\theta$ or $(1-\theta)^2$, and in some cases an inflection is observed which correlates with population of multiple states. Figure 4 shows $S(\theta, T_s)$ for N_2 on W(100)[8,6] which is a situation where only a single high binding state (β_2) exists with a density of 0.5 of the W atom density in a c(2x2) configuration. Data shown vary from T_s= 200 to 900K, and solid and dashed curves represent fits using the kinetic (Eq. 8) and successive site (Eq. 14) models with dissociative adsorption into the chemisorbed state. Figure 5 shows a plot of S_0 and K (Eq. 9) versus $1/T_s$ for N_2 on W(100). It is evident that S_0 is nearly independent of T_s until 600K, after which it falls rapidly. The parameter K which from Eq. 9 should have an Arrhenius temperature dependence, fits fairly well to two straight lines,[9] breaking at ~ 600K. The interpretation of Eq. 9 yields an activation energy difference of $E_d^*-E_a$ of 3 kcal/mole. If one interprets the precursor as a physically adsorbed or γ state of N_2 which has a heat of adsorption of 10 kcal, then E_d^*=10 kcal/mole and

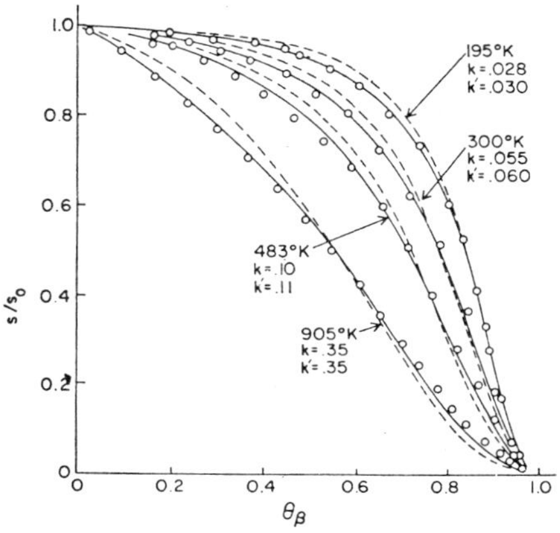

Figure 4. S/S_0 versus θ for $\beta_2 N_2$ on W(100) at temperatures indicated. Fits to models are indicated by solid and dashed lines.

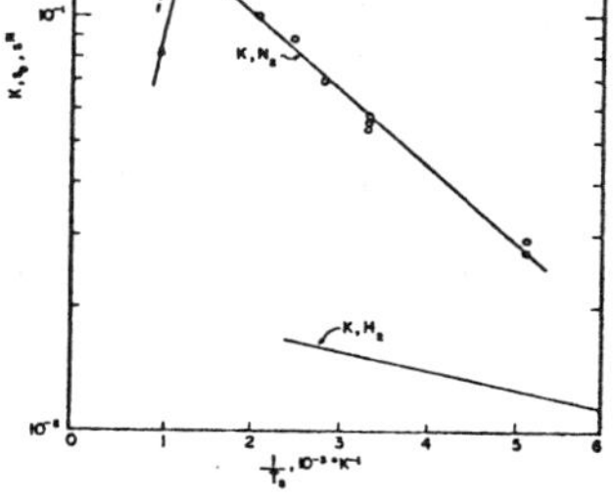

Figure 5. Arrhenius plots of S_0 and K for N on W(100).

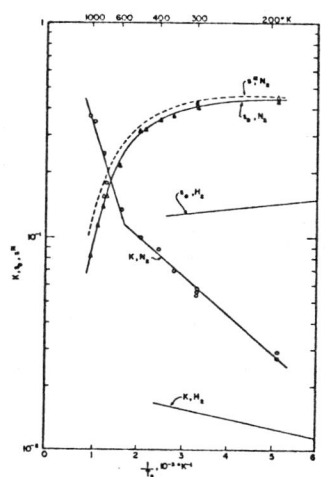

E_a = 7 kcal/mole as sketched in Figure 6.

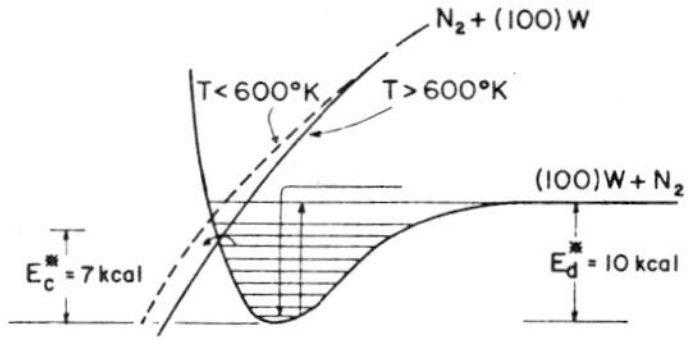

Figure 6. One-dimensional potential curves estimated from activation energy of Figure 5 and Eq. 9.

Dynamics of adsorption. The angle of incidence dependence of S should yield information on energy transfer between incident gas molecules and the surface for adsorption, and for situations with precursors, this probably involves thermal accommodation into the precursor as an intermediate step in chemisorption. This can only be measured with molecular beams, and Figure 7 shows some of the results[10] for S_0 as a function of angle from the surface normal ϕ.

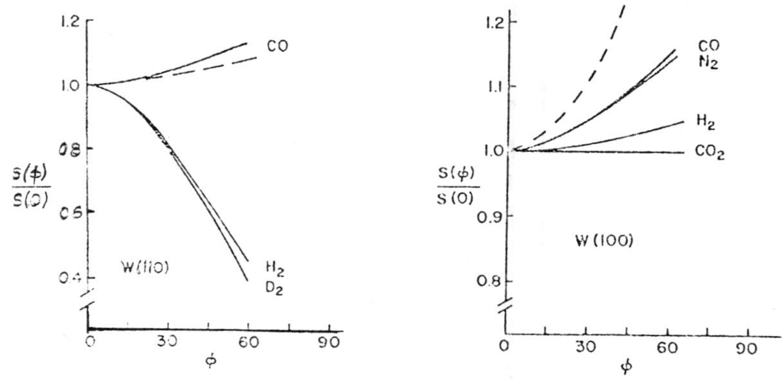

Figure 7. Variation of S_0 with angle of incidence on W(100) and W(100) for both gas and surface at 300K.

Results can be correlated with simple models of energy transfer, but there is a need for detailed experimental and theoretical work on the dynamics of adsorption kinetics with intermediate states.

DESORPTION

The role of precursors in desorption has been considered only recently, and the experimental situation is less clear than in adsorption. Shannabarger[11] first considered precursors in desorption, and several authors have developed rate expressions using various models.[12,13]

If the pressure is negligable, Eq. 1 becomes

$$A_g \xleftarrow{k_d^*} A^* \underset{k_d}{\overset{k_a}{\rightleftharpoons}} A_s \qquad (15)$$

and the steady state approximation yields a desorption rate

$$r_d = k_d^*\theta^* = \frac{k_d^* k_d \theta}{k_d^* + k_a(1-\theta)} \qquad (16)$$

If $k_d^* >> k_a$ we obtain a "normal" desorption rate expression, although the activation energy in k_d is less than the heat of desorption, and the approximation implies the absence of equilibrium between θ and θ^*.

If $k_a >> k_d^*$ this becomes

$$r_a \approx \frac{k_a^* k_a \theta}{k_a(1-\theta)} \qquad (17)$$

which gives an activation energy equal to the heat of adsorption but the coverage dependence has a factor of $(1-\theta)^{-1}$ not expected for single state first order desorption kinetics. This effect could be observed as an anomalously high desorption rate as the coverage approached saturation. This may be observed as high coverage "tails" in the temperature programmed desorption or as a pre-exponential factor which increases with coverage.

Desorption with empty and full site precursors

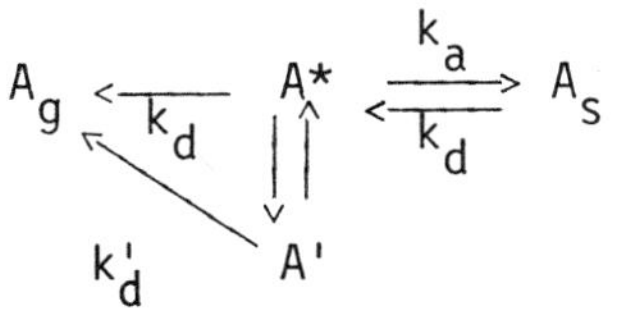

(18)

can be shown to yield a desorption rate expression

$$r_d \simeq \frac{\theta^2}{1-\theta} \tag{19}$$

for properly chosen values of the (many) rate parameters.

The Principle of Detailed Balance implies a relation between adsorption and desorption kinetics. The chemisorbed state assumed in all models obeys a Langmuir isotherm

$$\theta = \frac{K_e P}{1+K_e P} \tag{20}$$

where K_e is the adsorption equilibrium constant

$$K_e = K_{oe} e^{\Delta Ha/RT} \tag{21}$$

The simple precursor models of Eqs. 8 and 13 with S independent of θ are identical with the desorption model of Eqs. 15 and 16, and the $(1-\theta)^{-1}$ factor must be present in the desorption expression if $(1-\theta)$ is not present in the sticking coefficient. Similarly, detailed balance requires that if $r_d \sim \theta^2/(1-\theta)$, then $S \sim \theta$ for the model of Eq. 18.

Detailed balance then says that precursors should only be evident in desorption if they exist in adsorption. Experimentally this is difficult to demonstrate because temperatures of the two experiments are necessarily different. However in systems where a Langmuir model of adsorption obtains, desorption should also be simple. Similarly, it is doubtful if an expression such as Eq. 19 should be valid unless for the same system $S \sim \theta$.

As with adsorption kinetics, molecular beam experiments should reveal details of the process not accessible to angle averaged results. It is observed the H_2 desorption[14,15] from Cu and Ni is highly peaked in the forward direction and that the gas kinetic energy is higher than expected in this direction. No models[15] appear to be entirely successful in explaining these experiments. However they certainly demonstrate that simple one-dimensional pictures of desorption are inadequate.

REACTION

Reaction kinetics have usually been considered in terms of one-state Langmiur-Hinshelwood mechanisms, although of course a precursor state may be involved here also. For example, a unimolecular reaction A $\longrightarrow$ B may occur through the precursor intermediate indicated in Eq. 1. If the rate of reaction is slow and A_g, A^*, and A_s are in equilibrium, then it is easy to show that the reaction rate r_R is given by

$$r_R = k_R \overset{*}{A} = \frac{k_R\theta}{1-\theta} = \frac{k_R K^* P_A}{1+K^* P_A} . \qquad (22)$$

This gives a $(1-\theta)^{-1}$ factor in the rate measured as a function of coverage. When measured as a function of pressure P_A, the rate has the form of Langmuir-Hinshelwood kinetics but now K^* is the equilibrium constant relating P_A and θ_A^* , that is, it contains the heat of adsorption of the <u>precursor state</u> rather than the chemisorbed state.

This predicts the rather surprising result that reaction kinetics may be independent of the coverage or properties of the chemisorbed state if reaction proceeds through a precursor intermediate.

SUMMARY

Although is is probable that precursor intermediates play an important role in most surface processes, quantitative description of these states remains elusive, both experimentally and theoretically. We have merely discussed the models which have been proposed and some of the experiments which indicate the existence of precursors. However the current situation is highly unsatisfactory in that the models are extremely primitive and most experiments may be inadequate for modelling.

Only as details of the potential surfaces and dynamics of energy exchange formulated at surfaces are known can one expect to be able to describe these processes quantitatively. Similarly, experiments in which reactant molecules are state selected or product states are identified will be necessary to obtain data with which to compare theory.

REFERENCES CITED

1. J.B. Taylor and I. Langmuir, Phys. Rev. 44, 23 (1933).
2. J.A. Becker in Structure & Properties of Solid Surfaces, Eds. R. Gomer and C.S. Smith (Univ. of Chicago Press, 1952) p. 459.
3. G. Ehrlich, J. Phys. Chem. 59, 173 (1955).
4. P.J. Kisliuk, J. Phys. Chem. Solids 3, 95 (1957); 5, 5 (1958).
5. C. Kohrt and R. Gomer, J. Chem. Phys. 52, 3283 (1970).
6. D.A. King and M.G. Wells, Surface Science 23, 120 (1971).
7. P.W. Tamm and L.D. Schmidt, J. Chem. Phys. 52, 1150 (1970); 55, 4253 (1971).
8. L.R. Clavenna and L.D. Schmidt, Surface Science 22, 365 (1970).
9. L.D. Schmidt in Adsorption Desorption Phenomena, Ed. F. Rica (Cambridge Press 1972).
10. C.S. Steinbruchel and L.D. Schmidt, Phys. Rev. Lett. 32, 594 (1974).
11. M.R. Shannabarger, Surface Science 44, 297 (1974).
12. D.A. King, Surface Science 64, 43 (1977).
13. R. Gorte and L.D. Schmidt, Surface Science 76, 559 (1978).
14. R.E. Stickney et al., Surface Science 26, 522 (1971); 29, 590 (1972); 50, 263 (1975).
15. G. Comsa, R. David, and K.D. Rendulic, Phys. Rev. Lett. 38, 775 (1977); Chem. Phys. Lett. 49, 512 (1977).
16. L.D. Schmidt, to be published.

HETEROGENEOUSLY CATALYZED OSCILLATORY REACTIONS*

H. Suhl and R. E. Lagos**
Department of Physics and
Institute for Pure and Applied Physical Sciences
University of California, San Diego, La Jolla, CA 92093

ABSTRACT

Conditions are derived for setting a given catalyzed reaction into oscillation by manipulating the thermal contact with the heat bath. Also, expressions are given for the changes in selectivity of a catalyst in a small-amplitude oscillatory state compared with the selectivity in the time independent state.

Oscillatory chemical reactions are reactions that proceed non-uniformly in time, in space, or in both, in spite of ostensibly uniform external conditions, such as reactant supply, temperature, etc. They have stimulated a great many investigations ranging from detailed studies[1] of the simplest situations (e.g., monomolecular decomposition) to highly speculative proposals[2] (e.g., morphogenesis). Intermediate between these two extremes one finds the study of simple mathematical models (such as the "Brusselator" and the "Oregonator") where authors suggest that they may simulate, to some extent at least, certain highly complex realistic cases. In the same category one also finds ambitious attempts to treat these phenomena as a form of symmetry breaking, a far-from-equilibrium analogue of the familiar phase transitions of statistical mechanics.

The mathematical description of all these phenomena is hampered by the fact that their rate equations are intrinsically non-linear, at least when written in terms of variables having any reasonable immediacy of physical interpretation. General theories of non-linear differential equations are not available at this time, and therefore the analysis of any particular oscillatory reaction, however successful, does little to illuminate the whole class of phenomena. In particular, we do not know necessary and sufficient conditions that a particular reaction, described by a set of non-linear rate equations

---

*Supported in part by NSF grant #DMR77-24957.
**Partial fulfillment of requirements for Ph.D. degree.

ISSN:0094-243X/80/610097-12$1.50

$$\frac{d\vec{x}}{dt} = \vec{F}(\vec{x}) \qquad \text{(i)}$$

must satisfy in order to oscillate. (In this contribution, we limit the discussion entirely to the case of purely temporal oscillations, assuming spatial uniformity throughout.) Here $\vec{x}$ is the minimal set of variables (concentrations, etc.) needed to describe the reaction, and $\vec{F}$ is a non-linear vector function of $\vec{x}$. (We do know that certain features, such as the presence of an autocatalytic step may favor oscillations.[3])

Here we shall be concerned with deriving general conditions for oscillations when it is possible to manipulate one particular physical process effecting equation (i): the rate at which the temperature of the reacting mass can relax to the temperature of the heat bath. This rate is presumably rather readily controlled in the case of heterogeneous catalysis. Aside from establishing conditions, we shall also describe a possible use of the phenomenon: the manipulation of the selectivity of a catalyst. Under given external conditions it turns out that in the oscillatory state, the time-averaged product-yields in the various output channels differ from the corresponding yields in the steady state. We shall derive expressions for the incremental changes strictly valid only for small-amplitude oscillations, but the <u>relative</u> incremental changes hold for a significant range of amplitudes.

These developments were suggested by an analysis[4] of a particular oscillatory reaction: carbon monoxide over platinum. In ref. 4 the experiments of Dauchet and van Cakenberghe[5] were analyzed, adopting these authors' suggested mechanism for the oscillations. They assumed that the oxidation was a Langmuir Hinschelwood process and argued as follows: Dissociative adsorption of O_2 on platinum is a highly activated process. Beginning at a high temperature, with the surface covered mainly by oxygen, not much of the (exothermic) CO_2-production occurs and the surface cools down, favoring CO adsorption over O_2 adsorption. With more CO available, the reaction proceeds faster, raising the surface temperature. This in turn favors greater O-coverage, and the cycle begins anew. Note that this mechanism could not function if the thermal contact with the heatbath were perfect, i.e., under isothermal conditions.

The analysis in ref. 4 confirmed this picture. At constant partial pressures of O_2 and CO, as well as essentially instantaneous removal of the CO_2 produced, the vector $\vec{x}$ in equation (i) has only three components:

$$\vec{x} = (\theta_1, \theta_2, z)$$

where θ_1 and θ_2 are the fractional surface coverages of O and CO, respectively, and $z = (T - T_B)/T_B$ is the fractional deviation from the bath temperature. The three components of F are

$$\left.\begin{aligned} F_{\theta_1} &= k_o p_{o_2}(1 - \theta_1 - \theta_2)^2 - k_{\ell h}\theta_1\theta_2 \\ F_{\theta_2} &= k_{co} p_{co}(1 - \theta_1 - \theta_2) - k_{\ell h}\theta_1\theta_2 \\ F_z &= -\gamma z + h_1\dot{\theta}_1 + h_2\dot{\theta}_2 + k_3\theta_1\theta_2 \end{aligned}\right\} \qquad \text{(ii)}$$

where p_{O_2} and p_{co} measure the O_2 and CO partial pressures, $k_{\ell h}$ is the Langmuir Hinschelwood rate of formation of CO_2, k_o and k_{co} the adsorption rates of O_2 and CO, and γ is the rate of reversion of surface temperature to the bath temperature. h_1 and h_2 are proportional to the heats of formation of adsorbed O, respectively, and h_3 measures the heat released by the reaction of CO and O on the surface. All reverse reactions are neglected.

THE STEADY STATE

First consider the case $\gamma = \infty$ (i.e., $T = T_B$ at all times). The equations

$$\vec{F} = 0$$

give the critical points of the system (i). These are at most four in number: Total oxygen coverage $\theta_1 = 1$, $\theta_2 = 0$; total CO coverage $\theta_1 = 0$, $\theta_2 = 1$, and two more $(\theta_1^{\pm}, \theta_2^{\pm})$, with intermediate coverages. Only two of these four points are stable, and they are shown in Fig. 1. They are $(\theta_1 = 0, \theta_2 = 1)$ and the "spiral" point, (θ_1^+, θ_2^+), say, in Fig. 1. The point θ_1^-, θ_2^- is a saddle. If the system starts out below the "separatrix" passing through that saddle, it will move to carbon-monoxide "poisoning," If it starts above the separatrix, it ends at (θ_1^+, θ_2^+).

Under some conditions (e.g., high CO partial measure) the two points $\theta_1^{\pm}$, $\theta_2^{\pm}$ cease to exist. In that case the system goes to $(\theta_1 = 0, \theta_2 = 1)$ no matter where it starts (Fig. 2).

When γ is not infinite but still large enough, steady states of essentially the same kind still exist, but the orbits are now three dimensional.

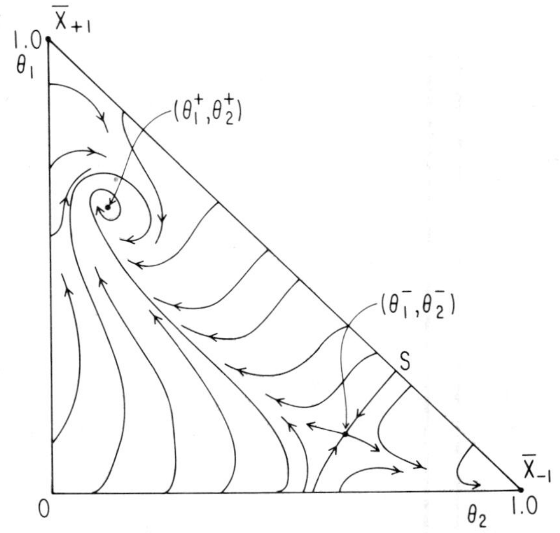

Fig. 1. Orbits in the phase plane for $\gamma = \infty$, under conditions such that $(\theta_1^{\pm}, \theta_2^{\pm})$ exist.

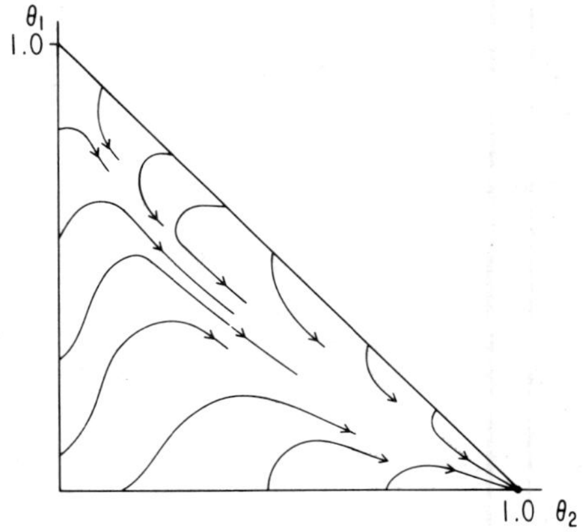

Fig. 2. Orbits in the phase plane for $\gamma = \infty$, under conditions such that $(\theta_1^{\pm}, \theta_2^{\pm})$ cease to exist.

OSCILLATORY STATES

The steady state (θ_1^+, θ_2^+) is stable in the sense of a spiral point only in a certain region of the six dimensional parameter space subtended by six of the seven variables $k_o p_{o_2}$, $k_{co} p_{co}$, $k_\ell h$, γ, h_1, h_2, h_3 (one of the seven may obviously be scaled out by choice of a time scale). Details of this region are given in ref. 4. On a certain hypersurface in this parameter space, the point (θ_1^+, θ_2^+) becomes a so-called center: a small deviation from this point neither grows nor decays, but circles indefinitely around θ_1^+, θ_2^+. When the parameters are situated on one side of the hypersurface, a deviation spirals inward; when they are situated on the other side, a deviation will spiral outwards, ultimately reaching a limit cycle. If the phasepoint starts outside the limit cycle (but not "too close to $\theta_1 = 0$, $\theta_2 = 1$, $z = 0$) it will spiral into the limit cycle. The situation is depicted in Figs. 3a, b and c. The passage of the point in parameter space from one side of the critical hypersurface to the other is called a Hopf bifurcation. In other problems, more elaborate kinds of bifurcation may occur, but we shall not consider these here.

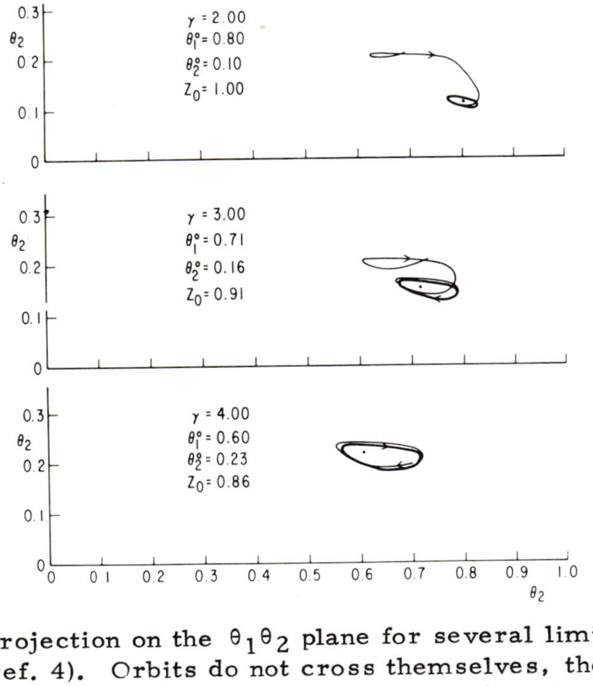

Fig. 3. Projection on the $\theta_1\theta_2$ plane for several limit cycles (ref. 4). Orbits do not cross themselves, the apparent crossing is due to the projection on the $\theta_1\theta_2$ plane.

QUASI-CHAOTIC MOTION

Oscillatory motion is generically connected to Fig. 1. For example, for suitable values the other variables, progressive reduction of the heat leak parameter γ will result in Hopf bifurcation and a limit cycle. To what kind of motion is Fig. 2 connected as γ is reduced? We note that for $\gamma = \infty$ (Fig. 2), the phasepoint seems to "remember" that there used to be a stable point for it to go to; unable to find that point, it is funneled towards ($\theta_1 = 0$, $\theta_2 = 1$). When γ is reduced and z is allowed to change, then under conditions that would favor a limit cycle when (θ_1^+, θ_2^+) exists, the motion becomes quasi-chaotic, with the phasepoint describing the elaborate orbit of Fig. 4 in search of the "lost" singular point. The closer the parameters are to the coalescence condition of $\theta_1^\pm, \theta_2^\pm$, the more elaborate the orbit. Of course, at exact coalescence the rate equation (i) breaks down and must be replaced by some equation for the probability distribution of (θ_1, θ_2, z), since fluctuation effects become important.

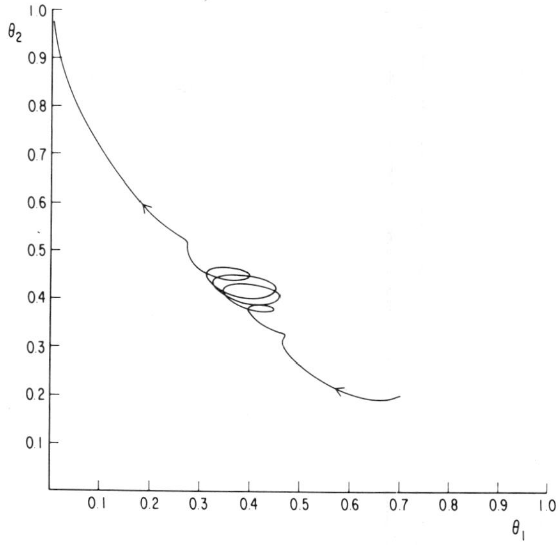

Fig. 4. Projection on the $\theta_1\theta_2$ plane of an orbit with parameters such that ($\theta_1^\pm$, $\theta_2^\pm$) cease to exist and $\gamma \neq \infty$ (ref. 4). Orbits do not cross themselves, the apparent crossing is due to the projection on the $\theta_1\theta_2$ plane.

CONDITIONS FOR HOPF BIFURCATION AND LIMIT CYCLE IN THE CASE OF AN ARBITRARY REACTION

Reverting to equation (i), it is clear how one searches for a limit cycle. One first finds the critical points, i.e., the solution $\vec{x}_o$ of $\vec{F}(\vec{x}) = 0$. One then seeks the characteristic time constants by solving the linearized problem

$$\delta \dot{x}_i = \sum_j F_{ij}(\vec{x}_o)\, \delta x_j$$

around each critical point, to determine its stability. Here

$$F_{ij}(\vec{x}_o) = \partial F_i / \partial X_j^o \quad .$$

Concentrating on a particular stable point, one then imposes the condition that it becomes a center, which requires that one pair of characteristic roots is purely imaginary, $\pm i\omega$.

In the general case, this is obviously quite an intricate problem. For the special case in which n-1 of the components of $\vec{x}$ are concentrations, and the remaining one is the temperature deviation z, a simple pair of conditions may be derived which involves no more than matrix inversion, once the critical point $\vec{x}_o$ has been found. This is because of the special structure of equation (i) in this case:

$$\begin{cases} \dot{\theta}_i = F_i(\theta_j, z) \\ \dot{z} = -\gamma z + \sum_{i=1}^{n-1} h_i \dot{\theta}_i \; {}^{*} \end{cases} \qquad i = 1, 2, \ldots, n-1 \quad \text{(iii)}$$

where we denote the concentrations by θ_i (for surface reactions they are coverages). $\left(\text{Evidently } F_n = \sum_{j-1}^{n-1} k_i F_i - \gamma z\right)$. The steady state is given by

---

*At first sight this seems to differ from the expressions (ii) for CO oxidation. The reason is that in (ii), the CO_2 was considered pumped off rapidly so that its concentration, θ_3 say, was zero. At a finite pump rate r, there will be an additional equation $\dot{\theta}_3 = k_{\ell h} \theta_1 \theta_2 - r$.

$$z = 0, \quad F_i(\theta^o, 0) = 0$$

The equations for the deviations $\delta\theta_i$, z from the stable point are

$$\delta\dot{\vec{\theta}} = \mathfrak{F}\,\delta\vec{\theta} + \vec{\ell}\,\delta z$$

$$\delta\dot{z} = -\gamma\,\delta z + \vec{k}\cdot\delta\dot{\vec{\theta}}$$

where the matrix

$$\mathfrak{F} = \left\| \frac{\partial F_i}{\partial \theta_j} \right\|_{\theta = \theta^o} \qquad i, j = 1, 2, \ldots, n-1$$

and where

$$\vec{\ell} = \partial\vec{F}/\partial z$$

and

$$\vec{h} = \{h_i\}$$

are n-component column vectors. $\vec{F}$ depends on z throughout the rates $\vec{k}$. For a time-variation of a form $\sim e^{\lambda t}$, it is then found that

$$\lambda + \gamma = \lambda\,\vec{h}'\,\frac{1}{\lambda I - \mathfrak{F}}\,\vec{\ell}$$

where $\vec{h}'$ is the transpose of $\vec{h}$ and I is the unit matrix. Bifurcation occurs when $\lambda = \pm i\omega$, ω real. Conditions for this are

$$\gamma = \vec{h}'\,\frac{\omega^2}{\mathfrak{F}^2 + \omega^2 I}\,\vec{\ell}\,; \qquad 1 = -\vec{h}'\,\frac{\mathfrak{F}}{\mathfrak{F}^2 + \omega^2 I}\,\vec{\ell}$$

To examine the content of these equations, suppose there are a total of N reactions in which the n reactants participate. Let W_α^+ be the "forward" rate of the α'th reaction, and W_α^- its backward rate. The W's are generally products of several of the θ's (depending on the order of the reaction). Then if $\nu_i^\alpha = (\nu_i^\alpha)^- - (\nu_i^\alpha)^+$ is the difference of the stoichiometric coefficients of the i'th reactant on the right and lefthand sides of the α'th reaction, we have

$$F_i = \sum_{\alpha=1}^{N} \nu_i (W_\alpha^+ - W_\alpha^-)$$

Noting that F_i depends on the temperature deviation only through the activation energies ϵ_α^+ and ϵ_α^- of the forward and reverse reactions, we immediately find

$$\ell_i = \frac{\partial F_i}{\partial z} = \sum_{\alpha=1}^{N} \nu_i^\alpha (\epsilon_\alpha^+ W_\alpha^+ - \epsilon_\alpha^- W_\alpha^-)$$

In the case of a very high frequency limit cycle at least, this enables us to draw some broad conclusions. For large ω, we must have

$$\gamma = \vec{h}' \cdot \vec{\ell}$$

$$\omega^2 = -\vec{h}' \mathfrak{F} \vec{\ell}$$

Suppose the reverse reaction rates are small. Then all matrix elements of $\mathfrak{F}$ are negative. Therefore, if all h_i are positive, the ℓ_i must be positive, i.e., the weighted excess of forward activation energies over reverse activation energies must be positive. Furthermore, $\vec{h}'$ or $\vec{\ell}$, or both, must be large. Since the elements of $\mathfrak{F}$ by hypothesis are much less than ω, γ must be of order ω^2/f, where f is a typical matrix element of $\mathfrak{F}$. Thus the heat leak to the bath must be large for a high frequency limit cycle.

For a low frequency limit cycle we must have

$$-h' \frac{1}{\mathfrak{F}} \ell \approx 1$$

and then $\omega^2 \propto \gamma$. No very general statement is now possible, because nothing very general can be inferred about the signs of the matrix elements of $(\mathfrak{F})^{-1}$ from the signs of the elements of $\mathfrak{F}$.

CHANGES IN THE SELECTIVITY OF A CATALYST IN THE OSCILLATORY STATE

We assume that the product formation occurs only as a bimolecular process. Then the formation rate of compound (ij) may be written

$$r_{ij}^{(1)} = k_{ij}^{(1)} \overline{\theta_i(t)\,\theta_j(t)}$$

if it occurs by the L. H. mechanism, and

$$r_{ij}^{(2)} = k_{ij}^{(2)} p_i \overline{\theta_j(t)} + k_{ji}^{(2)} p_j \overline{\theta_i(t)}$$

if it occurs by the Eley-Rideal process. (Note that many of the rate constants $k_{ij}^{(1)}$ and $k_{ij}^{(2)}$ will be zero. Note also that $k_{ij}^{(2)}$, in contrast with $k_{ij}^{(1)}$, is not symmetric in i and j.) The bar denotes the time average. The partial pressures p_i are assumed constant; if the experiment is such that they vary, $p_i \overline{\theta}_j$ must be replaced by $\overline{p_i \theta_j}$, etc. We write

$$\theta_i(t) = \overline{\theta_i(t)} + \delta\theta_i(t)$$

where $\overline{\delta\theta_i(t)} = 0$, and we further split $\overline{\theta}_i$ as follows

$$\overline{\theta_i(t)} = \theta_i^{o} + \delta\theta_i^{o}$$

where θ_i^{o} is the (unstable) critical point enclosed by the limit cycle (recall that we are considering Hopf bifurcation only).

The critical points of the set of equations (iii) are independent of the heat leak γ. Consider those critical points that are stable when $\gamma = \infty$, and confine attention to one particular one of these. Then the rates when $\gamma = \infty$ are steady state rates:

$$r_{ijs}^{(1)} = k_{ij}^{(1)} \theta_i^{o} \theta_j^{o}$$

$$r_{ijs}^{(2)} = k_{ij}^{(2)} p_i \theta_j^{o} + k_{ji}^{(2)} p_j \theta_i^{o}$$

Now let γ be reduced until the point θ_i^{o} goes unstable and a limit cycle occurs. Then, for L.H., the fractional increment in the rates

$$\frac{\Delta r_{ij}^{(1)}}{r_{ijs}^{(1)}} = \frac{\delta\theta_i^{o}}{\theta_i^{o}} + \frac{\delta\theta_j^{o}}{\theta_j^{o}} + \frac{\overline{\delta\theta_i(t)\,\delta\theta_j(t)}}{\theta_i^{o}\theta_j^{o}}$$

while for ER,

$$\frac{\Delta r_{ij}^{(2)}}{r_{ijs}^{(2)}} = \frac{\delta\theta_i^o}{\theta_i^o}$$

where it has been assumed that $k_{ji}^{(2)} \gg k_{ij}^{(2)}$. For small limit cycles, these can be evaluated in terms of the F-matrix. For small cycles, the various $\delta\theta_i^o$ are all proportional to one single scaling parameter A which may be called the amplitude of the limit cycle, and $\delta\theta_i$ is A times the i'th component of the eigenvector of the linearized equation at bifurcation. Because of the structure of equations (iii), it is particularly convenient to choose the temperature deviation itself, δz, as the scaling parameter. Writing $\delta z = \delta z\, e^{i\omega t} + \delta z^* e^{-i\omega t}$, $\delta\theta_i = \delta\theta_i e^{i\omega t} + \delta\theta_i^* e^{-i\omega t}$, it is then easily seen that

$$\overline{\delta\theta_i(t)\,\delta\theta_j(t)} = \delta\theta_i\delta\theta_j^* + \delta\theta_i\delta\theta_j^* + \delta\theta_i^*\delta\theta_j$$

$$= \left[\left(\frac{\mathfrak{F}}{\omega^2 + \mathfrak{F}^2}\ell\right)_i \left(\frac{\mathfrak{F}}{\omega^2 + \mathfrak{F}^2}\ell\right)_j + \omega^2\left(\frac{1}{\omega^2 + \mathfrak{F}^2}\ell\right)_i \left(\frac{1}{\omega^2 + \mathfrak{F}^2}\ell\right)_j\right]\delta z\,\delta z^*$$

This gives the sum-rule

$$\sum_{i,j} h_i\, \overline{\delta\theta_i(t)\,\delta\theta_j(t)\, h_j} = \left(1 + \frac{\gamma}{\omega}\right)\delta z^*\,\delta z \quad .$$

(Note that only for a small limit cycle can the higher harmonic content of $\delta\theta$ and δz be neglected.)

$\delta\theta_i^o$ is found from

$$0 = \overline{F_i(\theta(t),\, z(t))}$$

which yields, upon expansion

$$\delta\theta_i^o = \frac{1}{2}\sum_m (\mathfrak{F}^{-1})_{im}\left[\frac{\partial^2 F_m}{\partial\theta_j^o\,\partial\theta_k^o}\,\overline{\delta\theta_j(t)\,\delta\theta_k(t)} + \frac{\partial^2 F_m}{\partial z^2}\,\delta z^*\,\delta z\right]$$

This takes a particularly simple form if there is no dissociative adsorption, and no L.H. processes take place. Then the F_m are linear functions of the θ, and F is independent of the θ_i^0's. In that case

$$\delta\theta_i^0 = \frac{1}{2}\sum_m (F^{-1})_{im} \frac{\partial^2 F_m}{\partial z^2}\,\delta z^*\,\delta z$$

It is reasonable to expect that the ratios

$$\frac{\Delta r_{ij}}{\Delta r_{i'j'}}$$

which measure the relative changes in selectivity between channels ij and i'j', and which for small limit cycles are independent of amplitude to be only moderately sensitive to the actual value of a finite amplitude.

REFERENCES

1. P. Hugo, in <u>4th European Symposium on Chemical Reaction Engineering, Brussels 1968</u>, 1971, p. 459.

2. A. M. Turing, Phil. Trans. Roy. Soc. London B237 (1952), 37.

3. G. Nicolis and J. Portnow, Chem. Rev. 13 (1973), 365.

4. R. E. Lagos, B. C. Sales and H. Suhl, Surface Science 82 (1979), 525.

5. J. Dauchot and J. Van Cakenberghe, Nature Phys. Sic. 246 (1973), 61.

CLASSICAL TRAJECTORY STUDIES OF UNIMOLECULAR DYNAMICS

W.L. Hase
Department of Chemistry, Wayne State University
Detroit, Michigan 48202

ABSTRACT

The application of classical trajectories to the study of unimolecular dynamics is reviewed. A brief description is given of the methodology of the classical trajectory calculations. The earlier triatomic trajectory studies of Bunker are discussed, as well as studies of CH_3NC, H-C≡C-Cl, and C_2H_5 decomposition. Calculated lifetime and relative translational energy distributions are compared with experimental observations and predictions of the RRKM theory. The extension of gas-phase unimolecular trajectory studies to studies of gas-surface unimolecular reactions is also discussed.

I. INTRODUCTION

The first Monte Carlo trajectory study of a unimolecular reaction was performed by Bunker seventeen years ago.[1] Since that time the total number of unimolecular trajectory studies is still relatively small.[1-18] However, the activity within this field has increased substantially with the last few years, and it is expected to continue. This is because of the many unanswered questions which remain concerning the dynamics of unimolecular reactions.[19] Classical trajectories are the high energy and high particle mass limit of quantum-mechanical scattering.[20] A quantum-mechanical scattering calculation of a unimolecular reaction would involve an exceedingly large number of states, and such a calculation has not been performed. Thus, the appeal of classical trajectories.

An important question is the degree to which classical trajectories resemble the real, quantum-mechanical world. This question is answerable if potential energy surfaces are known exactly so that classical trajectory and experimental results can be compared directly, or if classical and quantum scattering results can be compared for the same potential energy surface. Neither of these tests have been made for a unimolecular reaction. However, they have been made for bimolecular A+BC → AB+C reactions. In general, these tests have shown that if tunnelling and interference effects are not in question classical trajectories make quantitative predictions about dynamical properties such as reactive cross sections, product energy distributions, and scattering angles.[21] For the $H+H_2$ reaction, good agreement is found between classical and quantum scattering values for the above dynamical properties.[22] Thus, classical trajectories are expected to give qualitative and in many cases quantitative results for unimolecular reactions.

The remainder of this paper is divided into three parts. In section II the methodology of unimolecular classical trajectory

ISSN:0094-243X/80/610109-27$1.50

calculations is briefly described. Section III consists of specific trajectory results, and conclusions reached from the trajectory calculations. The last section, IV, describes the extension of unimolecular trajectories to gas-surface unimolecular reactions.

II. CLASSICAL TRAJECTORY METHOD

A. Selecting the initial conditions

The procedure used in selecting the initial conditions will depend upon the experimental situation being simulated. We will first deal with cases where the initial states of the molecule are chosed randomly. Thermal excitation is thought to produce excited molecules with a near random distribution of initial states. The problem in random sampling is to prepare a uniform distribution of points in the phase space of the molecule, in a region bounded by the hypersurfaces on which H and H+dH are constant. Since the molecule contains more energy than required for dissociation, the extensions of the molecular phase space must be defined so that product states are not chosen in the selection procedure. For unimolecular reactions with a well-defined potential energy barrier along the reaction coordinate, the extensions of the molecular phase space should be defined so that only phase points up to the top of the barrier are selected. Defining the molecular phase space for a unimolecular reaction without a well-defined barrier, such as simple bond rupture, is more difficult.

In the development of activated complex theory for simple bond dissociation reactions, AB → A+B, a surface separating reactant AB and product A+B states was assumed to be orthogonal to the reaction coordinate, relative AB distance r_{AB}, and to pass through the top of the rotational barrier.[23] The rotational barrier arises from A-B orbital angular momentum, L,

$$V_{eff} = V(r_{AB}) + L^2/2\mu_{AB}r_{AB}^2 \quad . \tag{1}$$

Detailed trajectory studies by Bunker and Pattengill for triatomic models[3] have shown that the rotational barrier does not constitute a point of separation between reactant and product states. Their trajectory results indicated that the surface separating reactant and product states can be assumed to be orthogonal to r_{AB} but should be located at a point along r_{AB} where the number of states is minimized;[24] i.e.,

$$\partial W(E,r_{AB})/\partial r_{AB} = 0.0, \tag{2}$$

where $W(E,r_{AB})$ is the sum of vibrational states in the molecule for a specific internal energy E and at an A-B separation of r_{AB}. This criterion is the same as the minimum flux criterion discussed by Wigner,[25] Horiuti,[26] and Keck[27] in terms of variational reaction rate theory. For unimolecular reactions with well defined potential energy barriers Eq. (2) will locate the dividing surface at the top of the barrier. The value of the reaction coordinate chosen by Eq. (2) is called the critical value r^+ and the molecular structure at that point is referred to as the critical configuration. In general r^+ is a function of both vibrational and rotational energy. Examples are given in Fig. 1 and 2 for ethane dissociation;[28] i.e., $C_2H_6 \rightarrow 2CH_3$.

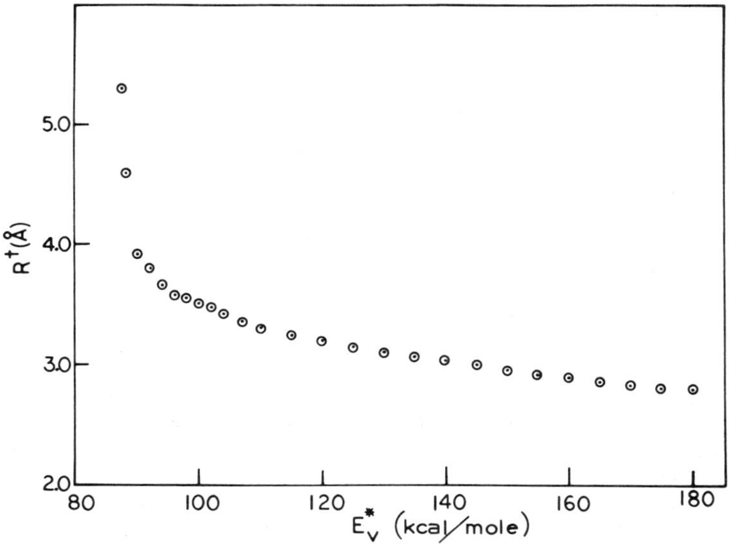

Figure 1. Plot of the critical value of the reaction coordinate, R^+, versus vibrational energy, $E_v{}^*$, for ethane dissociation.

Once the extensions of the molecular phase space have been evaluated a uniform distribution of points must be chosen. For up to about eight dimensions this can be done in an exact manner using extended rejection techniques.[1,29] As the number of dimensions increase one must use approximate methods for sampling the molecular phase space. This is because the number of orthants[30] is much greater than the total number of trajectories which can be economically computed. For example, methyl isocyanide (CH_3NC) has 2^{30} orthants, if center-of-mass motion is removed. Bunker and Hase[5] have described two different approximate sampling methods for situations where the total number of samplings is much less than the total number of orthants in the molecular phase space. They are orthant and progressive sampling.

The first step in orhtant sampling is to determine the extensions of the molecular phase space at a constant energy H(p,q). This is done by initially placing the atoms at their equilibrium positions

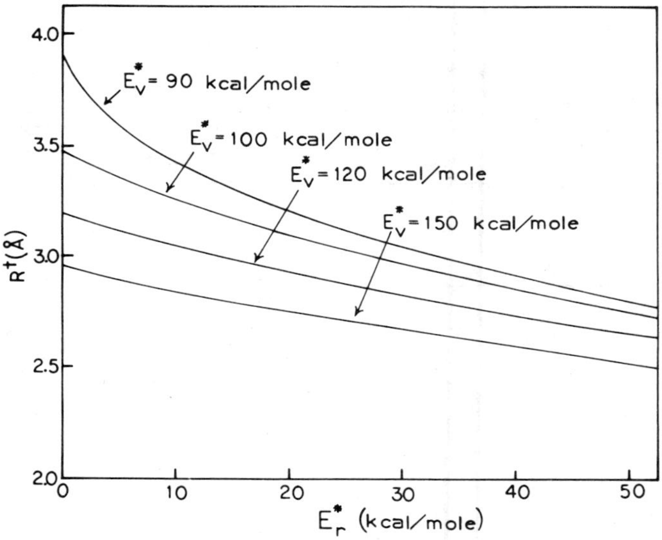

Figure 2. Effect of rotational energy, $E_r{}^*$, on R^+.

q_i°, with all conjugate momenta equal to zero. The q_i's and p_i's are then varied independently to find the maximum and minimum values for q_i and p_i given H(p,q). These limiting values for coordinates and momenta are labelled p_i^{max}, p_i^{min}, q_i^{max}, and q_i^{min}. Since the total energy of the molecule exceeds the unimolecular threshold(s), constraints must be placed on some of the q_i^{max} and q_i^{min} to eliminate the selection of product states. The second step in orthant sampling is to choose one of the 2^n orthants at random. Originally this was done by projecting a n-dimensional random unit vector with components x_i onto the hypersurface represented by (q_i^{max}, q_i^{min}, p_i^{max}, p_i^{min}). However, if the Hamiltonian is not diagonal in the momenta and coordinates, as is usually the case, the initial conditions within an orthant may not be random. As a result, the initial conditions may not give the correct relationship between kinetic and potential energy. A short-time sampling transient may also appear in the lifetime distribution.[5]

Recently, we modified the earlier orthant sampling procedure to produce initial states with the correct relationship between average potential ($\bar{V}$) and average kinetic ($\bar{T}$) energy.[18] For most situations the potential energy surface is strongly anharmonic so one cannot simply apply the virial theorem to obtain the correct relationship between the average values of potential and kinetic energy. However, this relationship may be evaluated by integrating bound trajectories for many cycles and recording the potential and kinetic energies at random time intervals. Interestingly, for every anharmonic potential energy surface we have studied $\bar{V} \simeq \bar{T}$ was found.

To attain the correct relationship between $\bar{V} \simeq \bar{T}$ for orthant sampling, the values for p_i^{max} and p_i^{min} are uniformly scaled by a parameter P_{scale}. The p_i's and q_i's are then chosen from

$$p_i = x_i \cdot (p_i^{max}, p_i^{min}) \cdot \text{scale}$$
$$q_i = x_{i+n/2} \cdot [(q_i^{max}, q_i^{min}) - q_i^o] + q_i^o, \quad (3)$$
$$i=1,n/2$$

where the x_i in Eq. (3) are all positive and $\sum_{i=1}^{n} x_i^2 \equiv 1.0$. Equal probabilities are assigned to the use of the maximum or minimum values for p_i and q_i in Eq. (3). Finally the values for p_i and $(q_i - q_i^o)$ are scaled uniformly so that the correct energy is attained.

Progressive sampling is a more sophiscated phase-space method with interesting properties. The starting point is any trajectory; e.g., initialized by orthant sampling. During integration, after about one-half of an average vibration period, a n-dimensional vector of coordinates and momenta is saved. Call this vector V_s. The key to the method is the not easily visualized fact that in spaces of high dimensionality, randomly oriented vectors are almost certain to be nearly orthogonal. This if we now choose a random vector V_r, values of the (V_s, V_r) angle will cluster just below 90°. We now bisect this angle

$$V_r' = V_s/|V_s| + V_r/|V_r| \quad (4)$$

and produce a new vector V_r' whose angle with V_s is near 45°, and by repetition, 22.5°, 11.25°, etc. In practice, a maximum angle Ψ is specified, and by repeated bisection V_s is displaced in a random direction between Ψ and 0.5Ψ. Scaling, like that used in orthant sampling, must then be applied. The next trajectory is then ready to start.

For Ψ=90°, progressive sampling reduces to orthant sampling. For Ψ very small, it starts trajectories in regions of phase space with high probability of traversal. Random sampling is approached, but at the expense of a requirement for very long selection chains in order to cover the whole space properly. For our calculations we have found Ψ-15° to be a reasonable choice.

In many experiments molecules are excited by nonrandom processes. For example;

chemical activation	$X+Y \rightarrow (XY)^*$
internal conversion	$S_1 \rightarrow S_o^*$
infrared laser	$nh\nu + XY \rightarrow (XY)^*$.

To simulate each of the above nonrandom processes requires a different sampling procedure. Chemical activation[31] will be discussed first. Here, the excitation is primarily due to the energy released as X and Y combine. If X is an atom and Y a diatom the sampling procedure developed for A+BC bimolecular reactions[20] can be used. If both X and Y are diatoms the sampling is similar to A+BC with only minor modifications. However, when X and/or Y are polyatomic the sampling is considerably more difficult. Here, sampling of the variables describing the X+Y relative spatial orientation and velocity is the same as for A+BC. The difficulty arises in choosing the rotational and vibrational states of X and Y. This is because rotational and vibrational energies are inseparable due to coriolis effects, and because the vibrational modes are not exactly orthogonal except at finite displacements. To date only approximate techniques have been used in sampling internal states of polyatomics. The assumptions of rotational and vibrational separability and of normal modes have been used in sampling internal states of the methyl radical (CH_3)[32] and of methane (CH_4).[33] For small amounts of vibrational excitation these assumptions seem to be valid. In a more recent trajectory calculation involving internal excitation of CO_2,[34] coriolis terms were included in a more rigorous treatment of the separability of vibration and rotation. For some calculations one may wish to choose random internal states of X and/or Y using orthant sampling.

Simulating the excitation of a molecule by $S_1 \rightarrow S_0$ internal conversion is an even more difficult problem than simulating chemical activation. For triatomic molecules one may sample the vibrational states in the higher electronic state (S_1) as they are prepared experimentally,[35] and then describe the $S_1 \rightarrow S_0$ electronic transition using either the surface hopping[36] or the semiclassical method.[37] Once S_0 is formed classical trajectories can be used to follow its decomposition. The surface hopping and semiclassical methods both become more approximate when applied to molecules containing more than 3-atoms because of the difficulty in rigorously treating the additional degrees of freedom. To treat the decomposition of chloroacetylene (H-C≡C-Cl) prepared by $S_1 \rightarrow S_0$ internal conversion,[12] Franck-Condon factors were computed for this transition using normal modes. The most probable distribution of normal mode vibrational states in S_0 was evaluated, and this distribution with random phases was then used in sampling the S_0 internal states. This procedure has significant approximations in that normal modes were used in computing the Franck-Condon factors and in sampling the internal states of S_0. A more rigorous treatment will require a better representation of internal states in highly excited molecules.

There are two techniques one can use to simulate the dissociation of molecules by a high power infrared laser. In the first, the theoretical model consists of classical trajectories moving on a potential energy surface with an external time-dependent force from the classical electromagnetic field. In this model the classical Hamiltonian is time dependent, since the molecule gains or loses energy from the external electromagnetic field. The Hamiltonian is

$$H(q,p,t) = H_o(q,p) + H_1(q,t) \tag{5}$$

where $H_o(q,p)$ is the Hamiltonian of the unperturbed molecule and $H_1(q,t)$ models the interaction of the molecule with the electromagnetic field. This theoretical model has been used in one unimolecular trajectory study.[13] Miller[38] has advanced a different method for treating both the electromagnetic field and the molecular system classically. In his theoretical model the equivalence of the radiation field to a set of harmonic oscillators is invoked, and the harmonic oscillators representing the field are treated as classical oscillators. The Hamiltonian for the complete system-radiation field, molecules, and their interaction is introduced, and it conserves the total energy of the complete system; i.e., molecules plus radiation field. Absorption, induced emission, and even spontaneous emission appear in a completely straightforward manner. To date this theoretical model has not been used in a unimolecular trajectory study.

B. Integrating the classical equations of motion

Hamilton's equations have been universally used in computing the trajectories. These are first order differential equations and are easily solved by numerical integration. In a conservative system the potential energy $V(q_i)$ is a function of coordinates alone.

$$P_i = \partial T/\partial \dot{q}_i \; ; \tag{6}$$

$$H = T(P,q) + V(q) \; ; \tag{7}$$

$$\dot{q}_i = \partial H/\partial P_i = \partial T/\partial P_i \; ; \tag{8}$$

$$\dot{P}_i = -\partial H/\partial q_i \; . \tag{9}$$

Eqs. (8) and (9) are numerically integrated in calculating the classical trajectory. The coordinate system one wishes to use is the one that minimizes the computational effort. Several aspects of the calculation must be considered. For many situations the coordinates used in selecting the initial conditions will have to be transformed to those used in the numerical integration. Coordinate transformations may also be required in analyzing the trajectory results. One would like to minimize the total number of coordinate transformations. In addition, the coordinates used for the numerical integrations should make Eqs. (8) and (9) as simple as possible.

Internal cartesian and reduced cartesian coordinates have been used in unimolecular trajectory studies. Internal coordinates have only been used for triatomic trajectories. This coordinate system reduces the number of coordinate tranformations, but results in complicated expressions for Eqs. (6)-(9). The Hamiltonian for an ABC molecule in internal coordinates with zero angular momentum is

$$H = (m_A+m_B)P_1^2/2m_Am_B + (m_B+m_C)P_2^2/2m_Bm_C$$

$$+P_1P_2\cos\alpha/m_B + [m_A(m_B+m_C)r_1^2 + m_C(m_A+m_B)r_2^2$$

$$-2m_A m_B r_1 r_2 \cos\alpha] P_\alpha^{\;2}/2m_A m_B m_C r_1^{\;2} r_2^{\;2}$$

$$-(P_1/r_2+P_2/r_1)P_\alpha \sin\alpha/m_B + V(r_1,r_2,\alpha), \qquad (10)$$

where the subscripts 1 and 2 belong to the AB and BC bonds respectively, and α is the ABC bond angle. For molecules containing four or more atoms the use of internal coordinates is not advised, since the kinetic energy in internal coordinates become extremely complicated.[39] In the reduced cartesian system the c.m. motion is suppressed and the number of coupled differential equations which must be numerically integrated is reduced by three. The Hamiltonian for an ABC molecule in reduced cartesians is

$$H = (m_A+m_B)(P_{x1}^2+P_{y1}^2+P_{z1}^2)2m_A m_B$$
$$+(m_C+m_B(P_{x2}^2+P_{y2}^2+P_{z2}^2)2m_C m_B$$
$$+(P_{x1}P_{x2}+P_{y1}P_{y2}+P_{z1}P_{z2})/m_B+V(x_1,y_1,z_1,x_2,y_2,z_2), \qquad (11)$$

where as in Eq. (10) the subscripts 1 and 2 belong to the AB and BC bonds respectively. Eq. (11) is easily extended to molecules containing four or more atoms.[40]

Integration algorithms which have been used in unimolecular trajectory studies are the 4th order Runge-Kutta-Gill, 6th order Adams-Moulton predictor-corrector, and 6th order hybrid Gear procedures.[20,41] The most popular procedure is to start the trajectory with the Runge-Kutta-Gill integrator and then shift to the faster but not self-starting Adams-Moulton or hybrid Gear routine. Studies have shown that there are no advantages in increasing the order of the Adams-Moulton integrator beyond 6th.[41,42] Numerical comparisons between the Adams-Moulton and Gear algorithms indicate that the Gear method is slightly more efficient, i.e., less than 20%. No advantage is found in using algorithms with variable integration stepsize. The effects are nil or even deleterious at times. Recently a new integration algorithm, with promising possibilities, has been developed.[43] However, it has not been used and tested in a unimolecular trajectory study.

C. Trajectory results

The properties of the classical trajectories which are of interest are:

1. The lifetime distribution. Lifetimes, τ, are obtained from the time it takes each reactive trajectory to reach the critical surface separating reactants and products. The observed lifetimes can be collected into a histogram, where the number of lifetimes N in each time division is normalized by $N_0\Delta t$ (N_0 is the total number of trajectories and Δt is the lifetime width). The lifetime distribution is then given by

$$\ln P(t) = \ln[N(t\rightarrow t+\Delta t)/N_0\Delta t] \; . \qquad (12)$$

If the lifetimes are random

$$\ln P(t) = \ln k - kt, \tag{13}$$

where k is the unimolecular rate constant.

2. Translational energy distribution between the dissociation fragments A and B. This is given by

$$E_{rel} = \tfrac{1}{2}\mu_{AB}\, V_{AB}^2 \ , \tag{14}$$

where μ_{AB} is the A+B reduced mass and V_{AB} is the relative velocity between the centers of mass.

3. The internal energies of the dissociation fragments. If the internal energies are too large, rotational and vibrational energies will not be separable nor will the vibrational energy will be separable into energies of specific normal modes. At low enough energies (tests are required) these separations are possible.

4. Bond energies. This is an approximate measure of the energy in a bond and is given by

$$E_{ij} = \tfrac{1}{2}\mu_{ij}|\dot{r}_{ij}|^2 + V(r_{ij}) \ , \tag{15}$$

where μ_{ij} is the reduced mass of the i and j atoms which constitute the bond, $|\dot{r}_{ij}|$ is the time derivative of the magnitude of the bond length, and $V(r_{ij})$ is the interaction potential between the bonds (e.g., a Morse function).

5. The orbital angular momentum of separation of A and B.

$$\vec{L}_{AB} = \mu_{AB}\vec{r}_{AB} \times \vec{V}_{AB} \quad . \tag{16}$$

6. The scattering angle. This quantity is only needed in simulating chemical activation and molecular beam experiments. It is found by taking the dot product between the initial and final relative velocity vectors.

7. Other properties. These include the angle between the initial and final orbital angular momentum vectors, and the product impact parameter b, found from

$$|\vec{L}_{AB}| = \mu_{AB} b |\vec{V}_{AB}| \tag{17}$$

at infinite separation.

D. Potential energy surface

When performing a unimolecular trajectory study an analytic representation of the potential energy surface is required.[20,44] This is because the classical trajectories will move over all regions of the potential energy surface. Different types of analytic potential energy surfaces have been used in unimolecular trajectory calculations, depending upon the type of study. Most studies have employed model potential energy surfaces. If the model is carefully chosen the results can be very insightful. They provide a representation of the unimolecular reaction at a microscopic level and can also be compared with predictions of statistical theories; e.g.,

RRKM and phase space theories. Recently, efforts have been made to derive "realistic" potential energy surfaces so that trajectory and experimental results can be directly compared.[11,14,18,44] However, these "realistic" potential energy surfaces are still partially models.

Recent advances in ab initio quantum chemical procedures have made it possible to calculate what are thought to be near quantitative potential energy surfaces.[45] For systems containing 3 atoms it is possible to calculate potential energies at a sufficient number of internuclear configurations to derive a complete potential energy surface. These points can then be fit to a particular functional form. If the quantum chemical surface does not agree with experimental observables it is often scaled to match the experimental heat of reaction, activation energy and vibrational frequencies. Thus, if scaling is required the surface is in some respects a model.

For unimolecular reactions containing more than 3 atoms it becomes extremely expensive to calculate a complete potential energy surface. What can be done for these cases, is to calculate a quantum mechanical surface for the degrees of freedom directly associated with the reaction and treat the remaining degrees of freedom by semi-empirical functions.

Probably in the next few years analytic potential energy functions will be derived from a combination of experimental and theoretical data. In addition to experimental heats of reaction and activation energies, spectroscopically derived quadratic, cubic and quartic force constants should also prove valuable in deriving analytic surfaces.

III. TRAJECTORY RESULTS

The early trajectory calculations by Bunker[1,2] suggested that there may be molecules for which internal energy is redistributed slower than the molecule decomposes. Bunker investigated triatomic models (A-B-C) with a potential function of the form

$$V_{ABC} = D_{AB}\{1-\exp[-\beta_{AB}(r_{AB}-r^{\circ}_{AB})]\}^2 + D_{BC}\{1-\exp(-\beta_{BC}(r_{BC}-r^{\circ}_{B})_C]\}^2$$

$$+\tfrac{1}{2}k_{\theta}(\theta-\theta^{\circ})^2, \qquad (18)$$

where θ is the A-B-C angle. Bond energies, force constants and eqiulibrium geometries were chosen to fit experimental data. For some models k_{θ} was written as a function of r_{AB} and r_{BC}. Initial states were chosen randomly at fixed total energies. Models with equivalent or nearly equivalent masses like O_3 and N_2O have random lifetime distributions, i.e., $P(t) = k\exp(-kt)$. However, triatomic H-C-C type models which resemble hydrocarbons were found to have nonrandom lifetime distributions. This is true even if rotation is included. Examples of random and nonrandom distributions are given in Fig. 3. As discussed by Bunker, the shapes of the nonrandom distributions, suggests that for these models passage through the dividing surface separating reactants and products is not the rate-limiting step in the dissociation. An internal bottleneck appears to be present This will be discussed in more detail below.

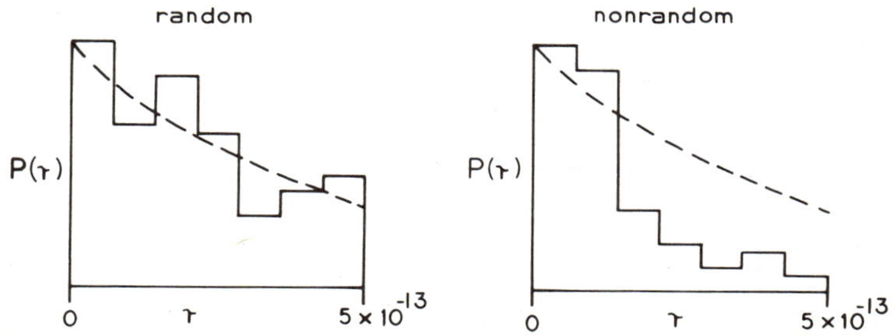

Fig. 3. Examples of trajectory, random and nonrandom, lifetime distributions.

Regardless of the nature of the lifetime distribution, the unimolecular rate constant can be equated to the intercept of the trajectory lifetime distribution. Bunker and Pattengill[3] found that if anharmonicity was accounted for and the critical configuration chosen by Eq. (2) the trajectory and RRKM rate constants agreed to within 50%.

The above trajectory calculations have provided a basis for later trajectory studies. In the remaining part of the section trajecotry studies in which the author has participated will be discussed. The unimolecular reactions which will be considered are methyl isocyanide isomerization ($CH_3NC \rightarrow CH_3CN$), chloroacetylene dissociation ($H{-}C{\equiv}C{-}Cl \rightarrow H{+}C{\equiv}C{-}Cl$ and $H{-}C{\equiv}C{+}Cl$) and ethyl radical dissociation ($C_2H_5 \rightarrow H{+}C_2H_4$).

CH_3NC. This study[4,5] consisted of approximate random sampling and of potential energy functions with varying degrees of anharmonicity. The simplest potential function was

$$V = \sum_{i=1}^{5} D_i\{1-\exp[-\beta_i(r_i-r_i^\circ)]\}^2 + \tfrac{1}{2}k_\phi \sum_{i=1}^{3} (\phi_i-\phi_i^\circ)^2 + \tfrac{1}{2}kx \sum_{i=1}^{3} (x_i-x_i^\circ)^2 + \sum_{i=0}^{5} \alpha_i \cos i\theta. \qquad (19)$$

It has five Morse bonds corresponding to the N≡C and three C-H stretches, and a stretch from the methyl C to the center of the NC group. Harmonic forces resist the distortion of the H-C-H angles χ and the H-C-(NC center) angles ϕ. The five term Fourier cosine series expresses the variation of potential energy along the reaction coordinate, which is the angle θ between NC and the bond from the NC center to the methyl C;

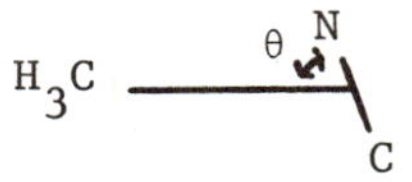

The five Fourier coefficients were chosen to give an activation energy and heat of reaction in agreement with experiment,[46] to match the C-N≡C(CH_3NC) and C-C≡N(CH_3CN) bending frequencies, and to place the isomerization barrier at $\theta=90°$.[47] The most complicated potential included quartic bending anharmonicity

$$f[\sum_{i=1}^{3} (\phi_i-\phi_i^\circ)^4 + \sum_{i=1}^{3} (\chi_i-\chi_i^\circ)^4, \tag{20}$$

and repulsion between the H's of CH_3 and the N and C of (NC)

$$c\sum_{i=1}^{3} [(r_{iN}-r_z)^2 + (r_{iC}-r_z)]^2 \tag{21}$$

with the assumption that $r_{iN} > r_z$ or $r_{iC} > r_z$ leads to a zero contribution from the corresponding squared term. The repulsions introduce a CH_3-(NC) torsional potential.[47a]

Random initial conditions for CH_3NC were chosen at total internal energies of 200,100 and 70 kcal/mole using orthant and progressive sampling, and a third approximate sampling procedure called mode sampling. A thermal (500°K) rotational energy distribution was also put on CH_3NC by the sampling procedures. The trajectories were integrated to maximum times of 4.0×10^{-13} sec. The primary result of these trajectory calculations were lifetime distributions. A comparison was made between the total number of reactive events and the number predicted by the RRKM theory. The ratio of these can be approximately equated to that of trajectory to RRKM rate constants.

For all cases the trajectory rate constants for the three sampling methods were found to differ with each other and with the RRKM prediction; i.e., the trajectory rate constants were always less than those of the RRKM theory. The obvious conclusion was that the reaction coordinate for CH_3NC is badly coupled to the remaining vibrational modes, and therefore the Monte Carlo rate constants are the intramolecular relaxation rates. This is the same result as Bunker found for his H-C-C models.[1,2] The trajectory results for CH_3NC were seemingly at odds with the thermal isomerization of CH_3NC which is well described by the RRKM theory.[46] However, recent experimental results[48] have suggested that intermolecular collisions may be inducing internal energy redistribution, and in the absence of collisions internal energy redistribution, may be slow as suggested by the trajectories.

H-C≡C-Cl. Photophysical experimental studies involving chloroacetylene[49] inspired this trajectory study.[7,12,50] It was suggested that internal energy did not redistribute within highly excited chloroacetylene on a 10^{-9} sec. time scale. To investigate this proposal three different types of calculations were performed: (1) orthant and progressive random sampling at 200,175 and 150 kcal/mole; (2) a nonrandom orthant type sampling at 200, 175 and 150 kcal/mole; and (3) nonrandom excitation of approximate normal modes at 135 kcal/mole. A simple potential energy function was assumed

$$V = \sum_{i=1}^{3} D_i\{1-\exp[-\beta_i(r_i-r_i^\circ)]\}^2 + \tfrac{1}{2}\sum_{i=1}^{2} k\theta_i(\theta_i-\theta_i^\circ)^2, \qquad (22)$$

where

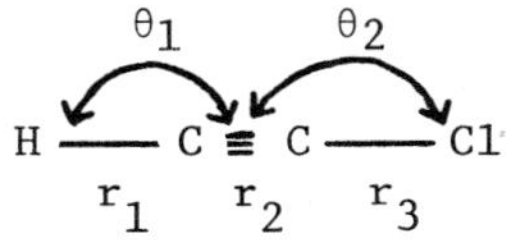

.

If energy redistribution is rapid for this model it should be even more rapid for a more anharmonic model.

The trajectory lifetime distributions for H-C≡C-Cl are random and the trajectory and RRKM unimolecular rate constants are in very good agreement. An example of the results is given in Table 1.

Table I. Trajectory and RRKM Rate Constants for H-C≡C-Cl Decomposition at 200 kcal/mole

Path	Rate Constants (sec^{-1}) RRKM	Trajectory
H-C Rupture	7.0×10^{11}	$4.9\pm0.7\times10^{11}$ [a]
		$5.8\pm0.8\times10^{11}$ [b]
Cl-C Rupture	4.9×10^{11}	$6.8\pm0.9\times10^{11}$ [a]
		$7.1\pm0.9\times10^{11}$ [b]

a. The rate constant is calculated using orthant sampling

b. The rate constant is calculated using progressive sampling.

In nonrandom orthant sampling only excitation in either the H-C≡C or C≡C-Cl part of the molecule was simulated. This was accomplished by sampling only 12 dimensions of the 18-dimensional phase space. Trajectory results are shown in Fig. 4 for excitation at 175 and 150 kcal/mole with the H-C≡C modes initially excited. The nonrandom distributions were fit using the following two state mechanism:

$$\text{H-C≡C-Cl}^{\dagger} \xrightarrow{k_{NR}} \text{H+C≡C-Cl}$$

$$\text{H-C≡C-Cl}^{\dagger} \xrightarrow{\lambda} \text{H-C≡C-Cl}^{*}$$

$$\text{H-C}\equiv\text{C-Cl*} \xrightarrow{k} \text{H-C}\equiv\text{C+Cl}$$

$$\text{H-C}\equiv\text{C-Cl*} \xrightarrow{k} \text{H-C}\equiv\text{C+Cl}$$

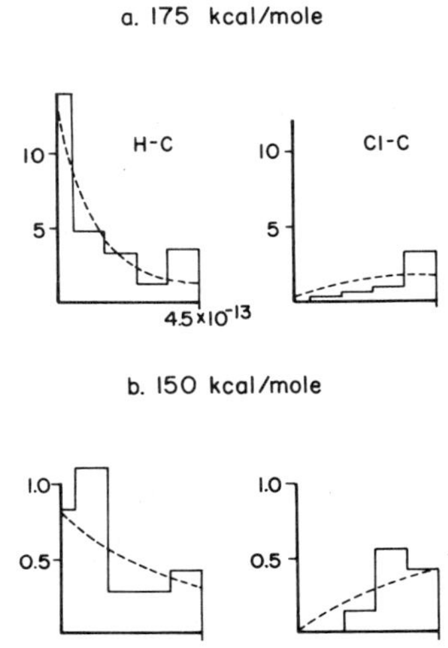

Figure 4. Histograms of $P(t)x10^{11}$ vs t for H-C≡C-Cl excited nonrandomly, i.e., the H-C≡C modes are initially excited. Dashed lines are fits of the two-state mechanism. $\lambda=2.5x10^{12}sec^{-1}$ and $8.0x10^{12}$ at 150 and 175 kcal/mole, respectively.

The nonrandomly prepared state is designated †, from which decomposition by C-H bond rupture can occur. The relaxation rate constant λ describes the rate at which † is transformed to a random distribution *. The state * then decomposes according to the RRKM rate constant k. k_{NR} is determined from the C-H bond rupture lifetime distribution. Values for λ varied from $4\text{-}11x10^{11}$ sec^{-1} at 200 kcal/mole to $1\text{-}4x10^{11}$ sec^{-1} at 150 kcal/mole.

The above nonrandom excitation trajectories were also analyzed to study the relationship between unimolecular lifetime and product relative translational energy distributions. As shown in Fig. 4 the lifetimes are strongly nonstatistical due to the nature of the initial excitation. In contrast the product relative translational energy distributions are statistical. This is shown in Fig. 5 where the Cl-C≡C modes are initially excited at 200 kcal/mole. Trajectories with lifetimes in ranges of $0\text{-}1.5x10^{-13}$ and $1.5\text{-}4.5x10^{-13}$ sec. both have statistical relative translational energy distributions. This result suggests that statistical relative translational energy distributions are not a diagnostic test for efficient intramolecular vibrational energy redistribution.

Finally, H-C≡C-Cl was prepared nonrandomly at 135 kcal/mole using three different "normal mode" energy distributions. These trajectories also showed efficient intramolecular energy redistribution. Plots of the average energy in the HC internal coordinate [Eq.(4)] is shown in Fig. 6 for the three different initial distributions. It is seen that the HC internal coordinate nearly redistributes its initial energy within 10^{-12} sec.

Though the above trajectory results are at odds with the interpretation of the choloracetylene photophysical experiments, they do not disagree with the experimental results per se. Interpretations of the experimental results are possible which agree with the trajectory results.[12]

$\underline{C_2H_5}$. This study has consisted of three parts: (1) quantum chemical

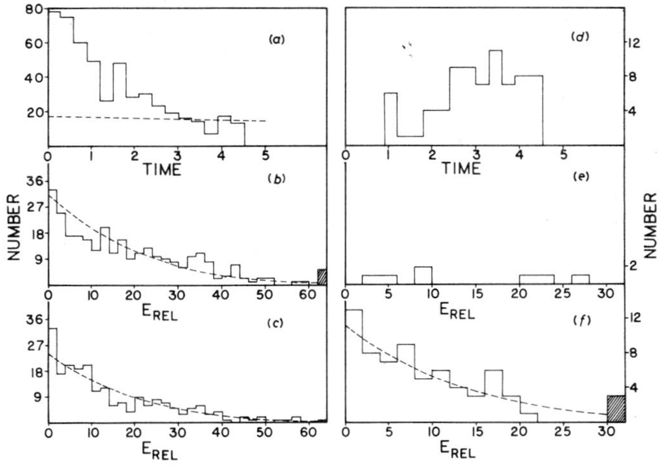

Figure 5. Lifetime and translational energy distributions for H-C≡C-Cl dissociation at 200 kcal/mole. Cl-C≡C modes are initially excited. (a) Lifetime distribution for C-Cl bond rupture. Time is in units of 10^{-13} sec. (b) HCC-Cl relative translational energy distribution for trajectories that have lifetime in range of 0-1.5x10^{-13} sec. E_{rel} is in units of kcal/mole. (c) same as (b) but for trajectories that have lifetimes in the range of 1.5-4.5x10^{-13} sec. (d),(e) and (f) are the same as (a),(b) and (c) except for C-H bond rupture. The dashed lines are the RRKM predictions. Events represented by shaded boxes have values of E_{rel} greater than 60 kcal/mole.

calculations of the $C_2H_5 \rightleftarrows H+C_2H_4$ potential energy surface;[51,52] (2) development of an analytic function for the potential energy surface;[51,52] and (3) the trajectory calculations.[18] Before we began the quantum calculations we had formulated a general analytic function for the potential energy surface. Our idea was to represent the asymptotic limits of the potential energy surface, C_2H_5 and $H+C_2H_4$, by modified valence force fields and then to connect these force fields by switching functions. Parameters for the switching functions would be determined from the quantum calculations. The two modifications in the C_2H_5 and $H+C_2H_4$ valence force fields were the representation of bond stretching potentials by Morse functions, and the expression of the CH_3 and CH_2 groups equilibrium angles as a

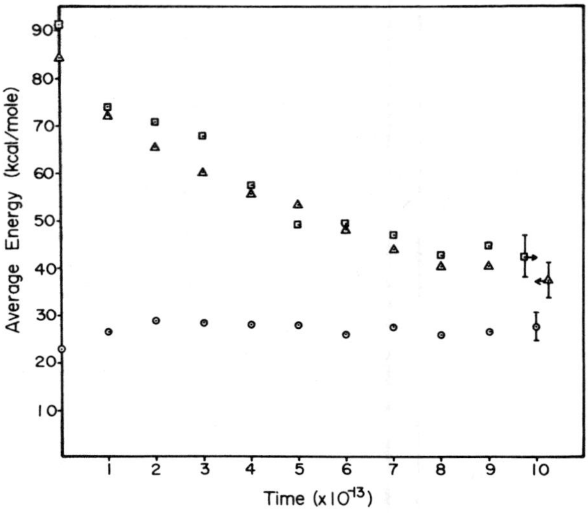

Figure 6. Plots of average energy in the HC internal coordinate versus time for three different initial energy distributions in H-C≡C-Cl; given by **O**, **□**, Δ. The error bars are representative of those for all the points.

function of CC distance (this relationship was found from our ab initio calculations).

Previous quantum chemical calculations suggested that an ab initio SCF calculation should yield a very good estimate of the potential energy surface between the C_2H_5 and $H+C_2H_4$ asymptotic limits.[52] Our ab initio SCF calculations were performed using the unrestricted STO-3G and 4-31G basis sets. We determined the geometries and force constants along the $H+C_2H_4 \rightleftarrows$ minimum energy path, the angular width of the $H+C_2H_4$ entrance channel, and the reaction path for H-atom migration. The extended 4-31G basis set was only used to calculate potential energies along the $H+C_2H_5 \rightleftarrows C_2H_5$ minimum energy path. It gives a reaction endothermicity and a barrier for C_2H_5 decomposition of 40.0 and 43.5 kcal/mole, respectively. Both numbers are in excellent agreement with experiment.[52]

A depiction of some of the internal coordinates used in the potential energy function is given in Fig. 7. Also included in the potential energy function are the dihedral angle τ between the H_1H_3C and CH_2H_4 planes, and the CH distance r_1, r_2, r_3 and r_4. Only one of the hydrogens (label H*) can dissociate along the minimum energy path and migrate between the carbon atoms. It is possible to write a symmetric analytic potential energy function if H-atom migration is not allowed.[51] We have not formulated a symmetric function that includes H-atom migration.

TABLE II. Comparison of Calculated and Fitted Geometries Along the $H+C_2H_4$ Minimum Energy Path[a]

	Δ_o		χ_o		ϕ_o		θ_o		R_o	
r_{CH}^*	Fit	STO-3G	Fit	STO-3G	Fit	STO-3G	Fit	STO-3G	Fit	STO-3G
2.5	116.0		87.6		121.8		100.6		1.34	
2.08	116.0	116.3	89.3	88.2	121.2	121.3	103.5	103.5	1.37	1.37
1.9	115.7		91.5		120.5		104.7		1.39	
1.7	114.1	114.2	96.6	97.3	118.8	117.8	105.6	107.3	1.43	1.425
1.6	112.6		100.1		117.2		106.2		1.45	
1.5	110.6		103.5		115.1		107.7		1.47	
1.4	108.8	110.1	106.0	105.3	113.0	112.7	109.6	110.3	1.49	1.495
1.3	107.9		107.3		111.7		110.8		1.50	
1.2	107.7		107.7		111.2		111.1		1.51	
1.08	107.8	108.2	107.8	107.6	111.1	110.8	111.1	111.6	1.51	1.51

[a]The angles are for the CH_3 group. Angles are in degrees and distances are in angstroms.

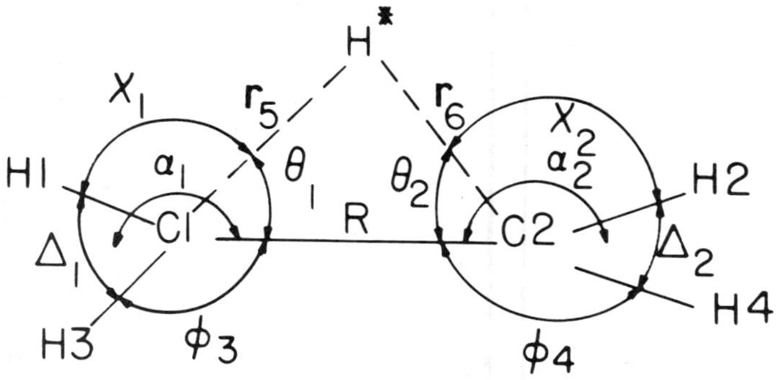

Figure 7. Internal coordinates used in the $C_2H_5 \rightleftarrows H+C_2H_4$ analytic potential energy function.

The complete analytic potential function is quite complicated and, here, we will only give a few of its salient features. A comparison of the ab initio and fitted geometries along the $H+C_2H_4$ minimum energy path is given in Table II. The fit to the 4-31G minimum energy path potential energies is shown in Fig. 8.

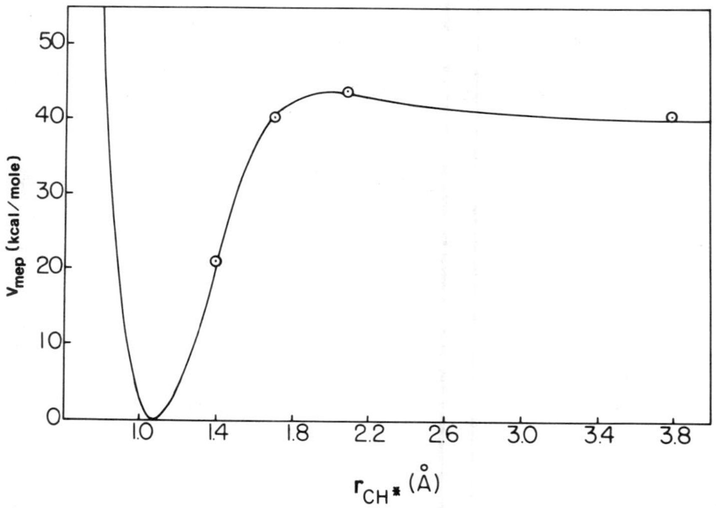

Figure 8. Potential energy along the minimum energy path (MEP) for H-atom addition to ethylene. The circles are 4-31G potential energies and the solid line is the fit of the analytic potential energy function.

The minimum energy path potential is given by

$$V_{mep} = D_H\{1-\exp[-\beta_H(g-r_o)]\}^2 - (D_{cd}-D_{cs})S_c, \qquad (23)$$

D_{cd} and D_{cs} are the CC double and single bond strengths and S_c is a switching function which depends upon the CH* distance variable g

$$S_c = \{1-\exp[-C_c(g-r_c)]\}^2 \qquad g \geq r_c$$
$$S_c = 0.0 \qquad g<r_c. \qquad (24)$$

g is a function of r_5, r_6 and R. Fits to the <u>ab initio</u> f_χ and f_θ quadratic force constants along the minimum energy path are shown in Fig. 9. These force constants were fit to Gaussian functions

$$f(r)=f_o e^{-a(r-1.0\text{Å})^2} \qquad r\geq 1.0\text{Å}$$
$$f(r)=f_o \qquad r<1.0\text{Å} \qquad (25)$$

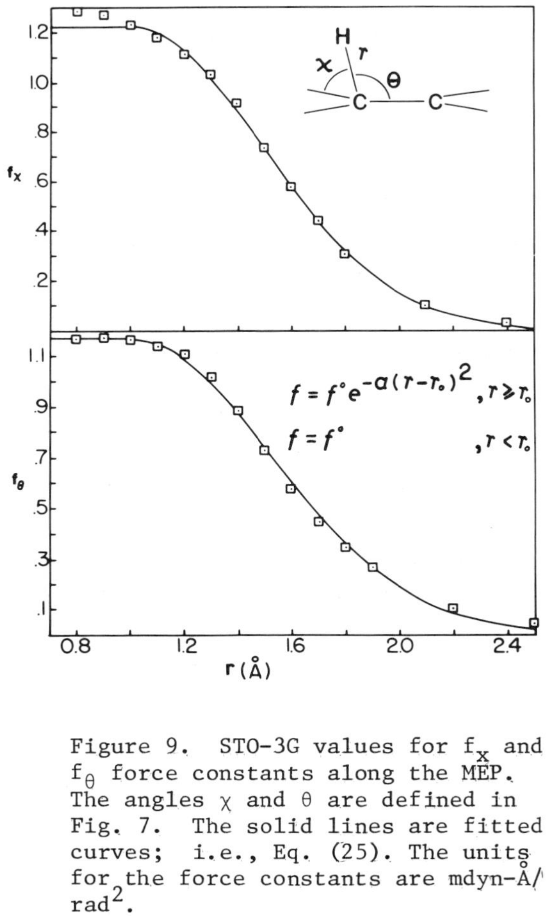

Figure 9. STO-3G values for f_x and f_θ force constants along the MEP. The angles χ and θ are defined in Fig. 7. The solid lines are fitted curves; i.e., Eq. (25). The units for the force constants are mdyn-Å/rad^2.

Contour plots of the analytic potential energy function are shown in Fig. 10 for R vs r_{CH*} and in Fig. 11 for θ vs r_{CH*}. The letter χ locates the saddlepoint in Fig. 10. The letters w,x,y, and z in Fig. 11 locate the bottom of the C_2H_5 well, the H-atom migration saddlepoint, the barrier for H-atom attack midway between the carbon atoms, and the barrier for $C_2H_5 \rightarrow H+C_2H_4$ dissociation.

In the trajectory calculations C_2H_5 was excited randomly at 150 and 100 kcal/mole (the maximum rotational energy was 2% of the total energy). At 150 kcal/mole both orthant and progressive sampling were used, while at 100 kcal/mole we only used orthant sampling. Also at 150 kcal/mole we studied dissociation on a surface that had a 40.1 kcal/mole dissociation barrier instead of 43.5 kcal/mole. The trajectory lifetime distributions are shown in Fig. 12. At 100 kcal/mole the lifetime distribution is nonrandom. The short time transient in the

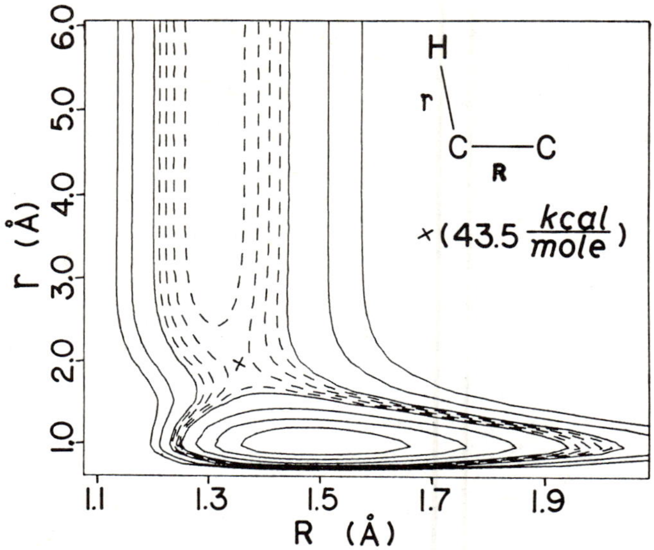

Figure 10. Potential energy surface in terms of the CC distance (R) and the CH* distance (r). The remaining internal coordinates have been optimized at each value of r and R.

progressive sampling lifetime distribution at 150 kcal/mole is an artifact resulting from the small number of chains of trajectories computed.

Relative H-C_2H_4 translational energies were computed at the dissociation barrier and at a large H-C_2H_4 separation (6.00Å); i.e., E_{rel} (barrier) and E_{rel} (final), respectively. We also recorded the shift in relative translational energy $\Delta E_{rel} = E_{rel}$(final)-E_{rel}(barrier). Calculated distributions of these three quantities are shown in Fig. 13. Average values of the distributions are given in Table III. The translational energy distributions are statistical at the barrier (one might have predicted the distribution to be nonstatistical at 100 kcal/mole since the lifetime distribution is nonrandom).

TABLE III. Average Translational Energies for $C_2H_5 \rightarrow H+C_2H_4$ Dissociation

Energy	E_o	Sampling	$\bar{E}_{rel}$(barrier)	$\bar{E}_{rel}$(final)	$\overline{\Delta E}_{rel}$
150	43.5	orthant	8.1±1.6	12.9±1.8	4.8±1.1
150	43.5	progressive	7.2±1.5	10.8±1.6	3.6±1.2
150	40.1	orthant	8.3±1.2	12.2±1.5	3.9±1.1
150	43.5	orthant	4.2±0.7	9.5±1.1	5.3±0.9

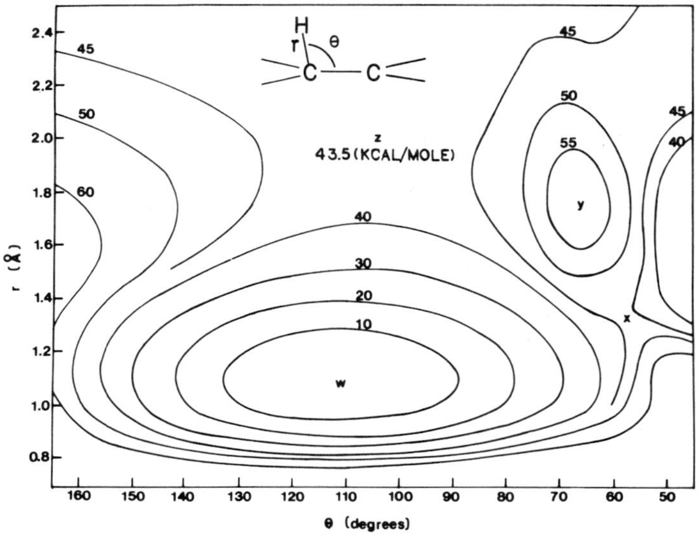

Figure 11. Potential energy surface in terms of the H*CC angle (θ) and the CH* distance (r). The remaining coordinates were given values corresponding to the MEP for H-atom addition to ethylene.

However, the final distributions are broader and shifted to higher values. The average values of $\overline{\Delta E}_{rel}$ and the shapes of the ΔE_{rel} distributions show that ΔE_{rel} can not be simply related to the minimum energy path potential energy release upon passing the dissociation barrier. The final translational energy distributions are similar to those observed for $C_2H_4F \rightarrow H{+}C_2H_3F$ dissociation.[53]

Because rotational excitation was so small in these calculations (a maximum of 2% of the total energy) angular momentum constraints are not significant. However, the product angular momentum distributions are still interesting. As shown in Fig. 14 the orbital angular momentum, $|L|$, and C_2H_4 rotational angular momentum, $|j|$, distributions are similar and tend to have larger average values than does the total angular momentum, $|J|$. Such distributions have been associated with repulsiveness in the dissociation event for A+BC reactions.[54]

A common arguement made in theoretical models of unimolecular reactions is that because of the lightness of the hydrogen atom the orbital angular momentum is not significant compared to the total angular momentum when a hydrogen atom dissociates. Therefore, for these cases the total angular momentum is assumed to be transformed into product rotational angular momentum. However, in these studies with small amounts of total angular momentum this arguement is not true. It would be interesting to see if it is true at higher total angular momentum.

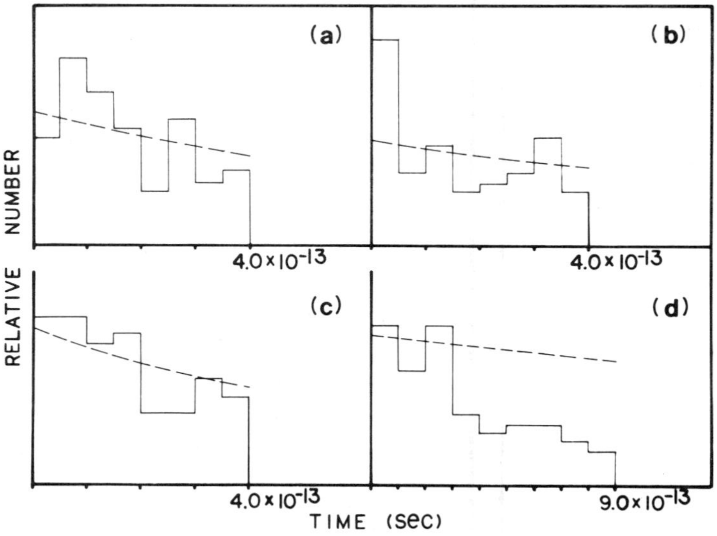

Figure 12. Lifetime distributions for $C_2H_5 \rightarrow H + C_2H_4$ dissociation: (a) orthant sampling at 150 kcal/mole; (b) progressive sampling at 150 kcal/mole; (c) same as (a) except the dissociation barrier equals 40.1 kcal/mole instead of 43.5 kcal/mole; and (d) orthant sampling at 100 kcal/mole.

There are several conclusions that can be reached from the above trajectory studies. They predict that there are molecules for which intramolecular vibrational energy distribution is slower than the statistical unimolecular rate constant. Such molecules are classified as intrinsically non-RRKM.[5] The trajectory calculations suggest that such molecules have a group of high frequency vibrations, such as C-H and O-H stretches, which do not couple strongly with the remaining modes. That non-RRKM behavior has not been observed in macroscopic unimolecular experiments (e.g., thermal dissociation and chemical activation) may be due to collision induced intramolecular energy transfer which obscures the slow intramolecular energy transfer in isolated molecules. Two recent experiments involving highly excited allyl isocyanide (CH_2CHCH_2NC)[55] and glyoxal ($C_2H_2O_2$)[56] have been interpreted in terms of incomplete intramolecular energy redistribution on a 10^{-6} sec time scale.

The possibility that internal energy redistribution may be inefficient in highly excited molecules is quite exciting. It suggests the presence of semi-periodic trajectories at energies much higher than the dissociation barrier. We have seen such trajectories in model H-C-C systems similar to those studied by Bunker.[1-3] The question is what fraction of the phase spece of a molecule consists of these semi-periodic trajectories. If it is large one would

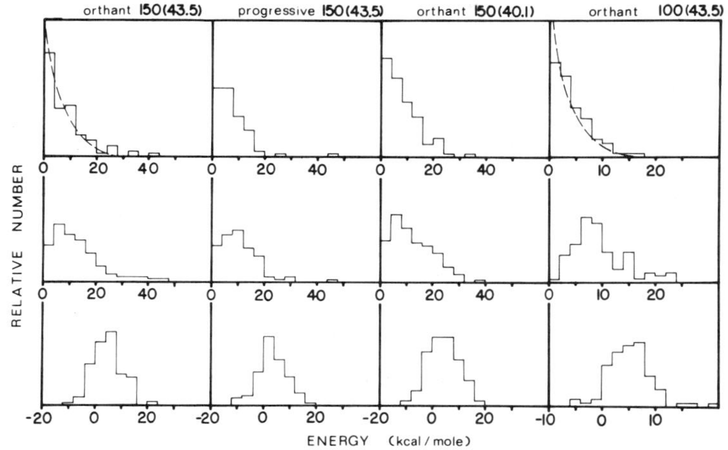

Figure 13. Translational energy distributions for $C_2H_5 \rightarrow H+C_2H_4$ dissociation. The top row contains the E_{rel} distributions at the exit-channel barrier (r_{CH*}=2.00Å), the middle row contains the E_{rel} distributions at large H-C_2H_4 separation (r_{CH*}=6.00Å), and the bottom row contains the ΔE_{rel} distributions. Orthant 150 (43.5) denotes orthant sampling at 150 kcal/mole and a dissociation barrier of 43.5 kcal/mole.

expect molecules to be intrinsically non-RRKM and also expect to be able to excite specific vibrational states in a highly excited molecule. These states would be described by a set of near orthogonal coordinates. However, these coordinates are probably not normal modes. Instead, for some cases they may be local modes.[57,58]

Slow intramolecular energy redistribution and intrinsic non-RRKM behavior suggests that the phase space for molecules may be intrinsically decomposable and, thus, the system non-ergodic. However, one should recognize that intrinsic non-RRKM behavior doe not necessarily infer non-ergodicity.[19]

The trajectory calculations show that statistical product translational energy distributions can be observed in the absence of complete intramolecular energy redistribution. Also, exit channel effects can yield non-statistical product energy distributions even though the distribution was statistical at the dissociation barrier. Thus, it does not appear to be possible to relate product translational energy distributions to intramolecular dynamics.

Future directions in classical trajectory sutdies of unimolecular reactions will involve comparison with exact and/or approximate quantum scattering calculations for small systems and extensions of the classical trajectory method to larger and more complex molecules.

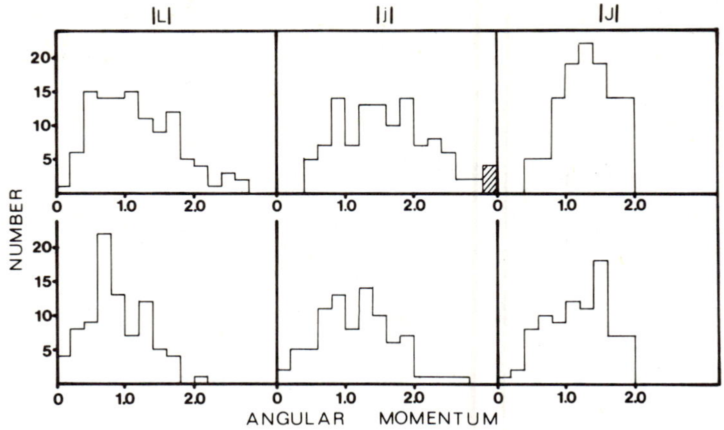

Figure 14. Angular momentum |L|, |j| and |J| distributions for $C_2H_5 \rightarrow H+C_2H_5$ dissociation. The top row is for orthant sampling at 150 kcal/mole and a dissociation barrier of 40.1 kcal/mole. The bottom row is for orthant sampling at 100 kcal/mole and a dissociation barrier of 43.5 kcal/mole.

Both of these types of studies will require accurate potential energy surfaces, for which quantum chemical calculations will be important. The results of the trajectory calculations for large molecules can be compared with theories that treat a small number of the modes dynamically and couple the remaining modes by approximate methods.[59,60] There will probably be more emphasis placed on comparing the trajectory results with experimental observations in the future.

IV. EXTENSIONS TO GAS-SURFACE DYNAMICS

The classical trajectory techniques used to study gas phase unimolecular dyanmics could easily be extended to unimolecular gas-surface reactions. One could study desorption from a solid,

$$XY\text{---(solid surface)} \longrightarrow XY+\text{(solid surface)},$$

unimolecular dissociation on a solid

$$XY\text{---(solid surface)} \longrightarrow X+Y+\text{(solid surface)}.$$

and clustering of metal atoms,

$$M_n + M \longrightarrow M_{n+1} \, .$$

A possible trajectory study would be to study H-atom reaction with ethylene absorbed on a solid surface,

$$H+\ C_2H_4\text{---(solid surface)}.$$

The potential energy surface could be derived by taking our surface for $H+C_2H_4$[52] and combining it with one for the solid surface with suitable interactions between C_2H_4 and the solid surface. From the trajectories the probability of the following processes could then be determined as a function of H-atom translational energy, C_2H_4 vibrational energy, and surface temperature:

(1) C_2H_5+(solid surface)

(2) C_2H_5---(solid surface)

(3) $H+C_2H_4$+(solid surface)

(4) $H+C_2H_4$---(solid surface)

(5) other dissociation processes; e.g., CH_3+CH_2+(solid surface).

Many different gas-surface reactions could be investigated. However, it will be best if there are dynamical experimental results available for comparison. In this regard, molecular beam studies should be very important.[61] The combination of classical trajectory calculations and dynamical experimental studies should prove very powerful in unraveling the microscopic details of gas-surface unimolecular reactions.

ACKNOWLEDGEMENT

The author is grateful to the National Science Foundation for financial support.

REFERENCES

1. D.L. Bunker, J. Chem. Phys., 37, 393 (1962).
2. D.L. Bunker, J. Chem. Phys., 40, 1946 (1964).
3. D.L. Bunker and M. Pattengill, J. Chem. Phys., 48, 772 (1968).
4. H.H. Harris and D.L. Bunker, Chem. Phys. Lett., 11, 433 (1971).
5. D.L. Bunker and W.L. Hase, J. Chem. Phys., 59, 4621 (1973); ibid, 69, 4711 (1978).
6. P. Brumer and M. Karplus, Faraday Discuss. Chem. Soc., 55, 80 (1973).
7. W.L. Hase and D.F. Feng, J. Chem. Phys., 61, 4690 (1974); ibid, 64, 651 (1976).
8. J.D. McDonald and R.A. Marcus, J. Chem. Phys., 65, 2180 (1976).
9. D.E. Carter, J. Chem. Phys., 65, 2584 (1976).
10. S. Goursaud, M. Sizun, and F. Fiquet-Fayard, J. Chem. Phys., 65, 5453 (1976).
11. K.S. Sorbie and J.N. Murrell, Molec. Phys., 31, 905 (1976).
12. C.S. Sloane and W.L. Hase, J. Chem. Phys., 66, 1523 (1977).
13. D.W. Noid, M.L. Koszykowski, R.A. Marcus and J.D. McDonald, Chem. Phys. Lett., 51, 540 (1977).
14. S.R. Kinnersly and J.N. Murrell, Molec. Phys., 33, 1479 (1977).
15. E.R. Grant and D.L. Bunker, J. Chem. Phys., 68, 628 (1978);

J. Santamaria, D.L. Bunker and E.R. Grant, Chem. Phys. Lett., 56, 170 (1978).
16. S. Goursaud, M. Sizun and F. Fiquet-Fayard, J. Chem. Phys., 68, 4310 (1978).
17. D.L. Bunker, K.R. Wright, W.L. Hase and F.A. Houle, J. Phys. Chem., 83, 933 (1979).
18. W.L. Hase, R.J. Wolf and C.S. Sloane, J. Chem. Phys., 71, xxxx (1979).
19. W.L. Hase, in Dynamics of Molecular Collisions Part B, edited by W.H. Miller (Plenum, New York, 1976), p. 121.
20. The methodology of classical trajectory calculations is reviewed in D.L. Bunker, Meth. Comp. Phys., 10, 287 (1971); and R.N. Porter and L.M. Raff, in Dynamics of Molecular Collisions Part B, edited by W.H. Miller (Plenum, New York, 1976) p.1.
21. W.H. Miller, Disc. Faraday Soc., 55, 119 (1973).
22. G.C. Schatz and A. Kuppermann, J. Chem. Phys., 65, 4668 (1976).
23. S. Glasstone, K.J. Laidler and H. Eyring, The Theory of Rate Processes (McGraw-Hill, New York, 1941), p. 131.
24. Bunker and Pattengill actually minimized the density of states along the reaction coordinate where one degree of freedom is a translation. For most cases, including triatomics, there is no difference between this criterion and the one of minimum number of states. See B.C. Garrett and D.G. Truhlar, J. Chem. Phys., 70, 1593 (1979).
25. E. Wigner, J. Chem. Phys., 5, 720 (1937).
26. J. Horiuti, Bull. Chem. Soc. Jpn., 13, 210 (1938).
27. J.C. Keck, Adv. Chem. Phys., 13, 85 (1967).
28. W.L. Hase, J. Chem. Phys., 64, 2442 (1976).
29. Multidimensional selection procedure must be used since the initial parameters do not have independent probability distributions; i.e., the system is not separable.
30. Orthants are in n-dimensional space what quadrants and octants are in two and three dimensions.
31. B.S. Rabinovitch and M.C. Flowers, Quart. Rev., 18, 122 (1964).
32. S. Chapman and D.L. Bunker, J. Chem. Phys., 62, 2890 (1975).
33. L.M. Raff, J. Chem. Phys., 60, 2220 (1974).
34. H.H. Suzakawa, Jr., Max Wolfsberg, and D.L. Thompson, J. Chem. Phys., 68, 455 (1978).
35. For example by $S_0+h\upsilon \rightarrow S_1$ photoexcitation.
36. J.C. Tully and R.K. Preston, J. Chem. Phys., 55, 562 (1971).
37. W.H. Miller and T.F. George, J. Chem. Phys., 56, 5637 (1972); and A. Komornick, K. Morokuma and T.F. George, 67, 5012 (1977).
38. W.H. Miller, J. Chem. Phys., 69, 2188 (1978).
39. The kinetic energy in internal coordinates is given by $2T=P^{+}GP$. Elements for the G-matrix are given in E.B. Wilson, Jr., J.C. Decius and P.C. Cross, Molecular Vibrations (McGraw-Hill, New York, 1955), p. 303.
40. M.D. Pattengill, Ph.D. Thesis, University of California, Irvine, CA 1969 (University Microfilms, Ann Arbor, Michigan).
41. P. Brumer, J. Comput. Phys., 14, 391 (1974).

42. C. Parr, Ph.D. Thesis, California Institute of Technology, Pasadena, CA 1969.
43. L.F. Shampine and M.K. Gordon, Computer Solution of Ordinary Differential Equations. The Initial Value Problem (W.H. Freeman, San Francisco, 1975); and L.D. Thomas, W.P. Kraemer and G.H.F. Diercksen, Chem. Phys., 30, 33 (1978).
44. J.N. Murrell, in Gas Kinetics and Energy Transfer (Chemical Society, London, 1978) Vol. III, p. 200.
45. R.F.W. Bader and R.A. Gangi, in Theoretical Chemistry (Chemical Society, London, 1975) Vol. II, p. 1.
46. F.W. Schneider and B.S. Rabinovitch, J. Am. Chem. Soc., 84, 4215 (1962).
47. Recent *ab initio* calculations indicate that θ=100.8° at the isomerization barrier: (a) D.H. Liskow, C.F. Bender, and H.F. Schaefer III, J. Chem. Phys., 57, 4509 (1972); (b) L.T. Redmon, G.D. Purvis and R.J. Bartlett, J. Chem. Phys., 69. 5386 (1978).
48. K.V. Reddy and M.J. Berry, Chem. Phys. Lett., 52, 111 (1977).
49. K. Evans, D. Heller, S.A. Rice and R. Scheps, Chem. Soc. Faraday Trans II, 69, 856 (1973).
50. W.L. Hase, "On the Relationship between Unimolecular Lifetime and Relative Translational Energy Distributions", accepted for publication in Chemical Physics Letters.
51. C.S. Sloane and W.L. Hase, Faraday Discuss. Chem. Soc., 62, 210 (1977).
52. W.L. Hase, G. Mrowka, R.J. Brudzynski and C.S. Sloane, J. Chem. Phys., 69, 3548 (1978).
53. J.M. Farrar and Y.T. Lee, J. Chem. Phys., 65, 1414 (1976).
54. J.C. Polanyi and J.L. Schreiber, Faraday Discuss. Chem. Soc., 62, 267 (1977).
55. K.V. Reddy and M.J. Berry, "A Nonstatistical Unimolecular Chemical Reaction: Isomerization of State-Selected Allyl Isocyanide", submitted to Chemical Physics Letters.
56. R. Naaman, D.M. Lubman and R.M. Zare, "Vibrational Energy Redistribution in Glyoxal Following Internal Conversion", submitted to the Journal of Chemical Physics.
57. R.L. Swofford, M.E. Long and A.C. Albrecht, J. Chem. Phys., 63, 179 (1976).
58. B.P. Henry, Accounts Chem. Res., 10, 207 (1977).
59. H. Shugard, J.C. Tully and A. Nitzan, J. Chem. Phys., 69, 336 (1978).
60. G.C. Schatz, Chem. Phys., 31, 295 (1978).
61. R.L. Palmer and J.E. Smith, Jr., Catal. Rev., 12, 279 (1975).

DENSITY-FUNCTIONAL STUDIES OF CHEMISORPTION ON SIMPLE METALS

N. D. Lang
IBM Thomas J. Watson Research Center
Yorktown Heights, N.Y. 10598

ABSTRACT

The uniform-positive-background model can be used to describe successfully many of the ground-state properties of the surfaces of simple (s-p bonded) metals. The model provides therefore a representation that is useful in analyses of chemisorption on these surfaces. We discuss here studies of this type which use a Hartree-like formulation of the many-electron problem that includes exchange and correlation.

The properties of simple (s-p bonded) metals have often been studied using the very elementary uniform-background model, in which the ionic lattice is imagined to be smeared out into a homogeneous background of positive charge. The conduction electrons will then also be distributed with constant density (except for special cases such as the Wigner lattice). The surface analogue of this model is simply a semi-infinite uniform positive background plus the conduction electrons, which assume a constant density (equal to that of the background) well inside of the surface, and tail off in some to-be-determined fashion in the surface region (the vicinity of the edge of the background). The only parameter characterizing the model is then the background density $\bar{n}$, which is generally expressed in the form of a radius r_s, with $(4/3)\pi r_s^3 \equiv \bar{n}^{-1}$; r_s runs from about 2 to 6 bohrs for simple metals.

We determine the electron density distribution in this model by solving self-consistently a set of single-particle equations due to Kohn and Sham.[1] These equations resemble the Hartree equations, but the electrostatic potential is supplemented by an effective exchange-correlation potential. This set of equations represents an exact formulation of the many-body problem for the ground-state density distribution of an inhomogeneous system of electrons in a static external potential (due for example to a lattice of nuclei or, in our case, to the positive background). We use the so-called local-density approximation[1] for the exchange-correlation potential, in which this potential at each point in space is given by the exchange-correlation part of the chemical potential of a uniform electron gas whose density is equal to that of the actual

ISSN:0094-243X/80/610137-13$1.50

system at that point. The use of this approximation has given good results in the study of solids,[2] molecules[3] and atoms.[4] We neglect the presence of spin moments in our discussion (that is, we use the non-spin-polarized version of the Kohn-Sham equations).

Figure 1 gives the electron density distribution for our surface model, at two different values of r_s, one ($r_s=5$) corresponding roughly to potassium and the other ($r_s=2$) to aluminum.[5] The density tails off exponentially into the vacuum; in the metal, it exhibits quantum density (Friedel) oscillations, which have the characteristic wavelength of half the Fermi wavelength, and an amplitude that decreases asymptotically with the square of the distance from the surface. The work function can be computed easily once the electron density distribution is known; it is compared with experimentally determined values for polycrystalline simple metals in Fig. 2.[6] The agreement

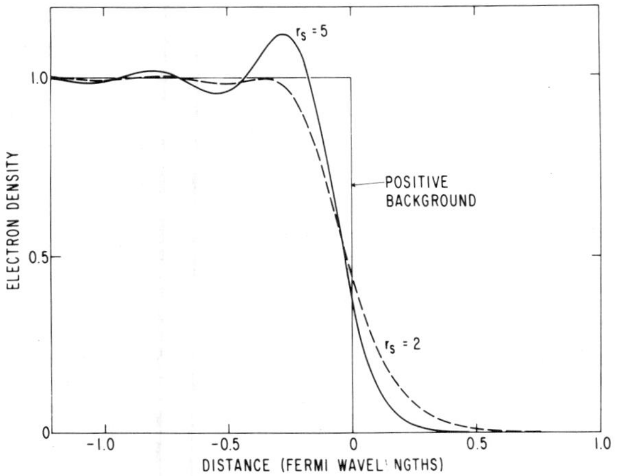

Fig. 1. Electron density in surface region of uniform-background model. (From Ref. 5.)

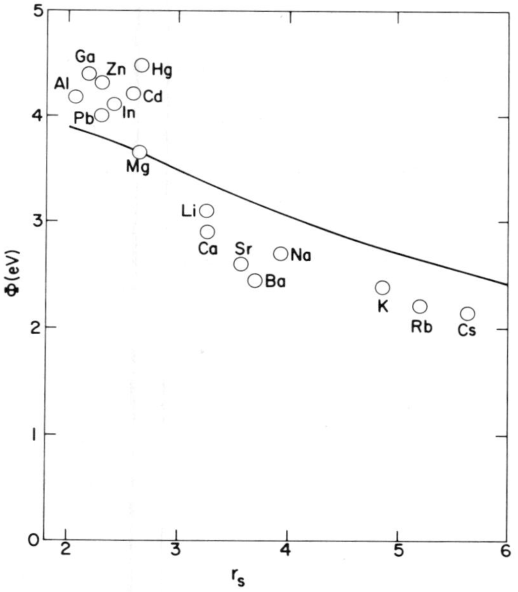

Fig. 2. Solid line: work function for the uniform-background model. Circles: measured work function for polycrystalline samples. (From Ref. 6.)

is quite reasonable, and is improved if the discrete ionic lattice of the metal is taken into account using first-order perturbation theory.[7] We note that the Thomas-Fermi treatment of this surface model[6] gives a work function that is identically zero for all values of r_s, indicating that this classic method is wholly inadequate in the present context.

We pass now in our discussion to an analysis of the ground-state properties of chemisorption systems, using the uniform-background model as a representation of the substrate. We discuss first a problem that was at one time a popular one for experimental study, namely the change in work function of a high-work-function substrate as alkali atoms are chemisorbed on its surface. It is found in particular that the work function drops rapidly as a function of alkali coverage, reaches a minimum, and then increases to roughly the value for the bulk alkali as the first full layer of chemisorbed atoms is completed. Now just as our substrate model consists of replacing each layer of substrate ions by a slab of positive charge (thereby building up the semi-infinite positive background), so also will we replace the layer of alkali ions by such a slab. This yields the positive-background configuration shown in Fig. 3; the electron distribution, also shown, is computed self-consistently just as in the case of the bare substrate. We fix the adsorbate (alkali) slab thickness to be the spacing of the closest-packed planes in the bulk alkali, and vary the slab density to simulate changes in coverage (since the alkalis are monovalent, the charge per unit area of the slab is equal to the number of adsorbed alkali atoms per unit area). For each value of the slab density, a self-consistent calculation is done, and the work function is obtained just as in the case of the bare surface.[8] Figure 4 shows the

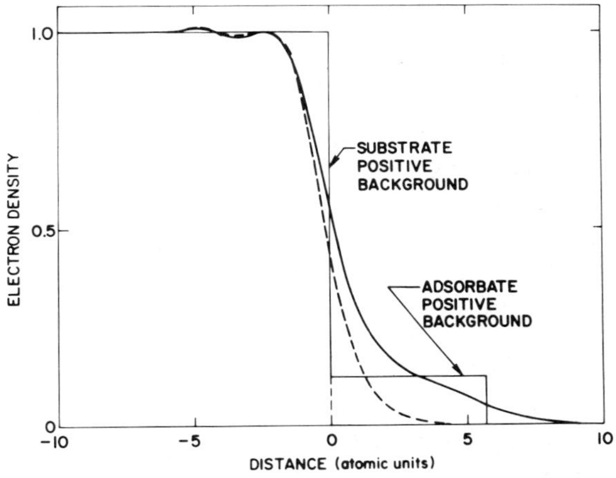

Fig. 3. Electron density distribution for uniform-background model of alkali adsorption. Solid line gives distribution for case of one full layer of adsorbed Na atoms; dashed line gives distribution for bare substrate ($r_s = 2$). (Atomic unit ≡ bohr.) (After Ref. 8.)

work function as a function of coverage for adsorbed sodium and cesium. A minimum is observed, just as in the experiment; Fig. 5 compares the minimum values with those obtained experimentally. The agreement is seen to be quite reasonable. A discussion of why in fact there is a minimum is given in Ref. 8.

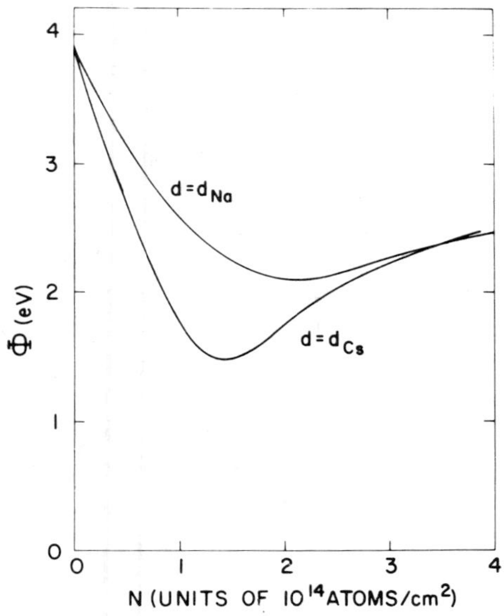

Fig. 4. Curves of work function vs. coverage for Na and Cs adsorption in the uniform-background model ($r_s = 2$ substrate). Values of the coverage at the minimum indicated here are smaller than those determined by adsorption on transition-metal substrates, whose work functions are much larger than that of the model substrate. The value of the work function at the minimum, however, is relatively independent of the substrate work function, and so a comparison can be made with experiments on transition metals (see Fig. 5). (After Ref. 8.)

Now let us consider the case of the chemisorption of a single atom on a metal surface, the case of more interest in a discussion of reactions at surfaces. We retain our simple substrate model, but treat the atom without any such approximation, beginning instead with a positive point charge of strength $Z|e|$ (i.e. the nuclear charge) at a distance d from the positive-background edge. Whereas the problems that we treated above had one-dimensional symmetry, this problem is cylindrically symmetric. The solution proceeds conveniently via an analysis of a Lippmann-Schwinger equation corresponding to each single-particle equation (for states degenerate with the conduction band of the substrate); the details of this treatment are given in Ref. 9. All of the core states of the adsorbed atom are computed in the full, non-spherical potential of the chemisorption system. The aim is to treat this simple model as carefully and completely as possible.

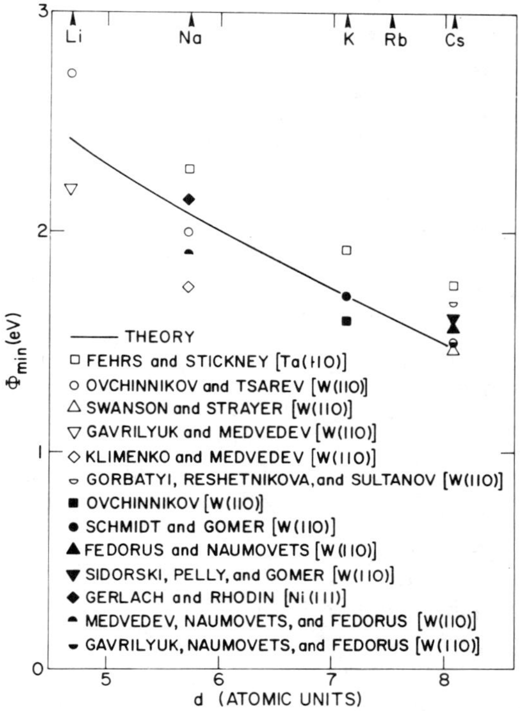

Fig. 5. Symbols: measured values of the minimum work function for alkali adsorption on close-packed metal substrates of initially high work function. Solid line: minimum value computed using the uniform-background model of alkali adsorption. (Atomic unit ≡ bohr.) (After Ref. 8.)

We show in Figs. 6 and 7 results for the chemisorption of Li, Si, and Cl, illustrating the three basic types of chemisorption bond--positive ionic, covalent, and negative ionic.[9] We consider a high-density substrate, with $r_s=2$, which is representative of a metal like aluminum ($r_s=2.07$). Figure 6 shows the difference in eigenstate density (of the Kohn-Sham single-particle equations) between the bare metal and the metal with the chemisorbed atom. The atom valence states which are (or become) degenerate with the metal conduction band broaden into resonances. The 2s resonance of chemisorbed Li is above the Fermi level, indicating that Li has lost its electron to the metal (though it will be seen in Fig. 7 that this electron doesn't go very far). The Cl 3p resonance is filled, indicating the acquisition of electronic charge. The Si 3p resonance straddles the Fermi level, which suggests covalent rather than ionic bonding.

The top row of Fig. 7 shows the contours of total electron density in the chemisorption system. Note that the contours return to their bare-metal form over a relatively short distance; this is a consequence of metallic screening. The second row shows difference plots: the difference between the total density in the chemisorption system and the superposition of bare metal and free-atom densities. Thus these plots

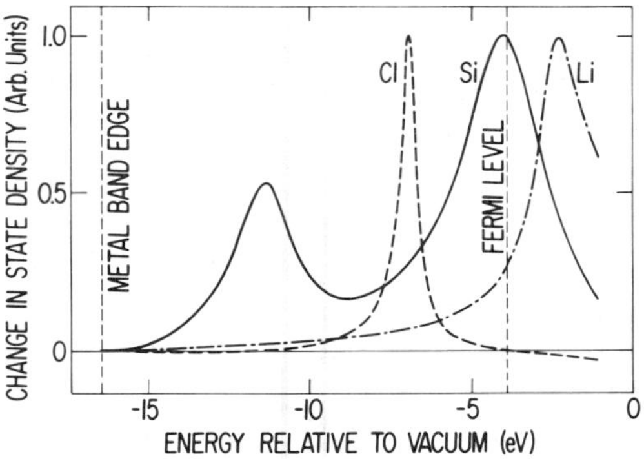

Fig. 6. Change in state density due to chemisorption of one atom on a uniform-background substrate ($r_s = 2$). Each curve corresponds to metal-adatom distance which minimizes the total energy. Note that the lower Si resonance corresponds to the 3*s* level of the atom; for Cl this is a discrete state below the bottom of the band (band edge). (From Ref. 9.)

show how the electronic charge rearranges itself to form the bond. The plot for Cl shows the case of a polarized negative ion. For Si, there is a charge accumulation in both bond and vacuum regions, which is characteristic of such plots[10] for covalently bonded diatomic molecules (when, as here, p states are involved). In the case of Li, it is not clear whether the plot represents a Li atom whose valence electron has been transferred to the metal, or whether it simply represents a very polarized Li atom. Figure 6 however resolves this problem, in the sense that the presence of an empty valence resonance for Li shows that the first picture is more appropriate.

Figure 8 shows the density difference between chemisorption system and bare metal, decomposed by energy, for four energies, in the case of Si.[9] Note that the lower part of a resonance tends to add charge to the bond region ("bonding"), while the upper part tends to subtract it ("antibonding"). The fact that the upper part of the 3p resonance is unoccupied leads to a net bonding interaction.

Given the eigenvalues and eigenfunctions of the single-particle Kohn-Sham equations that we have calculated, it is straightforward to calculate the binding energy of the chemisorbed atom, that is, the difference between the energy of bare metal plus free atom and the energy of the chemisorption system. A graph of this energy as a function of metal-adatom distance d is shown in Fig. 9 (dashed curve).[9] If we want to discuss the binding energy (and equilibrium distance) for chemisorption at different sites on metal surfaces, we must in some way re-introduce the discrete lattice structure of the substrate into our model. In the case of simple-metal substrates, we can represent the ion core potential at each lattice site by a pseudopotential; the re-introduction of the lattice

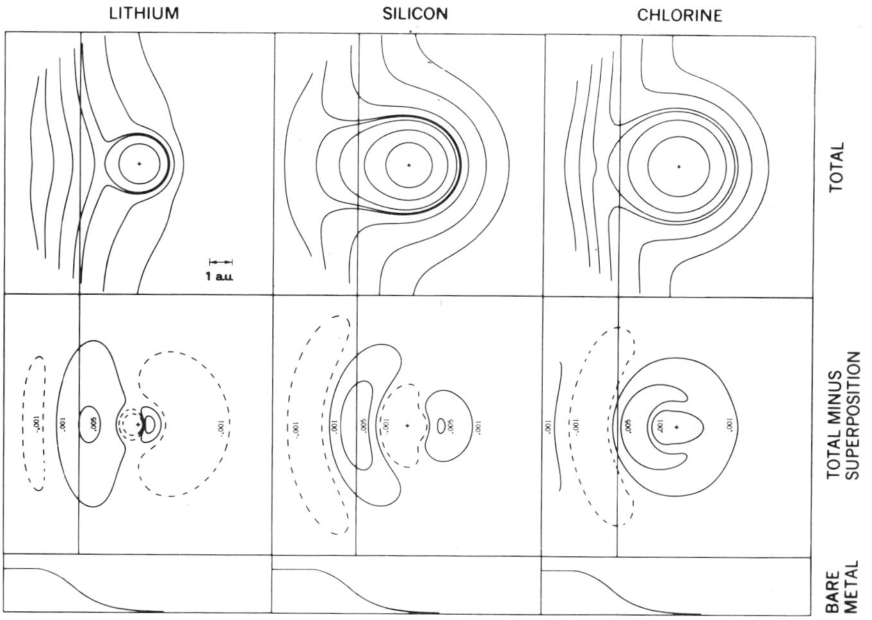

Fig. 7. Electron-density contours for chemisorption of one atom on a uniform-background substrate ($r_s = 2$). Metal-adatom distances shown minimize the total energy. Upper row: Contours of constant density in (any) plane normal to the metal surface containing the adatom nucleus (indicated by +). Metal is to the left-hand side; positive-background edge indicated by vertical line. Contours are not shown outside the inscribed circle of each square; contour values were selected to be visually informative. Center row: total electron density minus the superposition of atomic and bare-metal electron densities (electrons/bohr^3). The polarization of the core region, shown for Li, was deleted for Si and Cl because of its complexity. Bottom row: bare-metal electron-density profile (shown to establish physical distance scale). (From Ref. 9.)

therefore takes the form of adding a perturbing potential which is equal to the difference between the total lattice pseudopotential and the potential due to the semi-infinite uniform positive background. Since we wish only to consider changes of the total energy, we can use first-order perturbation theory. Such a treatment has been successfully applied to the study of work-function anisotropies and surface energies in simple metals.[5,7] Application to the present case yields curves such as those shown in Fig. 9 (solid curves) for Si chemisorbed on the (111) face of Al.[9] The most favorable adsorption site is seen to be the three-fold site.[11] As noted earlier, the positive background is to be thought of as arising from a symmetric smearing of the ionic charge on each lattice plane into a slab of positive charge; thus the first lattice plane of pseudopotentials is half an interplanar spacing inside the background edge. This fact, together with the equilibrium d value seen in Fig. 9 for adsorption in the three-fold site gives an Al-Si bond length of 2.6 Å (no experimental data is yet available).[12]

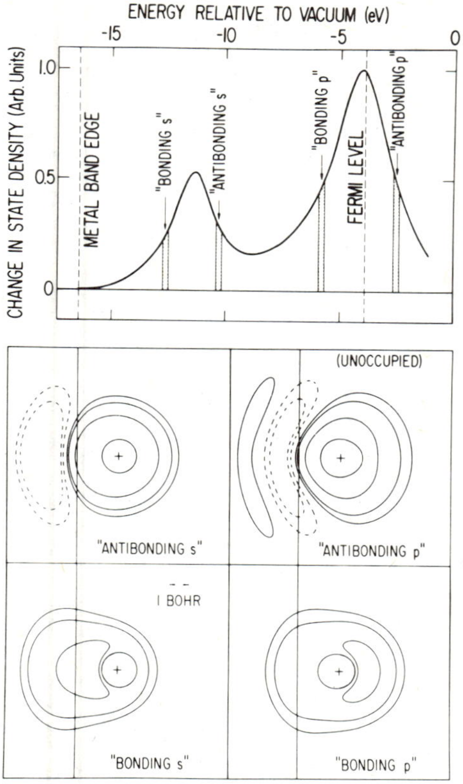

Fig. 8. Upper part of figure reproduces the state density curve from Fig. 6 for a Si atom chemisorbed at its equilibrium distance on a uniform-background substrate ($r_s = 2$). The two peaks correspond to the $3s$ and $3p$ atomic states. The lower part of the figure shows the density contours associated with the four shaded regions in the state density curve. (See caption of Fig. 7 for details of such contour maps.) Solid lines correspond to positive contour values, dashed lines to negative values. The same set of contour values is used for all four cases. The crescent-shaped contours in the two "bonding" maps correspond to maxima in the density. (Contours near the nucleus were deleted for clarity.) (From Ref. 9.)

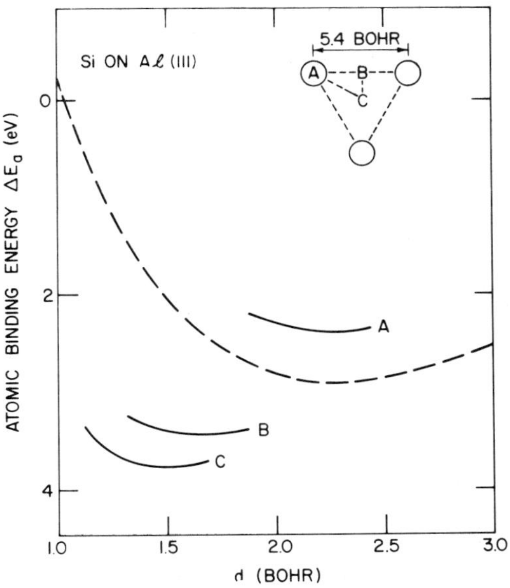

Fig. 9. Atomic binding energy as a function of distance *d* for a Si atom on an Al(111) substrate. Solid lines give results for the case in which the substrate is represented by the uniform-background model ($r_s = 2$) with a pseudopotential lattice included using first-order perturbation theory. Dashed line gives result for uniform-background model without lattice correction. Separation *d* is measured from the positive-background edge; the outermost plane of substrate nuclei lies 2.2 bohrs behind this edge in the present case. The calculated equilibrium position for adsorption in the centered site *C* (which is over a hole in the second layer of atoms) corresponds to an Al-Si bond length of 2.6 Å. (From Ref. 9.)

Even though the calculations provide equilibrium metal-adatom separations, the results of calculations performed for other separations yield information useful to discussing dynamics of atoms at surfaces, and provide insight into the distance dependence of the metal-adatom interaction. Figure 10 shows the distance dependence of the dipole moment for adsorbed Na, Si, and Cl atoms.[9] The slopes of the Na and Cl curves are consonant with their character as ionic (or largely ionic) adsorbates; and the essentially zero slope of the Si curve is consonant with the basically non-ionic character of its bonding. Figure 11 shows how the resonance for adsorbed hydrogen changes with distance.[9] At the largest distance, the metal-adatom interaction is not strong, and the resonance is fairly narrow. As the atom moves closer, the interaction becomes stronger, and the resonance widens. The resonance position tends to follow the metal double-layer potential however, and so as the atom moves in, the resonance shifts down to lower energies. This has the result that as the hydrogen is moved still closer (bottom panel of Fig. 11), the resonance becomes narrower again, because the density of metal states at the resonance energy is very small near the bottom of the metal conduction band. A similar effect is seen for the Si 3s resonance in Fig. 12.[9] This figure also indicates the distance dependence of the splitting between the $3p_z$ and $3p_{xy}$ states of Si (the sign of this splitting in fact changes at distances only slightly smaller than those shown).[13,14]

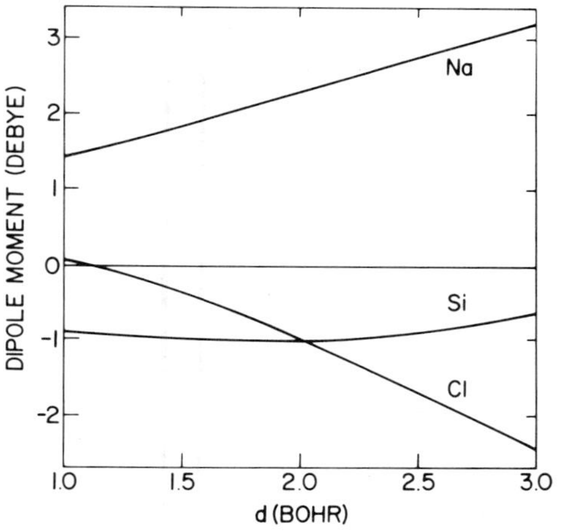

Fig. 10. Dipole moment as a function of metal-adatom distance *d* for Na, Si, or Cl atom chemisorbed on a uniform-background substrate ($r_s = 2$). The sign of the dipole moment is defined so that a negative moment corresponds to an increase in substrate work function. (From Ref. 9.)

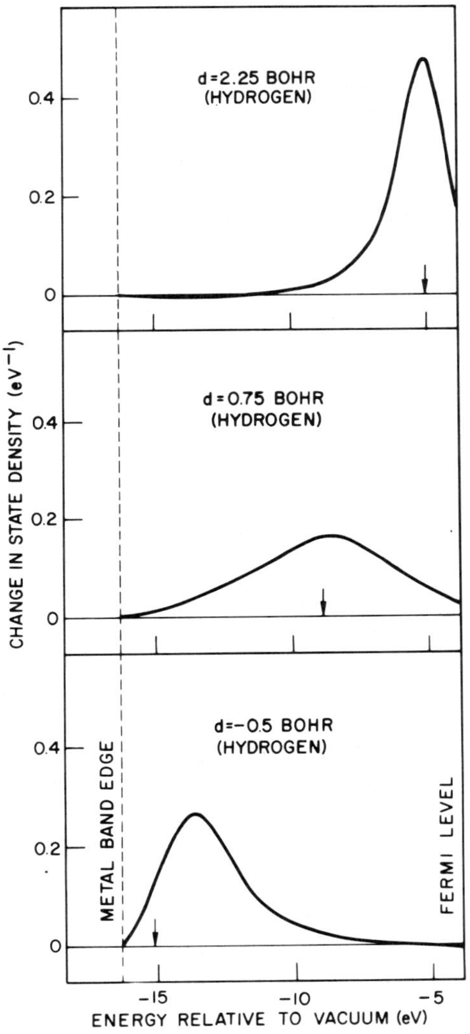

Fig. 11. State density change due to chemisorption of a hydrogen atom on a uniform-background substrate ($r_s = 2$). Curves are shown for three different metal-adatom distances d. (From Ref. 9.)

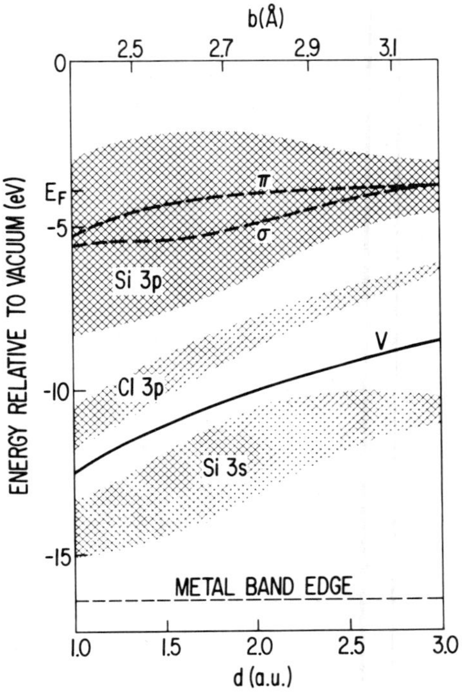

Fig. 12. Characteristics of Si and Cl state-density resonances (cf. Fig. 6) as a function of metal-adatom separation for adsorption on a uniform-background substrate ($r_s = 2$). Boundaries of shaded regions indicate half-maximum energies. Lower axis *d* is separation of adatom nucleus from positive-background edge (a.u. ≡ bohr). The plane through the outermost nuclei of the substrate represented by this background is half an interplanar spacing behind the background edge. Thus *d*, the crystal structure of the substrate, and the adsorption site determine an adatom to metal-atom bond length *b*. The upper axis provides *b* for a threefold site on a (111) surface of Al as an example. Dashed curves show peak positions for p_z (σ) and p_{xy} (π) components of Si-3*p* resonance. (Not shown for Cl.) Curve *V* gives effective one-electron potential of bare metal (displaced downward for pictorial reasons). (From Ref. 9.)

1. W. Kohn and L. J. Sham, Phys. Rev. 140, A1133 (1965).
2. V. L. Moruzzi, J. F. Janak, and A. R. Williams, Calculated Electronic Properties of Metals (Pergamon Press, New York, 1978).
3. O. Gunnarsson, J. Harris, and R. O. Jones, J. Chem. Phys. 67, 3970 (1977); J. Harris and R. O. Jones, J. Chem. Phys. 68, 1190 (1978); 70, 830 (1979); K. Kitaura, C. Satoko, and K. Morokuma, Chem. Phys. Lett. 65, 206 (1979).
4. O. Gunnarsson and B. I. Lundqvist, Phys. Rev. B 13, 4274 (1976).
5. N. D. Lang and W. Kohn, Phys. Rev. B 1, 4555 (1970).
6. N. D. Lang, in: Solid State Physics (F. Seitz, D. Turnbull, and H. Ehrenreich, eds.), Vol. 28 (Academic Press, New York 1973), p. 225.
7. N. D. Lang and W. Kohn, Phys. Rev. B 3, 1215 (1971).
8. N. D. Lang, Phys. Rev. B 4, 4234 (1971).
9. N. D. Lang and A. R. Williams, Phys. Rev. B 18, 616 (1978).
10. R. F. W. Bader, W. H. Henneker, and P. E. Cade, J. Chem. Phys. 46, 3341 (1967); R. F. W. Bader, I. Keaveny, and P. E. Cade, J. Chem. Phys. 47, 3381 (1967).
11. This is a three-fold site over a hole in the second layer of atoms.
12. Adding the covalent radius of Si and the metallic radius of Al gives this same bond length.
13. The treatment of chemisorption described above can be used to study core-level binding energies of adsorbed atoms. See N. D. Lang and A. R. Williams, Phys. Rev. B 16, 2408 (1977).
14. Work of a type similar to that in the above discussion, for chemisorbed H and H_2, has been done by H. Hjelmberg, Physica Scripta 18, 481 (1978); Surf. Sci. 81, 539 (1979); J. K. Nørskov, H. Hjelmberg, and B. I. Lundqvist, Solid State Commun. 28, 899 (1978); H. Hjelmberg, B. I. Lundqvist, and J. K. Nørskov, Physica Scripta (to be published).

TRENDS IN THE CHEMISORPTION ENERGY OF HYDROGEN AND OXYGEN ON TRANSITION METALS

C. M. Varma and A. J. Wilson*
Bell Laboratories
Murray Hill, NJ 07974

INTRODUCTION

One of the simplest properties characterizing the interaction of gases with solid surfaces is the chemisorption energy, ΔH_{ad}. It has been known for sometime now[1] that ΔH_{ad} for gases like H_2, N_2, O_2, etc. vary systematically across the transition metal series. The available data has recently been summarized by Toyashima and Samarjai.[2] ΔH_{ad} is maximum at the beginning of the series and decreases as the number of d-electrons decreases; the decrease is faster at the beginning of the series and saturates as we approach the end. Within the scatter of the data there is no discernible systematic variation going down from 3d to 5d metals.

Our aim here is to derive the physical basis for these trends by simple but well-defined methods and to express ΔH_{ad} in terms of the simple parameters of the adsorbed atoms and in terms of a few essential parameters characterizing the transition metals that have clear meaning in the theory of such solids.

TRANSITION METAL ELECTRONIC STRUCTURE

The systematics for ΔH_{ad} on transition metals suggest that it is governed by some general features of the transition metal electronic structure rather than by the details. The valence-electron structure of transition metals consists of the relatively tightly bound d-bands hybridized with nearly free-electron s-p bands. In our treatment we assume that the ad-atom bond is primarily with the d-orbitals and that the free s-p electrons serve primarily to renormalize like Coulomb repulsive energies,

* Permanent address: Cavendish Laboratory, Cambridge, England

ISSN:0094-243X/80/610151-09$1.50

etc., so that the ad-atom - d-orbital bond is most effective. This assumption is based on experience with transition metal alloys and our confidence in it is reinforced by the results of the present calculations.

The minimum essential parameters necessary to characterize the d-bands of a transition metal are

(i) the mean energy or the first moment of the density of states,

(ii) the band-width or the second moment of the density of states, and

(iii) the number of d-electrons or the Fermi-energy.

These parameters are obtained from bandstructure calculations and renormalized-atom calculations.[3]

The Fermi energy with respect to vacuum does not vary significantly across a given transition metal series. There is a significant variation in the bandwidth - as the number of d-electrons increases, the bandwidth shrinks. The most significant trend, however, is in the position of the mean energy of the band - it decreases rapidly with d-electron number. We note that the trends in ΔH_{bond} are similar to those in the mean energy. The physics of this correlation will be described in our concluding section.

PROCEDURE OF CALCULATIONS

Our starting point for a calculation of ΔH_{ad} is an expression for ΔH_{bond}, the bond-energy in terms of the change in the local density of states (LDOS) of the ad-atoms and the transition metal atoms, upon adsorption. Let $n_{i\alpha}(\varepsilon)$ be the LDOS of any orbital α on an atom i involved in the adsorption, before adsorption, and $\tilde{n}_{i\alpha}(\varepsilon)$, be the corresponding quantity after adsorption. Let $C_{i\alpha}$ and $\tilde{C}_{i\alpha}$ be their respective first moments. Then it can be shown in a self-consistent one electron-approximation (like the density-functional approximation) that

$$\Delta H_{bond} \simeq \sum_{i,\alpha} (E_{i\alpha} - \tilde{E}_{i\alpha}) - \tfrac{1}{2} \sum_{j} (D_{i\alpha,j\beta} + \tilde{D}_{i\alpha,j\beta}) \times \Delta N_{j\beta} \tag{1}$$

where

$$E_{i\alpha} = \int_{-\infty}^{\varepsilon_f} d\varepsilon(\varepsilon - C_{i\alpha}) n_{i\alpha}(\varepsilon) \quad , \tag{2}$$

$$E_i = \int_{-\infty}^{\tilde{\varepsilon}_f} d\varepsilon(\varepsilon - \tilde{C}_{i\alpha}) \tilde{n}_{i\alpha}(\varepsilon) \quad , \tag{3}$$

$$D_{i\alpha,j\beta} = C_{i\alpha} - C_{j\beta} \quad , \tag{4}$$

$$\tilde{D}_{i\alpha,j\beta} = \tilde{C}_{i\alpha} - \tilde{C}_{j\beta} \quad , \tag{5}$$

and $\Delta N_{j\beta} = \tilde{N}_{j\beta} - N_{j\beta}$ is the change in the total charge in $j\beta$.

The first term in Eq. (1) is evidently a co-valent or metallic type bonding term and the second an ionic type term. Note that for $\tilde{n}_{i\alpha} = n_{i\alpha}$, i.e., for the ad-atom far from the metal surface $D = \tilde{D} = (\mu - \varepsilon_f)$, where μ is the chemical potential of the ad-atom. In this case $\Delta H_{bond} \simeq (\mu - \varepsilon_f)\Delta N$.

To calculate ΔH_{bond} with Eq. (1), we need to specify the LDS of the ad-atom and of the transition metal atoms before and after adsorption.

As already mentioned, we do not believe a detailed description of the LDOS is needed to explain the empirical results. We determine the general features of the density of states from their first few moments evaluated in the tight binding approximation. The further constraints on the LDS are that their first moment be self-consistent with the charge transfer and the associated Coulomb interactions. An additional important constraint is that the system obey charge neutrality locally. We shall take locally to mean the ad-atom and its immediate neighboring transition metal atoms. Local charge neutrality (or Friedel sum-rule) is enforced in general through the long-range Coulomb interaction. We parametrize these interactions here by an ad-atom - nearest neighbor transition metal atoms repulsion parameter U_{am} and determine it by enforcing local charge neutrality.

The onsite screened Coulomb repulsion parameter, U_m, is known to have the value 3 to 4 eV. The repulsion parameter on the ad-atom, U_a, is determined from the ionization energy, the affinity energy and from the image forces. Its value for Oxygen and Hydrogen at a distance of about 1Å from the surface is about 5 eV.

The ad-atom - surface metal atom transfer integral V has been calculated using atomic orbitals and atomic potentials for metals for which the equilibrium position of the ad-atom on the metal surface is known. What is important is that it is more or less a constant across the series. Since the experimental data we are comparing our results to pertains to polycrystalline surfaces, we have taken V to be an adjustable parameter, but constant across the series. The parameter used is about 3/2 times the parameter calculated for the 100 surface of the metals, with the ad-atom atop the center with four-fold co-ordination.

We briefly summarize the procedure of the calculations. The details will be published elsewhere. First the parameters of the surface atom (charge on the surface atom, the mean energy and the second moment of LDS on it) are determined insofar as they differ from the bulk. The calculated bulk parameters and U_m are used in this calculation. Next the LDS on the ad-atom and surface atoms is self-consistently determined using the free-atom energy level, V, U_a, U_m, and with U_{am} such that local charge neutrality is obtained. Features of the LDS are determined by fixing the first three moments and finding a suitable functional form of LDS incorporating these moments. (We have also performed much more detailed calculation of the LDS using the continued fraction method with results similar to those from the much simpler method outlined here.) The resulting LDS is used in Eq. (1) to calculate ΔH_{bond}.

RESULTS AND DISCUSSIONS

In Fig. (1), we present our calculated results for the binding energy of Oxygen on 4d-metals, using the parameters specified on the figure. The experimental results are also shown, as is the calculated ΔN_a, the d-electron charge transfer to the oxygen atom and the parameter U_{am} required to achieve local charge

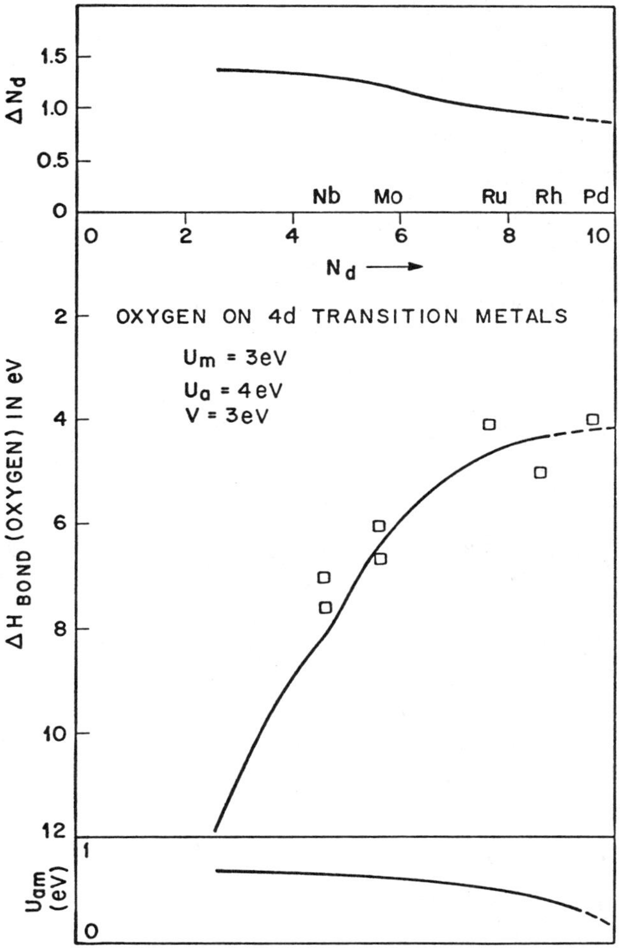

Fig. 1. Calculated and experimentally obtained binding energy for Oxygen on 4d-transition metal surfaces. Also given is the (d-electron) charge transfer to Oxygen and the metal atom-ad-atom Coulomb parameter required to obtain local charge neutrality.

neutrality. U_{am} is of order ½ eV and varies smoothly across the series. ΔN_a is of order 1 electron and also varies smoothly. We envisage that the overflow s-electron charge on the pure metal surface moves back on to the surface transition metal atoms to screen the d-electron charge flow and give us renormalized repulsion parameters, U_m, etc.

In Fig. (2), we compare the results obtained by varying the various parameters for oxygen on 4d-series. A larger U_m or U_a leads to an overall reduction in the binding energy without affecting the general trend. A larger V leads to an increased binding energy again without affecting the general trend.

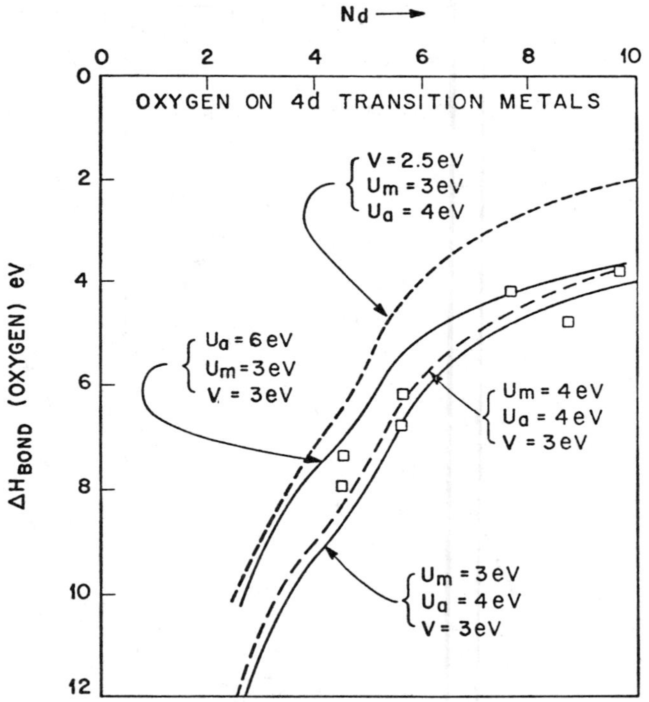

Fig. 2. Calculations for Oxygen binding energy on 4d-transition metal surface for various parameters. The experimental points are also shown.

In Fig. (3), we present results for hydrogen adsorbed on the 4d-series metals together with the experimental results.

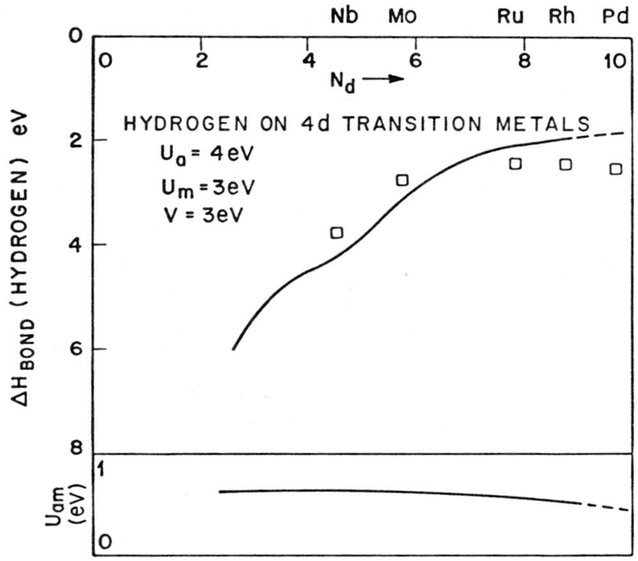

Fig. 3. Calculated and experimentally obtained binding energy for Hydrogen on 4d-transition metal surfaces. Also given is the (d-electron) charge transfer to Hydrogen and the metal atom-ad atom Coulomb parameter required to obtain local charge neutrality.

Our results for Oxygen and Hydrogen on 3d-metals and for Nitrogen compare equally well with the experimental results. For Nitrogen we get about ½ electron larger charge transfer than oxygen and somewhat larger binding energy.

Having obtained the systematics of the chemisorption energies of simple atoms on transition metals from a relatively simple calculation, we are able now to comment on the physics of the problem. From an examination of the variation of transition metal parameters and the experimental results, one can make the empirical observation that the transition metal parameter primarily determining the trend is the mean energy of the d-band. The reason is the following: the change in the density of states near the top of the d-band or the bottom is nearly the same as that near the center of the d-band if

$$\frac{W^2V^2}{2[(\varepsilon_a-C)^2 + 4V^2)^2} << 1 \qquad (6)$$

as may be verified by second-order perturbation theory. This condition is always met in the problems under consideration. If the variation in LDS are similar in all energy ranges of the d-bands, the characteristic d-band parameter playing the crucial role may be taken to be the mean energy of the d-band. It is for the same reason that our simple characterization of the d-band has worked: detailed features of the band are unimportant if there is similar variation everywhere. In any event this assumption is checked in detail in another paper, where detailed LDOS are calculated.

If the important metallic parameter determining the change in LDS is C, two points follow. As C approaches the partially filled level ε_a the binding energy due to interaction through V and the charge transfer to the lower energy state decreases. This may be verified by diagonalizing the matrix of a level at ε_a connected to the degenerate levels at C through V and calculating the new energy through filling the lowest levels. The difference in energy before and after adsorption will be $[\frac{1}{4}(\varepsilon_a-C)^2 + nV^2]^{\frac{1}{2}}$, where n is the degeneracy at C.

A slightly better calculation would be to do the above not with C, but with C_f, the mean energy of the filled part of the band. C_f has a trend in the transition metal series similar to (but slowly varying than) C. Also as C approaches ε_a the antibinding levels formed of the interaction between ad-atom and metal increasingly get populated. This further reinforces the trend.

The variation in the bandwidth, W, across the series is next in importance (after the mean d-electron energy, C or C_f), in determining the trend in the binding energy. Our conclusion from the calculations is that other parameters remaining the same, a smaller W gives a larger binding energy. The leveling off of the binding energy curve to the right of the series (even though C rapidly falls) is due to the decrease in W. This is due to two effects: C remaining the same, C_f is higher for a small W; also the alteration in the local density of states depends (besides ε_a-C) on V/W, being larger for larger V/W.

ΔH_{bond} for Hydrogen is smaller, primarily because there is only one unfilled level as opposed to two for Oxygen. Correspondingly the variation across the series is also smaller.

We have also compared the position of the atomic chemisorption level obtained in the calculations with experimental photoemission results, where available and obtained good agreement.

We have tried in this work and partially succeeded in understanding the binding energy trends of simple atoms on transition metals. Through a simple physically motivated method, we have tried to highlight the essential parameters determining the binding energy. The principal question left undecided is the precise value of the parameters employed. These can only be obtained through enormously difficult calculations on transition metal surfaces with modifications of the methods of the type used, say by Lang and Williams[4] for atoms on jellium. We do not, however, expect any significant changes in the physical picture presented here.

REFERENCES

1. See for instance, J. M. Thomas and W. J. Thomas, Introduction to the Principles of Heterogeneous Catalysis (1967), Academic Press (New York), Chapter 2.
2. I. A. Toyashima and G. A. Somarjai, Catal. Rev. **19**, 105 (1979); G. A. Somarjai, Catal. Rev. **18**, 173 (1978).
3. L. Hodges, R. E. Watson and H. Ehrenreich, Phys. Rev. **B5**, 3953 (1972), and further calculations by R. E. Watson (to be published).
4. N. D. Lang and A. R. Williams, Phys. Rev. **B18**, 616 (1978).

THERMAL DESORPTION AND DISSOCIATION CATALYZED BY A SOLID SURFACE*

Uzi Landman and G.S. De, School of Physics, Georgia Institute of Technology, Atlanta, Georgia 30332

M. Rasolt, Solid State Division, Oak Ridge National Laboratory, Oak Ridge, Tennessee 37836

INTRODUCTION

One of the major objectives of surface science is the understanding of the fundamental processes involved in chemical reactions catalyzed by solid surfaces. A catalyzed surface reaction may conveniently be described by a sequence of elementary kinetic steps[1,2] such as adsorption (which may proceed via stages involving precursor states, physisorption etc.); diffusion[3,4] (on the surface and in certain cases through the bulk); reaction (dissociation, association, rearrangement etc.) and desorption. The traditional thermochemical analysis of reactions involving the balance of heats of the above distinct processes, such as heats of adsorption, desorption, dissociation and bond formation, provides criteria for the occurrence of certain reaction paths. However, such analysis, while determining the direction of the process in the reactants-products space does not provide detailed information about the reaction rates and the fundamental microscopic processes which govern them. The employment of specific catalytic agents in a chemical reaction provides the means for modifying reaction paths and rates. In addition through the use of the proper catalyst a particular reaction channel, selected from the manifold of thermodynamically allowed ones, may be enhanced. In order to enable to choose specific catalysts with the objective of optimizing the yield of a desired reaction product it is necessary to develop an understanding of the microscopic interaction processes and to identify the physical parameters of the catalytic system (catalyst, reactants, and ambient conditions) which govern the outcome and rate of the reaction.

The evolution of a chemical reaction is conveniently described by rate data, i.e., plots of the decay or growth of the concentrations of the reactants and products, respectively. When treated on a phenomenological level[5,6], a reaction scheme is postulated and described by a set of simultaneous rate equations. In this class of studies the rate constants are regarded as parameters to be determined by fitting the solutions of the equations to the experimental data. The functional form of the solutions to the kinetic decay and growth equations serves to catagorize reactions by their orders with respect to reactant concentrations (i.e., the exponents of the concentrations as they appear in the rate equations). However, it should be emphasized that since rate equations form the most contracted description of a chemical reaction there need not be a simple correspondence between the order and the physical mechanism of the reaction.

Understanding of chemical reactions on a more fundamental level is provided by statistical mechanical approaches[8-17], in which rate

ISSN:0094-243X/80/610161-19$1.50

constants are evaluated using partition functions[6], stochastic methods[15] or "phase-space theories" and molecular dynamics[16,17] (i.e., classical trajectory analysis). The latter methods describe the classical evolution of the system on its potential energy surface and provide information about collision dynamics, energy transfer and rates. While most popular in gas-phase kinetics, these powerful statistical methods have not been thoroughly investigated in the context of surface reactions.

The most refined treatment of reaction kinetics is by microscopic models[18-29] in which the underlying physical processes, such as coupling, excitation and energy transfer, are investigated. It is via investigations on this level that one may expect to unravel the correlations between catalytic reaction kinetics and characteristic parameters of the solid substrate (geometric structure, vibrational and electronic spectra), properties of the reactants (electronic, vibrational) and ambient conditions (pressure, temperature and external fields). While the importance of the structural, electronic and vibronic factors in catalysis has been recognized[30] detailed microscopic calculations of catalytic reaction rates are scarce. However, we believe, that current advents of theoretical methods coupled with experimental spectroscopic techniques[31,32,33,34] (such as ultra-violet photoemission, Auger chemical shifts, electron loss spectroscopies, and surface sensitive infrared techniques) and kinetic measurements (in particular molecular beams techniques[35,36]) provide the impetus for an increased activity in the above approach.

Of paramount importance for the understanding of reaction dynamics on a microscopic level are the mechanisms of coupling and energy exchange and redistribution between the reactants[37]. In the case of gas phase reactions certain general conclusions with regard to the relative importance of translational, vibrational and rotational energies in surmounting the reaction barrier have been drawn. For example, when the barrier is encountered early in the approach of the reactants translational energy is the key factor. While for reactions in which the potential energy barrier for reaction occurs when the reactants have come to close proximity, the vibrational energy content is dominant in dictating the probability of reaction. A heterogeneous surface catalytic reaction presents a considerably more complicated situation than a gas phase reaction. Metal substrates, for example, exhibit a wide variation in activity and specificity. These substrates possess a quasicontinuous spectrum of electronic excitations filling states up to the Fermi level and a spectrum of vibrational excitations. The lattice vibrations (phonons) as well as elementary excitations of the highly polarizable conduction electrons (e.g., electron-hole pairs) provide efficient mechanisms for energy transfer and dissipation. In addition impurities and structural imperfections modify the spectrum of elementary excitations and at the same time they are known to influence the course of surface reactions with respect to their efficiency, selectivity and specificity. Consequently, it is of interest to explore the coupling and excitation mechanisms between molecular species and metallic surfaces and to formulate the evolution of the excitations leading to bond formation (sticking), bond rupture (desorption and dissociation), rearrangement and migration. In this study we focus on thermally induced bond rupture elementary reactions, i.e., dissociation and desorption[38]. Moreover, we investigate coupling to lattice vibrations (phonons) although similar considerations could be applied to electron-hole pair excitations in the quasi-boson representation. The work reported herein is the result of a collaborative effort with Drs. G. S. De (Georgia Institute of Technology) and M. Rasolt (Oak Ridge National Laboratory).

In the following an outline of the method of calculation and sample results are presented. For a more detailed, comprehensive presentation, the reader is referred to a recent publication by the above authors[39].

To facilitate our discussion we specify the following ingredients of the theory:

a) coupling of an atom or a molecule to the surface
b) substrate induced thermal energy transfer and excitation
c) temporal evolution of the system, i.e., time-evolution and calculation of reactional probabilities and rates.

These elements are considered in turn in the following sections.

II. HAMILTONIAN AND COUPLINGS

(i) Hamiltonian

The first step in our formulation is a statement of the Hamiltonian of the system. The total Hamiltonian of the adsorption system may be written as

$$H = T_e + T_N + V(\underset{\sim}{r},\underset{\sim}{R}) \qquad (2.1)$$

where T_e and T_N represent kinetic energies of electrons and nuclei of the system (molecule and substrate) and $V(\underset{\sim}{r},\underset{\sim}{R})$, various contributions to the potential energy, ($\underset{\sim}{r} \equiv \{\underset{\sim}{r}_M,\underset{\sim}{r}_s\}$, $\underset{\sim}{R} \equiv \{\underset{\sim}{R}_M,\underset{\sim}{R}_s\}$, where $\underset{\sim}{r}_M$ and $\underset{\sim}{r}_s$ are the electronic coordinates of the adsorbed molecule and solid respectively, and $\underset{\sim}{R}_M,\underset{\sim}{R}_s$ the corresponding nuclear coordinates). In the adiabatic approximation the total wavefunction is taken as

$$\psi(\underset{\sim}{r},\underset{\sim}{R}) = \phi(\underset{\sim}{r},\underset{\sim}{R})\chi_N(\underset{\sim}{R}) \ . \qquad (2.2)$$

The electronic wavefunction satisfies the equation

$$[T_e + V(\underset{\sim}{r},\underset{\sim}{R})]\ \phi(\underset{\sim}{r},\underset{\sim}{R}) = \varepsilon_e(\underset{\sim}{R})\phi(\underset{\sim}{r},\underset{\sim}{R}) \qquad (2.3)$$

solved for fixed $\underset{\sim}{R}$, where the direct interaction between nuclei is included in $V(\underset{\sim}{r},\underset{\sim}{R})$. The equation for the nuclear motion can be found variationally[41]

$$\left\{ - \sum_{\mu=1}^{N} \frac{1}{2M_\mu} \nabla_\mu^2 + [\varepsilon_e(\underset{\sim}{R}) + V'(\underset{\sim}{R})] \right\} \chi_N(\underset{\sim}{R}) = \varepsilon_N \chi_N(\underset{\sim}{R}) \qquad (2.4)$$

where

$$V'(\underset{\sim}{R}) = \sum_{\mu=1}^{N} \int d^3\underset{\sim}{r}\ \phi^*(-\frac{1}{2M_\mu}\nabla_\mu^2)\phi \quad , \tag{2.5}$$

and the summations are over all the nuclei. The effective potential (expression in square brackets in Eq. (2.4)) for the nuclear motion is dominated by the electronic energy $\varepsilon_e(\underset{\sim}{R})$ and the term $V'(\underset{\sim}{R})$ is small. To go beyond the adiabatic approximation requires the evaluation of the non-adiabatic couplings $-(h^2/M)\langle n|\nabla_{\underset{\sim}{R}}|m\rangle\nabla_{\underset{\sim}{R}}$ and $-(h^2/2M)\langle n|\nabla^2_{\underset{\sim}{R}}|m\rangle$, where $|n\rangle$ and $|m\rangle$ are adiabatic electronic states depending upon the nuclear coordinates $\underset{\sim}{R}$. While it is possible to include non-adiabatic contributions using a generalized Green's function method[42], we do not include them in the following.

At this stage the electronic energy $\varepsilon_e(\underset{\sim}{R})$ may be modeled in the following manner. First one identifies in $\varepsilon_e(\underset{\sim}{R})$ those components which correspond to intramolecular bonds and to binding between atoms of the molecule and a localized region in the solid. The electronic interactions for a fixed configuration of nuclear molecular coordinates, $\{\underset{\sim}{R}_M\}$, and solid nuclear coordinates, $\{\underset{\sim}{R}_S\}$, determine the adsorption potential between the solid and the adsorbate. We separate the interaction into two parts: one in which the solid is kept stationary and the other where the solid is allowed to vibrate. It is via the latter contribution that an energy exchange between the molecule and solid (which may eventually yield desorption or dissociation) becomes possible. The first contribution to these "bond-potential-energies" may then be modeled by some analytical potential formulae such as a harmonic well, a Morse potential or other suggested potential formulae. The rest of $\varepsilon_e(\underset{\sim}{R})$, i.e., that part which can be identified with the solid coordinates may then be replaced by a certain model of the solid (remember that internuclear interactions were included in $\varepsilon_e(\underset{\sim}{R})$).

The corresponding nuclear motion equations (2.4) may then be solved with the above mentioned model replacement for $\varepsilon_e(\underset{\sim}{R})$, and their solutions provide the vibrational spectrum for the modeled system.

(ii) Coupling

In this section we derive, under certain approximations, an expression for the coupling between a point charge (of charge + Z_A^*e, where Z_A^* is the effective charge of the ion) adsorbed at a distance z_1 from a metal surface and the <u>fluctuating</u> part of the metal substrate. As discussed previously, in the model which we develop the role of the electronic (including direct nuclear interactions) energy is to establish a bounded molecule-solid system characterized by, for example, a Morse potential with an equilibrium distance d_1 of the atomic constituents from the surface, with an associated manifold of vibrational levels.

To make the calculation tractable and yet preserving the essential physical features we model the surface in the following manner.

We consider an electron gas bounded by an infinite potential barrier (Fig. 1). The static ions are then placed within this potential and

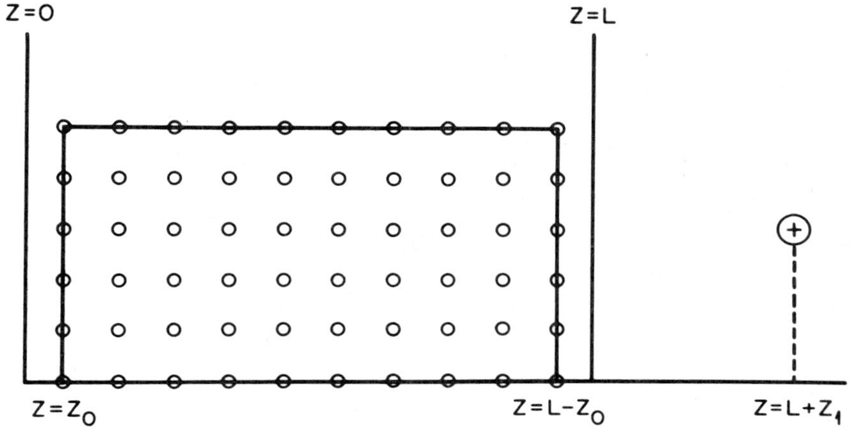

Fig. 1. The model adsorption system. The electron gas is bounded in a slab of linear dimension L. The first plane of substrate ions is located at $z = L-z_0$ with $z_0 \to 0$. The adsorbed ion is located at $z = L+z_1$.

the first plane is positioned at a distance $L - z_0$ with $z_0 \to 0$ from the potential barrier, where L is the linear dimension of the slab.

We now relax the static ions and allow them to fluctuate by emitting phonons. Denoting such a density fluctuation of the ions as $\delta n_b(\underset{\sim}{r})$ and the electronic response to such a fluctuation as $\delta n_e(\underset{\sim}{r})$, the coupling may be written as

$$v(\underset{\sim}{r}) = \delta u_b(\underset{\sim}{r}) + \delta u_e(\underset{\sim}{r}) , \tag{2.6}$$

where $\delta u_b(\underset{\sim}{r})$ and $\delta u_e(\underset{\sim}{r})$ are the fluctuating ionic and electronic interaction potentials with the adsorbed charge $(+Z_A^* e)$ positioned at $\underset{\sim}{r}$, i.e.,

$$\delta u_b(\underset{\sim}{r}) = Z_A^* Z_s e^2 \int \frac{d\underset{\sim}{r}' \, \delta n_b(\underset{\sim}{r}')}{|\underset{\sim}{r}-\underset{\sim}{r}'|} ,$$

$$\delta u_e(\underset{\sim}{r}) = - Z_A^* Z_s e^2 \int \frac{d\underset{\sim}{r}' \, \delta n_e(\underset{\sim}{r}')}{|\underset{\sim}{r}-\underset{\sim}{r}'|} , \tag{2.7}$$

with Z_A^* and Z_s and atomic charges (possibly screened) of the adatom and metal ions, respectively. The position vector $\underset{\sim}{r}$ of the point charge is set equal to $z_1 + L$ in Eqs. (2.6) and (2.7), where L is the thickness of the sample (Fig. 1). We choose next a wave-vector $\underset{\sim}{q}$ of the fluctuating ionic background and express the single Fourier component of the density fluctuation as

$$\delta n_b(z,\underset{\sim}{r}_{\parallel}) = \delta n_b(q_\perp,\underset{\sim}{q}_{\parallel})\, e^{i(q_\perp z+\underset{\sim}{q}_{\parallel}\cdot\underset{\sim}{r}_{\parallel})} + c.c. \quad . \tag{2.8}$$

In the following we omit writing explicitly the parallel components $\underset{\sim}{r}_{\parallel}$, $\underset{\sim}{q}_{\parallel}$ from δn_b due to translational invariance in planes parallel to the surface, and denote $\delta n_b(z) \equiv \delta n_b(z,\underset{\sim}{r}_{\parallel})$ and $\delta n_b(q_\perp) \equiv \delta n_b(q_\perp, \underset{\sim}{q}_{\parallel})$. It is convenient to define an extension of $\delta n_b(z)$ (see below) by an even function $\delta n_b'(z)$ such that for z in the range $[-L,0]$, $\delta n_b'(z) = \delta n_b(-z)$. Treating the electrons semiclassically and solving the linearized static collisionless Boltzmann equation for the fluctuation in the electron Fermi-Dirac distribution driven by the self consistent field given by Eqs. (2.6) and (2.7) we obtain the following expression for the coupling $v(z_1+L)$[39].

$$v(z_1+L) \equiv v(z_1) = \frac{Z_A^* Z_s e^2\, 2\pi\, \delta n_b(q_\perp) e^{-q_\parallel z_1}}{(q_\perp^2+q_\parallel^2)\, D}\left[1 - \frac{\lambda_{TF}^2}{q_\perp^2+q_\parallel^2+\lambda_{TF}^2}\right] , \tag{2.9}$$

where λ_{TF} is the Thomas-Fermi wavelength of the electrons and D is given by

$$D = 1 - \frac{\lambda_{TF}^2}{2(q_\parallel^2+\lambda_{TF}^2)^{1/2}[q_\parallel+(q_\parallel^2+\lambda_{TF}^2)^{1/2}]} \quad . \tag{2.10}$$

Within the context of the infinite surface barrier as a model for the static semi-infinite crystal the generalization of Eq. (2.9) to include crystallinity of the substrate would involve simply treating $\delta n_b(q_\perp,\underset{\sim}{q}_{\parallel})$ in terms of the phonon crystal propagator. In the present calculation we, however, would use a continuum Debye model of the solid.

In a continuum model of the solid the Fourier components of the positive background number density fluctuations, $\delta n_b(\underset{\sim}{q})$ see Eq. (2.8), are defined by the relation

$$\delta n_b(\underset{\sim}{r}) = \sum_{\underset{\sim}{q}} \delta n_b(\underset{\sim}{q})\, e^{iq\cdot\underset{\sim}{r}} \tag{2.11}$$

and $\delta n_b(\underset{\sim}{r})$ is given as[43]

$$\delta n_b(\underset{\sim}{r}) = -\, Z_s\, n_o\, \underset{\sim}{\nabla}\cdot\underset{\sim}{D} \tag{2.12}$$

where n_o is the ionic number density of the solid and $\underset{\sim}{D}$ is the displacement of the background from its equilibrium position. From Eqs. (2.11) and (2.12) it follows that

$$\delta n_b(q_\perp,\underset{\sim}{q}_{\parallel}) \equiv \delta n_b(q_\perp) = Z_s n_o \sqrt{\frac{\hbar}{2M_s n_o \Omega\omega_{\underset{\sim}{q}}}}\; q(b_{\underset{\sim}{q}}+b_{-\underset{\sim}{q}}^+) \quad , \tag{2.13}$$

where M_s is the atomic mass of a solid atom, Ω is the volume of the solid and $b_{\underset{\sim}{q}}(b^+_{\underset{\sim}{q}})$ are the annihilation (creation) operators of a phonon of wave vector $\underset{\sim}{q}$ and frequency $\omega_{\underset{\sim}{q}}$. We now express the instantaneous position of the adsorbed atom z_1 as $z_1 = d_1 + u_1$, where d_1 is the equilibrium distance of the atom from the surface, and expand the exponent, $\exp(-q_{\parallel} z_1)$, in Eq. (2.9) to obtain

$$v(z_1) = v_o(\underset{\sim}{q};d_1) + \delta v_1(\underset{\sim}{q};u_1) + \delta v_2(\underset{\sim}{q};u_1) + \ldots , \qquad (2.14)$$

$$\delta v_1(\underset{\sim}{q};u_1) = g^{(1)}_{\underset{\sim}{q}} \; u_1(b_{\underset{\sim}{q}} + b^+_{-\underset{\sim}{q}}) , \qquad (2.14a)$$

$$\delta v_2(\underset{\sim}{q},u) = g^{(2)}_{\underset{\sim}{q}} \; u_1^2(b_{\underset{\sim}{q}} + b^+_{-\underset{\sim}{q}}) , \qquad (2.14b)$$

where $\delta v_1(\underset{\sim}{q};u_1)$ is the potential corresponding to bilinear coupling between the vibrations of the atom and the phonons of the solid. Assuming for simplicity an acoustic continuum model for the solid i.e., $\omega_q = sq$ where s is the sound velocity, $g^{(1)}_{\underset{\sim}{q}}$ in Eq. (2.14) is given by

$$g^{(1)}_q = -F \frac{q^{1/2} q_{\parallel}}{q^2+\lambda^2_{TF}} \left[1 + \frac{\lambda^2_{TF}}{2(q^2_{\parallel}+\lambda^2_{TF})^{1/2}[q_{\parallel}+(q^2_{\parallel}+\lambda^2_{TF})^{1/2}]-\lambda^2_{TF}} \right] e^{-q_{\parallel} d_1} \qquad (2.15a)$$

where

$$F = 2\pi \, Z_s Z^*_A \, e^2 \left(\frac{\hbar n_o}{2M_s \Omega s} \right)^{1/2} . \qquad (2.15b)$$

Our results can be summarized by writing the Hamiltonian for our model system as

$$H = \varepsilon^o_e + H^o_v + H' , \qquad (2.16a)$$

$$H' = \sum_i v(z_i) , \qquad (2.16b)$$

where ε^o_e is the electronic energy (including direct nuclear-nuclear interactions) for equilibrium nuclear positions, H^o_v is the zeroth-order vibrational Hamiltonian for the adsorption system, and the last term Eq. (2.16b) in Eq. (2.16a) contains couplings between the vibrations of intramolecular and chemisorptive bonds and fluctuations in the solid.

(iii) Transition Rate

The transition rate between vibrational levels of the adsorbed system due to the coupling operator H' (Eqs. 2.16) can be evaluated to lowest order in perturbation theory using the golden rule formula[44]

$$W_{v\to v'} = \frac{2\pi}{h} \sum_{n_q, n_q'} p(n_q) |\langle v', n_q' | H' | v, n_q \rangle|^2 \, \delta(\varepsilon_{v'} - \varepsilon_v + \varepsilon_{n_q'} - \varepsilon_{n_q}) \ , \tag{2.17}$$

where we sum over phonon final states n_q' and average over phonon initial states n_q using the probability distribution $p(n_q)$.

In order to evaluate $W_{v\to v'}$ we need to specify the manner in which we model the vibrational spectrum of the adsorbed molecule.

(i) In the crudest approximation the vibrations are modeled by harmonic oscillators, truncated at the appropriate predissociation levels (see Section III). Keeping only terms up to bilinear coupling (δv_1, see Eq. 2.14) allows only for single quantum transition ($v \to v\pm1$), accompanied by the absorption (emission) of a phonon. An expression for $W^{(h)}_{v\to v+1}$ was derived (see reference 39).

(ii) As an improved model of the bond potential energy between an adsorbed atom and the surface we consider the Morse potential[45]

$$\varepsilon_e(z-d_1) = D_e \{[1 - \exp[-\beta(z-d_1)]]\}^2 - D_e \ , \tag{2.18}$$

where D_e is the dissociation energy referred to the minimum, d_1 the equilibrium distance of the adsorbed atom from the surface and the parameter β determines the width of the potential. One often defines the anharmonicity parameter $x_e = \frac{h\omega_o}{4D_e}$ where ω_o is the vibrational frequency for infinitesimal amplitudes (x_e is often determined empirically). Using previously derived expressions[46,47] for the matrix elements of the deviation from equilibrium, u, the transition rates $W^{(M)}_{v\to v+1}$ and $W^{(M)}_{v\to v+2}$ can be derived[39].

III. EVALUATION OF FIRST PASSAGE TIMES FOR THE TRUNCATED HARMONIC OSCILLATOR AND MORSE-POTENTIAL MODELS

Having obtained explicit expressions for the couplings between the adsorbate and the substrate and for the rates of transitions between vibrational levels of the binding potential, induced by the couplings, we turn next to the temporal evolution of the excitations. Since for most systems of interest the allowed quanta of excitation, dictated by the characteristics of the phonon spectrum of the substrate are much smaller than the barrier for bond rupture, an incoherent multiphonon mechanism is formulated. This, however, is applicable to systems in which the spacings between vibrational levels of the potential associated with the reaction coordinate do not exceed the maximum phonon frequencies. When the above is not satisfied coupling may occur through a mode other than the bond-rupture reaction coordinate which serves as a "door-way" state (see Section IV).

Consider an oscillator system with $x_n(t)$ the distribution describing the population of vibronic levels n at time t. The time evolution of this distribution is governed under certain approximations by a master equation[48]

$$-\frac{dx_n}{dt} = \sum_{\nu=0}^{N+1} W_{\nu n}x_n - \sum_{m=0}^{N} W_{mn}x_m \qquad n = 0, 1, \ldots, N \qquad (3.1)$$

where N is the predissociation level, W_{nm} is the transition probability per unit time from m to n. In the above equation second-order terms due to recombination are neglected. The initial distribution (t=0) is normalized according to

$$\sum_{n=0}^{N} x_n(o) = 1 , \qquad (3.2)$$

and the $x_n(0)$'s are given by a Boltzmann distribution at temperature T, i.e.,

$$x_n(o) = e^{-\beta\varepsilon_n} \Big/ \sum_{n=0}^{N} e^{-\beta\varepsilon_n} . \qquad (3.3)$$

For the calculation of the reaction rate we will be interested in the mean time for the system specified above, to pass the N-th level for the first time - i.e., the mean first passage time, $\bar{t}$. The distribution of first passage times P(t) is given by[48]

$$P(t) = -\frac{d}{dt}\sum_{n=0}^{N} x_n(t) \qquad (3.4)$$

and $\bar{t}$ is the first moment of P(t), i.e.,

$$\bar{t} = \int_0^\infty tP(t)dt . \qquad (3.5)$$

An expression for the mean first passage time for an initial population distribution $x_n(o) = \delta_{n,o}$ for a truncated harmonic oscillator system and transitions between neighboring levels only, was first given by Montroll and Shuler.[48] This has been generalized by Kim[49] for the Boltzmann initial distribution (Eq. 3.3) for both a truncated harmonic and Morse oscillators with nearest and next nearest neighbor transitions.

(i) For the truncated harmonic oscillator the result is[49]

$$\bar{t}^{(h)} = \frac{(v+1)}{W^{(h)}_{v\to v+1}(1-e^{-\theta})} \sum_{j=1}^{N+1} j^{-1}(e^{j\theta}-1)(1-e^{-j\theta}) \qquad (3.6)$$

with $\theta = \hbar\omega_0/kT$, where ω_0 is the harmonic oscillator frequency.

(ii) To obtain an expression for the mean first passage time out of a Morse potential well, $\bar{t}^{(M)}$, with transitions between nearest and next nearest levels, we adopt the methods developed by Kim. The generalization of Kim's result [Eqs. (6.15) and (6.16) of reference 49] amount to taking into account that in our case the exchange of excitation is with a solid characterized by thermal occupation numbers $\bar{n}_{\omega_q} = [\exp(\hbar\omega_q/kT)-1]^{-1}$. Consequently contributions corresponding to transitions between the Morse potential levels must be weighted appropriately.

In the stochastic formulation of nonequilibrium kinetics which we have employed the reaction rate, R, is given by the inverse of the mean-first-passage time $\bar{t}$ (see also discussion in Chapter IV). Results for the rates of desorption of Potassium and Xenon from a tungsten substrate are shown in Figs. 2-4. In Fig. 2 results for the two models of the binding potential (harmonic - solid line, and Morse - dashed line) using experimentally[50] suggested values for the desorption energy $D_e = 2640$ meV, equilibrium distance $d_1 = 2.38$ Å and fractional charge $Z_A^* = 0.27$ e are compared, along with the experimentally obtained[34] rate given by $R = 10^{12.8} \exp(-D_e/kT)$, (open circles). It is evident that the results for the Morse potential and truncated harmonic oscillator models yield both an Arrhenius-like straight line in the semilogarithmic plot of R vs. 1/T. The pre-exponential factors, however, differ markedly with the Morse potential in agreement with experiment. It should be noted that the differences between the two models are less pronounced than those which were exhibited in the transition rates. Similar results, with a somewhat less pronounced difference between the two models and in agreement with experiment[51] are shown for Xenon desorption from Tungsten (with the parameters given in the figure captions) in Figs. 3 and 4. In Fig. 4 the rather weak dependence of the results for the rate of desorption on the equilibrium distance parameter is exhibited.

IV. DISCUSSION

From the results presented in Figs. 2-4 it is apparent that the model provides a rather adequate description of desorption for both weak (Xe/W) and strong (K/W) chemisorption systems. The principal merit of this model is that it exhibits explicitly the dependencies on various microscopic quantities characteristic to the substrate and adatom. Due to the complexity of the problem our model relies on a number of simplifying assumptions certain of which we enumerate below: (a) the substrate was modeled as a continuum structureless solid, (b) surface phonons have been ignored, (c) the electron response to fluctuations of the ionic charge was calculated semi-classically (quantum interference effects neglected) and with specular boundary conditions imposed, (d) electronic band-structure effects were ignored which implies weak electron-ion coupling in the substrate, (e) bilinear coupling between the nonstationary adatom and substrate was used in the numerical examples and transition

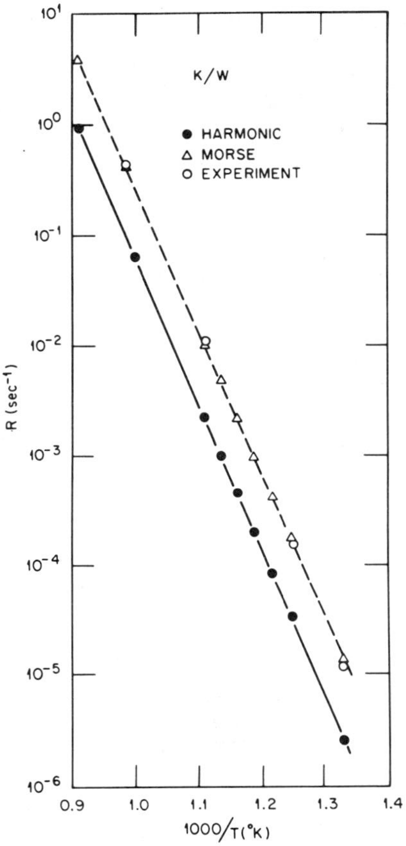

Fig. 2. Semilogarithmic plots of desorption rates versus inverse temperature for the system Potassium adsorbed on Tungsten. The desorption energy D_e was taken as 2.64 eV after reference 50. The parameters used in the calculation: fractional charge on the Potassium $Z_A^* = 0.27$ e and equilibrium distance of the adsorbate from the substrate, $d_1 = 2.38$ Å, were chosen after reference 50; the Debye temperature of the substrate was taken as $\theta_D = 220^oK$, the electron number density of the substrate was $n_e = 38x10^{22}$ cm^{-3} and $x_e = 6.5x10^{-4}$; the vibrational quantum in the harmonic well was taken as $\hbar\omega_o = 13.7$ meV (after L. M. Kahn and S. C. Ying, Solid State Commun. 16, 799 (1975)). The experimental points (open circles) were taken after reference 50, $R(T) = 10^{12.8} \exp(-D_e/kT)$. Both the Morse potential (dashed) and truncated harmonic (solid) yield linear relationships in the plot of ln R(T) vs. inverse temperature, parallel to one another (same activation energy for desorption) but with different intercepts (frequency factors). The results based on the Morse potential description of the chemisorptive bond are in bettwe agreement with the experimentally deduced results than those derived from a truncated harmonic potential.

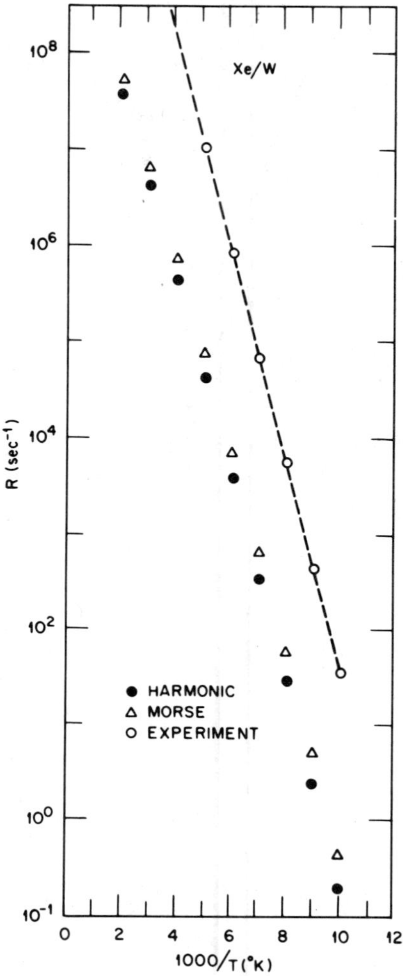

Fig. 3. Semilogarithmic plots of desorption rates R(T) versus inverse temperature for the system Xenon adsorbed on Tungsten. The characteristic parameters are: d_1 = 2.0 Å, $\hbar\omega_0$ = 3 meV; Z_A^* = 0.04e; the desorption energy, D_e, was taken as 217 meV after reference 51. The experimental points (open circles) were calculated from the rate expression given in the above reference $R(T) = 10^{12} \exp(-D_e/kT)$. Results obtained by using Morse (triangles) and truncated harmonic potentials (dots) are shown.

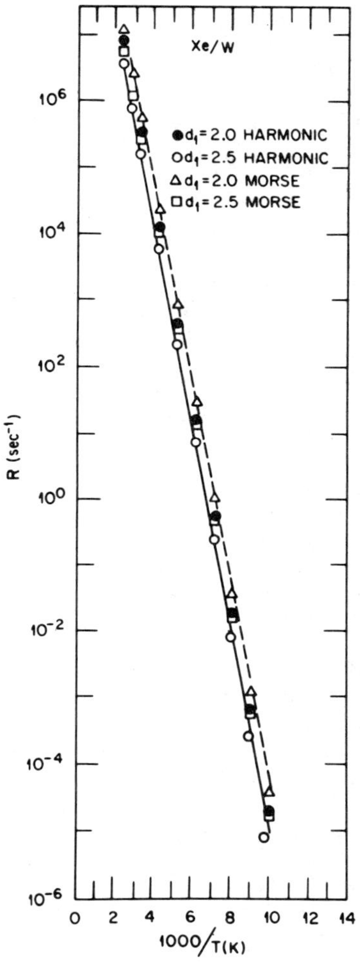

Fig. 4. Semilogarithmic plots of desorption rates, R(T), versus versus inverse temperature for Xe adsorbed on Tungsten. Characteristic parameters used; D_e = 300 meV; Z_A^* = $0.04e^{14}$; $h\omega_o$ = 3.0 meV; x_e = $1.25x10^{-3}$. The substrate parameters are those given in Fig. 2. Results are shown for both truncated harmonic (solid and open dots) and Morse (triangles and squares) potentials. The apparent slight sensitivity of the rates to the equilibrium distance of the adsorbed atom from the surface is shown.

rates due to these couplings were calculated using the Fermi golden rule, (f) an immobile adsorbate was assumed. The inclusion of adsorbate migration on the surface will add an extropy correction to the rate expression. While further improvements within this model are possible, the present study allows for a first evaluation of the sensitiviey of desorption kinetics to the various microscopic parameters.

The standard approaches to reaction kinetics, such as Absolute Rate Theory[6] (ART) and the various statistical methods[16] (e.g., RRKM) rely upon certain criteria of applicability. The main requirement of the above is that the initial and final (or transition complex) states are uncorrelated.[6] As discussed originally by Kramers[18] and further investigated recently,[19,20,52] the applicability of ART is related to the strength of the fluctuating part of the coupling (friction in the nomenclature of the above studies) between the adsorbate and the adparticle. The analytical results obtained[18] in the limits of small and large coupling support the assertion that the applicability of ART is limited to an intermediate regime of the coupling strength. In this regime the coupling is strong enough as to replenish instantaneously the equilibrium Maxmellian tail of particle momenta, necessary for surmounting the reaction barrier, and thus the rate becomes independent of the coupling. Outside this regime the Arrehnius behavior of the rate constant gets modified by multiplicative factors which vary with temperature. Our calculation of $\bar{t}$ in principle assumes the weak coupling regime since our transition rates $W_{v\to v+1}$ are treated to lowest order in perturbation theory. This assumption of weak coupling has internal consistency in that our low order treatment does yield good agreement with experiment. While clearly the interplay between the temperature dependence of $W_{v\to v+1}$ and the usual statistical occupations (the sum over j in Eq. 3.6) is a complicated one and cannot rigorously reduce to the Arrhenius-like form our numerical results give a measure for the weak deviation from such a behavior. This conclusion cannot be inferred directly from the evaluated transition rates alone but requires an analysis of the rates. In this context it is important, however, to notice the dependence of the results on the model potential used (truncated harmonic vs. Morse potentials), and that the difference in rates of desorption corresponding to the two model potentials is smaller than that exhibited in the associated level transition probabilities.

The stochastic treatment of the time evolution of the vibrational excitations leading to desorption which we have used, is a convenient formulation of non-equilibrium kinetic processes. In the equilibrium theory the rate constant depends only upon transitions which couple bound vibrational levels directly to the dissociated state,[48,49] N+1, and similar to the basic assumption of ART there is no dependence on the details of the excitation mechanism. The first passage-time, $\bar{t}$, calculated via the stochastic non-equilibrium formulation does not in general equal the reciprocal of the equilibrium rate constant.[49] One limit in which the above equality holds is when the energy required for a transition is large compared to the available thermal energy. This condition was not obeyed in our cases. Thus, it was necessary to investigate the full stochastic behavior.

The order of coupling in our model should also be commented on. While we have used bilinear coupling, a generalization would in principle allow treatment of higher order coupling terms. In this context we could argue that by a proper transformation of the coordinates $\{R_S\}$ and $\{R_M\}$ the bilinear coupling term could be removed and the frequencies of the substrate-adatom system renormalized accordingly. The coupling enters now through the new frequencies in these transformed canonical coordinates. It is now, in principle, possible to calculate desorption rates with the simple assumption of a Boltzmann occupation of these new levels, and with proper retransformation of our coordinates to define the stage of dissociation. Such a calculation is expected to yield similar results to ours (particularly in the weak coupling limit) but it is rather complex and has not as yet been carried out for the model systems discussed in this study.

A detailed investigation of reaction mechanisms requires an analysis of the reaction products. Indeed, a well established practice in gas-phase kinetic studies is to perform state selective measurements, i.e., identification of products, their center of mass translational energies and excitation of internal degrees of freedom. This mode of investigation is not common practice in current studies of surface reactions. In most, if not all, studies to date the experimental information consists only of mass, and in certain cases angular, distribution of the products.

Once the energy of the bond to be ruptured achieves an energy above the dissociation energy, the system may evolve via several channels. For example, it may couple to a translational continuum (desorption) or to states corresponding to diffusion along the surface. Formally, the problem is similar to that encountered in the study of autoionization[53], predissociation[53-56], radiationless transitions[57-59] and photodissociation[60-64]. Employing formalisms developed for the above problems we have derived elsewhere[39] expressions for the probabilities of decay via coupling to multichannel final states. Of particular importance is the diffusion channel. Associated with the activated diffusion of the adsorbate is an activation entropy which would modify the value of the rate constant. (See reference 1.)

Finally we comment on the application of our model to adsorbed molecular species. These systems possess additional degrees of freedom certain of which are of bond-stretching character and others which describe bond-bending, wagging, etc. While the energies typical to molecular bond-stretching modes might (and often do) exceed in magnitude those of single phonon excitation by the solid, the energies associated with the non-stretching modes are smaller. For example, the lowest vibrational level spacings in H-H, C-H, C=0, C=C and C-C stretches are 500, 360, 210, 200 and 110 meV, respectively, while the level spacings in non-stretching modes are often 10-50 meV. Consequently it is suggested that the latter modes through their coupling to the vibrations of the solid may be excited up to high levels via an incoherent multiphonon mechanism similar to that used in the present investigation and subsequently couple via anharmonicity to the high-lying, densely spaced levels of the bond-stretching modes. Once these high-lying levels have been populated, the excitation may propagate further via direct coupling to the substrate, eventually leading to fragmentation (see Fig. 5). In other words the non-stretching modes may serve as door-way states towards bond rupture via intramolecular energy redistribution.[64-65] In addition localized high-frequency impurity (in particular light impurities) modes may participate actively in the excitation mechanism.

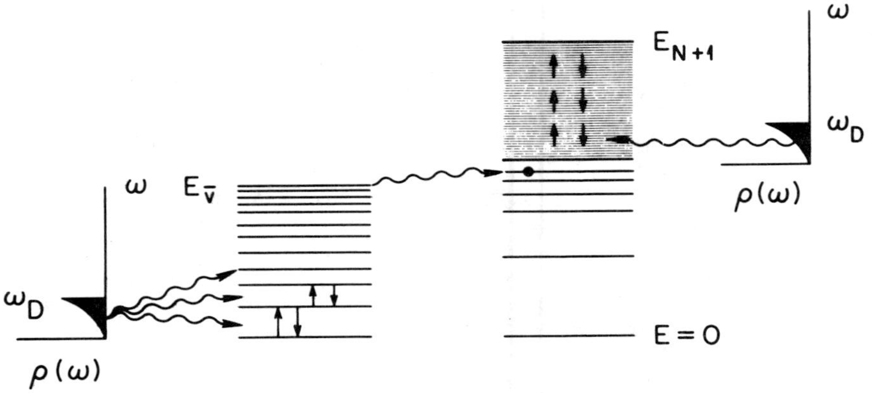

Fig. 5. Schematic picture of the door-way state model for thermal surface desorption or dissociation reaction mechanism. A characteristic Debye phonon density of states, $\rho(\omega)$ is shown on the left and right. Excitation of a low-frequency, door-way, mode of vibration (typically a nonstretching mode) occurs via an incoherent multiphonon mechanism. Upon achieving the level $E_{\bar{v}}$ the excitation is transferred to the high-lying levels of a stretching mode (or combination of such modes) via anharmonic coupling. Further excitation in the dense vibrational manifold corresponding to the bond-rupture coordinate can occur via direct incoherent multiphonon excitations induced by thermal coupling to the substrate. The predissociation level is denoted by E_{N+1}. Having achieved this level the reaction proceeds through coupling to possible final state channels such as dissociation, desorption or migration.

REFERENCES

*Work supported by DOE Contract No. EG-77-S-05-5489.

1. R. J. Madix, this volume.
2. L. D. Schmidt, this volume.
3. M. F. Shlesinger and U. Landman, this volume.
4. R. Gomer, this volume.
5. M. Boudart, in Physical Chemistry: An Advanced Treatise, Vol. VII, ed. by H. Eyring (Academic Press, New York, 1975) Chap. 7; K. J. Laidler, Reaction Kinetics, Vols. 1, 2 (Pergamon Press, New York, 1963).
6. S. Gladsstone, K. J. Laidler, H. Eyrig, The Theory of Rate Processes (McGraw-Hill, New York, 1941).
7. H. Suhl and R. E. Lagos, this volume.
8. L. S. Kassel, The Kinetics of Homogeneous Gas Reactions (The Chemical Catalog Co., New York, 1932).
9. H. S. Johnston, Gas Phase Reaction Rate Theory (Ronald Press, New York, 1966).
10. D. O. Hayward, B. M. Trapnell, Chemisorption (Butterworths, London, 1964).
11. R. Gomer, "Chemisorption" in Adv. Sol. Stat. Phys. 30, 94 (1975).
12. C. G. Goymour and D. A. King, J. Chem. Soc. Faraday I, 68, 280 (1972).
13. A. K. Mazumdar and H. W. Wassmuth, Surface Sci. 30, 617 (1972).
14. M. J. Dresser, T. E. Madey and J. T. Yates, Jr., Surface Sci. 42, 533 (1974).
15. E. W. Montroll, this volume; see also reference 3.
16. P. J. Robinson and K. A. Holbrook, Unimolecular Reactions Wiley-Interscience, London, 1972); W. Frost, Theory of Unimolecular Reactions (Academic Press, New York, 1973).
17. D. L. Bunker, Theory of Elementary Gas Reaction Rates (Pergamon Press, New York, 1966); W. L. Hase in Dynamics of Molecular Collisions, Part B, ed. by W. H. Miller (Plenum Press, New York, 1976), and W. L. Hase, this volume.
18. H. A. Kramers, Physica 7, 284 (1940).
19. (a) H. Suhl, J. H. Smith, P. Kumar, Phys. Rev. Letters 25, 1442 (1970); (b) E. G. d'Agliano, W. L. Schaich, P. Kumar, and H. Suhl in Collective Properties of Physical Systems, 24th Nobel Symposium 1973 (Academic Press, New York, 1974) p. 200;

(c) E. G. d'Agliano, P. Kumar, W. Schaich and H. Suhl, Phys. Rev. B 11, 2122 (1973).
20. P. B. Visscher, Phys. Rev. B 13, 3272 (1976); Phys. Rev. B 14, 347 (1976).
21. G. Iche and P. Noziers, J. Phys. 37, 1313 (1976).
22. N. B. Slater, Theory of Unimolecular Reactions (Methuen, London, 1959).
23. N. Rosen, J. Chem. Phys. 1, 319 (1933).
24. O. K. Rice, J. Chem. Phys. 55, 439 (1971); ibid. 1, 625 (1933).
25. P. J. Pagni and J. C. Keck, J. Chem. Phys. 58, 1162 (1973); R. M. Logan and J. C. Keck, J. Chem. Phys. 49, 860 (1968); P. J. Pagni, J. Chem. Phys. 58, 2940 (1973).
26. J. E. Lennard-Jones and C. Strachan, Proc. Roy. Soc. A150, 442 (1935); J. E. Lennard-Jones and A. Devonshire, Proc. Roy. Soc. A156, 6, 29 (1936).
27. B. Bendow and S. C. Ying, Phys. Rev. B 7, 622 (1973); S. C. Ying and B. Bendow, Phys. Rev. B 7, 637 (1973).
28. B. J. Garrison, D. J. Diestler, and S. A. Adelman, J. Chem. Phys. 67, 4317 (1977); B. J. Garrison and S. A. Adelman, J. Chem. Phys. 67, 2379 (1977).
29. B. I. Lundqvist, O. Gunnarsson, H. Hjelmberg and J. K. Nørskov, Surface Sci. 89, 196 (1979).
30. Physical Basis for Heterogeneous Catalysis, Eds. E. Drauglis and R. L. Jaffee (Plenum Press, N.Y., 1975).
31. T. N. Rhodin and D. L. Adams in Treatise on Solid State Chemistry, ed. by N. B. Hannay (Plenum Press, New York, 1976), Vol. 6AI.
32. H. D. Hagstrum and E. G. McRae in Treatise on Solid State Chemistry, ed. by N. B. Hannay (Plenum Press, N.Y., 1976), Vol. 6AI.
33. E. W. Plummer et al., this volume.
34. C. R. Brundle, this volume.
35. R. J. Madix, this volume.
36. R. J. Madix in Physical Chemistry of Fast Reactions, ed. by. D. O. Hayward (Plenum Press, New York, 1977); see also this volume.
37. J. C. Polanyi, Acc. Chem. Research 5, 161 (1972).
38. For recent reviews see (a) D. Menzel in Topics in Applied Physics, Vol. 4, ed. by R. Gomer, (Springer, Berlin, 1975), p. 101; (b) T. E. Madey and J. T. Yates, Surface Sci. 63, 203 (1977); (c) L. A. Peterman in Progress in Surface Science, Vol. 3, ed. by S. G. Davison (Pergamon Press, New York, 1972); (d) D. A. King, Surface Sci. 47, 384 (1975); (e) D. Menzel in The Physical Basis for Heterogeneous Catalysis, ed. by E. Dragulis and R. I. Jaffee (Plenum Press, New York, 1975), p. 437.
39. G. S. De, U. Landman and M. Rasolt, Phys. Rev. B (April 15, 1980).
40. M. Born and K. Huang, Dynamical Theory of Crystal Lattices (Oxford University Press, London, 1954).
41. H. C. Longuet-Higgins, Proc. Phys. Soc. 60, 270 (1948).
42. G. P. Brivo and T. B. Grimley, Surface Sci. 89, 226 (1979).
43. A. L. Fetter and J. D. Walecka, Quantum Theory of Many-Particle Systems (McGraw-Hill, New York, 1971), pp. 390-399.
44. L. I. Schiff, Quantum Mechanics (McGraw-Hill, New York, 1955).

45. P. M. Morse, Phys. Rev. 34, 57 (1929).
46. H. S. Heaps and G. Herzberg, Z. fur Physik 133, 48 (1952).
47. R. Herman and K. E. Shuler, J. Chem. Phys. 21, 373 (1953); 22, 954 (1954).
48. E. W. Montroll and K. E. Shuler, Adv. Chem. Phys. 1, 361 (1958).
49. S. K. Kim, J. Chem. Phys. 28, 1057 (1958).
50. L. Schmidt and R. Gomer, J. Chem. Phys. 42, 3573 (1965).
51. G. Ehrlich in Advances in Catalysis, Vol. 14 (Academic Press, New York, 1963), p. 255.
52. J. L. Skinner and P. G. Wolyness, J. Chem. Phys. 69, 2143 (1978).
53. U. Fano, Phys. Rev. 124, 1866 (1961).
54. O. K. Rice, Phys. Rev. 35, 1551 (1930).
55. F. Mies and M. A. Krauss, J. Chem. Phys. 45, 4455 (1966); F. Mies, ibid. 51, 798 (1969).
56. R. W. Numrich and K. G. Kay, J. Chem. Phys. 70, 4343 (1979), and references cited therein.
57. M. Bixon and J. Jortner, J. Chem. Phys. 48, 715 (1968).
58. M. Bixon and J. Jortner, J. Chem. Phys. 50, 4061 (1969); ibid. 48, 715 (1968).
59. J. Jortner and S. Mukamel, in The World of Quantum Chemistry, ed. by R. Daudel and B. Pullman (Reidel, Amsterdam, 1973).
60. S. A. Rice, I. McLaughlin, and J. Jortner, J. Chem. Phys. 49, 5756 (1968).
61. K. G. Kay and S. A. Rice, J. Chem. Phys. 57, 3041 (1972).
62. E. J. Heller and S. A. Rice, J. Chem. Phys. 61, 936 (1974).
63. D. F. Heller, M. L. Elert, and W. M. Gelbart, J. Chem. Phys. 69, 4061 (1978).
64. S. Mukamel and J. Jortner, J. Chem. Phys. 65, 5204 (1976).
65. A door-way state mechanism has been suggested in studies of multiphonon molecular fragmentation; for a recent review, see N. Blombergen and E. Yablonovitch, Phys. Today 31, 23 (1978).

INTERATOMIC FORCES AND SURFACE PROCESSES

William L. Clinton*
Georgetown University, Washington, D.C. 20057

ABSTRACT

I discuss some of the qualitative aspects of surface processes that can be uncovered with the use of exact theorems involving interatomic forces. In particular the Hellmann-Feynman and virial theorems are applied to: the forces determining the energy and angular distributions in electron stimulated desorption; various atom-surface interactions including diffraction, knock-out reactions, and precursor barriers to chemisorption; the three-body overlap force and its connection with the indirect inter-adsorbate interaction; ion neutralization processes at metal surfaces. In conclusion, some brief remarks are made about the problems involved in going from interatomic interactions to cross sections and activation energies and finally to reaction rates.

I. INTRODUCTION

Many aspects of atomic and molecular processes occurring at or on surfaces could be qualitatively and quantitatively understood if the various interatomic forces could be accurately calculated. This, of course, is one of the goals of modern many electron theories of surfaces and surface adsorbates.[1] But a bona fide, fully self-consistent calculation of the electronic energy of a poly-atomic interaction occurring on a surface is well beyond the current state-of-the-art. In this paper, I would like to present an intermediate approach to the problem of calculating interatomic forces at or on surfaces. It is based on the use of independent methods for the calculation of the total electronic energy and its derivatives namely the electrostatic theorem[2] and the virial theorem.[3] These theorems have the advantage of providing a well-defined formulism for calculating the energy and its atomic position vector gradients using only the one electron density matrix. The disadvantage, of course, is that they provide no systematic way of calculating either the electron density or more generally the one electron density matrix and unless the latter is exact or variationally optimal, the results of an energy calculation will not necessarily correspond to the more standard method of calculating the electronic energy. In this paper, however, I will try to show how to use these theorems to extract at least qualitative, and hopefully even more quantitative information, about interatomic forces from model electron densities.

---

*On leave of absence to DOE, Division of Materials Sciences until 8/15/79.

ISSN:0094-243X/80/610181-15$1.50

In particular, I will discuss applications to the following surface processes: (a) Electron Stimulated Desorption (ESD) - estimates of the magnitude and direction of the forces on desorbing atoms and ions will be made; (b) Atomic Beam Diffraction - a possible method for calculating the diffraction pattern using the same model electron density as in X-ray diffraction will be given; (c) Atomic Beam Induced Desorption - discussion of the possibility of observing atomic and/or molecular beam knockout reactions and the calculation of the cross section using interatomic force models will be given; (d) The Theory of the Precursor State - I will give a simple model force calculation of a pseudo-atom approach to an understanding of the precursor state; (e) The Indirect Inter-Adsorbate Interaction - the long range Friedel oscillations will be discussed using a simple model of the atom-metal overlap force; (f) Ion Neutralization - a knowledge of the final state forces in an ion neutralization process is necessary in order to calculate the cross section.

Finally, I will discuss the possibility of making contributions to real kinetic processes. That is, if activation energies and reactive cross sections can be calculated, then rate constants can be estimated and surface chemical kinetics will have at least a tenuous link to microscopic theory. I will show in this paper that model calculations of forces and energy differences may well satisfy a need for intermediate qualitative determinations that cannot be achieved at present with bona fide many electron theories.

A. Hellman-Feynman or Electrostatic Theorem

It has been known for some time that the gradient of the electronic energy $\vec{\nabla}E$ of any many electron, polyatomic system can be written solely in terms of the electron density. This is a special case of the Hellmann-Feynman theorem[2] which relates the energy derivative $\delta E/\delta\lambda$ to the expectation value of $\delta H/\delta\lambda$ and is a simple consequence of $H\Psi = E\Psi$ and $<\Psi|\Psi> = 1$. Specifically consider an N electron polyatomic system and let $n_e(\vec{r},\vec{R}_A\cdots)$ denote the electron density defined by

$$n_e(\vec{r},\vec{R}_A\cdots) \equiv N \int d^3r_2\cdots d^3r_n \; |\; \Psi_e(\vec{r},\vec{r}_2\cdots\vec{r}_n,\vec{R}_A\cdots)|^2 \qquad (1)$$

where Ψ_e is the N electron wave function. The total electronic energy $E_e(\vec{R}_A\cdots)$ can be written

$$E_e(\vec{R}_A\cdots) = T_e(\vec{R}_A\cdots) + V_{en}(\vec{R}_A\cdots) + V_{ee}(\vec{R}_A\cdots) \qquad (2)$$

where the terms are kinetic, nuclear attraction and electron repulsion energy respectively. The electrostatic theorem (or Hellmann-Feynman theorem for nuclear coordinates) has the form

$$\vec{\nabla}_A E_e(\vec{R}_A\cdots) = \int d^3r n_e(\vec{r},\vec{R}_A\cdots)\vec{\nabla}_A \left(\frac{-Z_A}{|\vec{r}-\vec{R}_A|}\right) \qquad (3)$$

and when the nuclear repulsion term $V_{nn} \equiv \Sigma_{A \neq A'} Z_A Z_{A'} / |\vec{R}_A - \vec{R}_{A'}|$ is added to $E_e(\vec{R}_A \cdots)$ we have the potential energy $U(\vec{R}_A \cdots) = E_e(\vec{R}_A \cdots) + V_{nn}$. Defining the total charge density $\rho_T(\vec{r},\vec{R}_A \cdots) \equiv n_e(\vec{r},\vec{R}_A \cdots) - \Sigma_{A'} Z_{A'} \rho(\vec{r}-\vec{R}_{A'})$ allows Eq. (3) to be expressed as

$$-\vec{F}_A \equiv \vec{\nabla}_A U(\vec{R}_A \cdots) = \int d^3 r \rho_T(\vec{r},\vec{R}_A \cdots) \vec{\nabla}_A \left(\frac{-Z_A}{|\vec{r}-\vec{R}_A|}\right) \tag{4}$$

where $\vec{F}_A$ is the net force on atom A. We notice now that an independent way of caluclating $\vec{\nabla}_A E_e$ would be from Eq. (2) whereby

$$\vec{\nabla}_A E_e = \vec{\nabla}_A T_e + \vec{\nabla}_A V_{en} + \vec{\nabla}_A V_{ee} \tag{5}$$

which when compared to Eq. (3) leads to the cancellation corollary

$$\vec{\nabla}_A T_e + \vec{\nabla}_A V_{ee} + \int d^3 r \, \vec{\nabla}_A n_e(\vec{r},\vec{R}_A \cdots) \, \Sigma_{A'} \frac{-Z_{A'}}{|\vec{r}-\vec{R}_{A'}|} = 0. \tag{6}$$

We conclude that there is a great deal of cancellation in the direct calculation of the force and Eq. (3) represents, in a sense, a most efficient force calculation.

We now briefly consider the conditions under which Eqs. (3) or (4) will be valid, i.e. those conditions under which Eq. (3) for example will agree with Eq. (5). Alternately, what are the conditions for the validity of the cancellation corollary Eq. (6)? Clearly, from what has already been said the main condition that will ensure the validity of Eq. (3) is; (a) N_e is derivable from an eigen function of H. But from this condition and the interpretation of quantum mechanics Eqs. (3) and (5) are also satisfied if; (b) N_e is an exact experimental electron density, e.g. as determined from accurate X-ray structure factors, Finally, it can be shown that Eqs. (3) and (5) will agree if; (c) N_e is derivable from a variationally optimal wave function. If any of these three conditions are met only approximately, then, of course, one expects a corresponding agreement between Eqs. (3) and (5) and it may be difficult to judge how accurate a calculation using Eqs. (3) or (4) will be on an absolute scale. In the later discussion of surface processes, we will emphasize the qualitative use of Eqs. (3) or (4) via the technique of modeling the electron density, or that part of it that is most important for a particular process. There is ample precedent for this approach in the area of molecular physics and one would expect that if the model replicates reality at all then qualitative conclusions should be valid. Recent calculations on jellium seem to support this idea.[4]

B. Virial Theorem

An even longer history is associated with the virial theorem having its roots in classical physics. For coulomb interactions alone, as is the case in a many electron system, this well-known theorem relates the average values of total potential and kinetic energies via $\langle V\rangle = -\langle T\rangle$. It was first recognized by Slater[3] that for a polyatomic system additional energy terms would be necessary when the system is displaced from equilibrium. In its more general form we obtain

$$\Sigma_{A'}\vec{R}_{A'}\cdot\vec{\nabla}_{A'}U(\vec{R}_{A'}\cdots) + U(\vec{R}_{A'}\cdots) = -T_e(\vec{R}_{A'}\cdots) \tag{7}$$

or the same equation with E_e replacing U since the nuclear repulsion term gives zero when operated on by $(\Sigma_{A'}\vec{R}_{A'}\cdot\vec{\nabla}_{A'} + 1)$.

We note that for any atomic configuration where the forces are all zero, the gradient terms vanish and the energy difference between such configurations is given by[5]

$$\Delta U \equiv U(2) - U(1) = -(T(2) - T(1) \equiv -\Delta T. \tag{8}$$

We will return to this energy difference corollary when discussing energy barriers.

We also make note of the force constant corollary[5]. For a diatomic molecule there is just one significant atomic position parameter R. Differentiating Eq. (7) and evaluating at R_e one obtains

$$K \equiv \frac{d^2U(R)}{dR^2}\bigg|_{R_e} = -\frac{1}{R_e}\frac{dT_e(R)}{dR}\bigg|_{R_e}. \tag{9}$$

Eq. (9) has been discussed in the diatomic chemical bonding context but its ramifications in adatom-surface interactions have not been considered. We will return to this result when discussing bonding criteria in a later section.

It was emphasized by Löwdin[3] that the virial theorem could always be satisfied if a scale factor for the electronic coordinates were chosen in a variationally optimal way. In this connection if the electronic variables are scaled with a scale factor $S(\vec{R}_A,\cdots)$ then $\Psi(\vec{r}_1\cdots\vec{R}_A\cdots) \rightarrow S^{3/2}\,\Psi(S\vec{r}_1\cdots, S\vec{R}_A\cdots)$ and

$$T_e(S,\vec{R}_A\cdots) = S^2(\vec{R}_A\cdots)\;T_e(,S\vec{R}_A\cdots). \tag{10}$$

This scaling theorem is a consequence of the degree of the kinetic energy operator and allows for easy calculation of $T_e(S)$ when S and T_e (1) are known.

A model calculation of the He-He interaction using only a scale factor in the atomic densities was used by the author with some modest success.[6] In this case combination of Eqs. (3) and (7) completely determined S(R) and the concomitant potential curve. This model may have some utility in atomic diffraction where little atomic distortion is expected.

II. APPLICATION TO SURFACE PROCESSES

In this section I will discuss several different kinds of processes that occur at or on a surface where one can extract at least some qualitative information about the interatomic forces and suggest ways of obtaining even quantitative and analytical results.

A. Electron Stimulated Desorption (ESD).[7]

As the name implies ESD involves using an electron beam to induce desorption of either ions or neutral species. In this technique the latter are observed and energy as well as angle analysis lead to information about the bonding of the adatom to the surface. Fig. 1 is intended to give a graphic description of one such process.

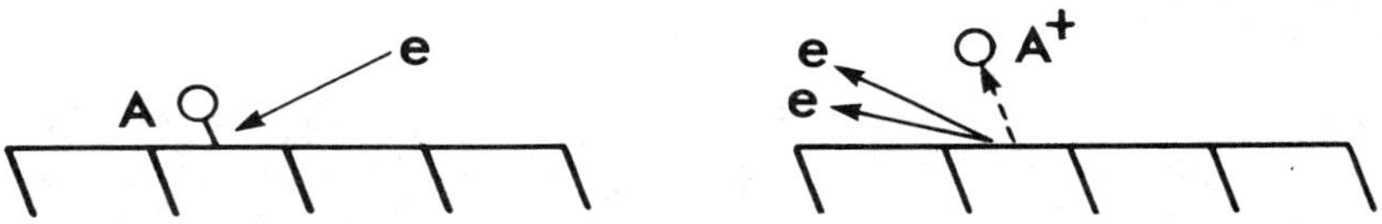

Fig. 1. Electron stimulated desorption of ions.

This process is believed to be Franck-Condon in nature which is corroborated by the observed energy distribution illustrated in Fig. 2.

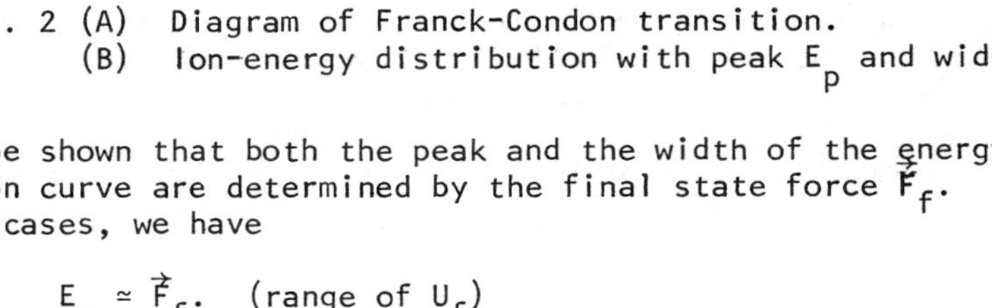

Fig. 2 (A) Diagram of Franck-Condon transition.
(B) Ion-energy distribution with peak E_p and width E_w.

It can be shown that both the peak and the width of the energy distribution curve are determined by the final state force $\vec{F}_f$. In the simplest cases, we have

$$E_p \simeq \vec{F}_f \cdot \text{ (range of } U_f) \tag{11}$$

$$E_w \simeq \vec{F}_f \cdot \text{ (vibrational amplitude in } U_o).$$

If one further simplifies the situation by taking a frozen molecular orbital viewpoint i.e. the excited orbitals in the ground state Ψ_o are used to calculate Ψ_f, then

$$\vec{F}_f = \int d^3r \, |\Psi_j(\vec{r},\vec{R}_A\cdots)|^2 \, \vec{\nabla}_A \left(\frac{-Z_A}{|\vec{r}-\vec{R}_A|} \right) \tag{12}$$

for ion desorption where Ψ_j is the initial state MO from which the electron was ionized. That is, the final state force is just that due to the hole left behind or produced in the orbital Ψ_j. This simple picture allows one to make estimates of both E_p and E_w. In addition, one can also extract some qualitative information from Eq. (12) the angular distributions discussed below.

Recent ESD experiments[9] on ion angular distributions (ESDIAD) have generated some very interesting new results about chemisorptive binding. One kind of observation with its chemical bond interpretation is represented in Fig. 3.

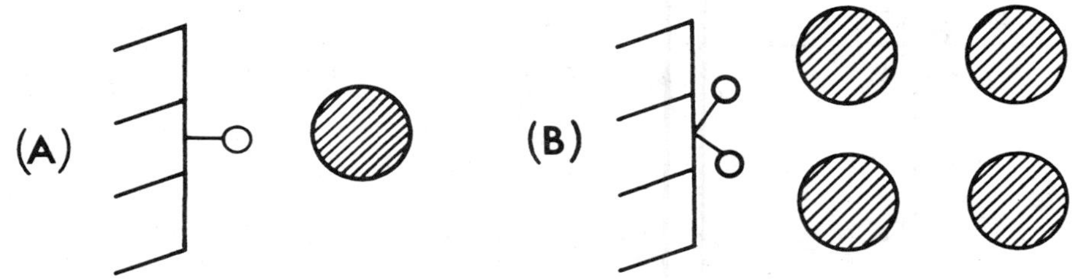

Fig. 3. ESDIAD: (A) Single spot pattern due to bonding normal to surface.
(B) Multiple spot pattern due to symmetrical bonding skew to surface.

Using Eq. (12), one can argue as follows: assuming for the moment that Ψ_j is a bonding orbital then $|\Psi_j|^2$ exerts a force tending to compress the bond (i.e. hold the atom to the surface). Therefore, when a hole is produced in Ψ_j the unbalanced force will be in the opposite direction i.e. along the bond but tending to eject the ion. This qualitative argument tends to support the intuitive chemical bond model of the process first put forward in reference (9).

Further developments of this model have established that, in addition to bond angles, even bending frequencies may be extracted from ESDIAD data.[10] To see this, consider the Franck-Condon expression for the cross section in the reflection approximation. We have

$$\sigma_D \simeq \sigma_{el} \, |\phi_\nu(\vec{R}_f)|^2 \, . \tag{13}$$

More precisely, Eq. (13) would include a neutralization factor which we will take to be isotropic. In this approximation the main contribution to the angular distribution of ions is from the ground vibrational wavefunction - i.e. from the Franck-Condon factor in the reflection approximation. This function is anisotropic in that it is essentially zero everywhere except in the direction of a chemical bond since $|\phi_\nu(\vec{R}_f)|^2$ is the probability density for observing an atom at the point $\vec{R}_f$. The position vector R_f is determined by the reflection approximation prescription of finding that point on $U_f(\vec{R})$ where $E_f = U_f$, E_f being the kinetic energy of the desorbed ion at a given angle. It is clear from Eq. (13) that bond angles, as well as bending vibrational frequencies, can in principle be extracted from ESDIAD data. In a very simplistic sense, one can argue that the center of an observed spot will be interpretable in terms of the bond angle of the chemisorption bond while the spread of beam intensity about the center will be a measure of the bending frequency. One very obvious way in which this simplistic view is modified is in the bond angle distortion due to the image potential acting on the desorbing ion. Corrections for this effect have been made.[10]

B. Atomic Beam Diffraction

Of somewhat recent vintage in the surface crystallography business is the method of atomic beam diffraction. Work on systems such as He-graphite[11], He-Silicon (100),[12] and He-scattering from surface charge density waves[13], has all been reported in just the past year. As a surface crystallography probe, the atomic beam has the advantage of being essentially completely surface sensitive. On the other hand, detection presents a more formidable problem. Nonetheless, results such as those taken from reference (12) and shown schematically in Fig. 4 show much promise for the future.

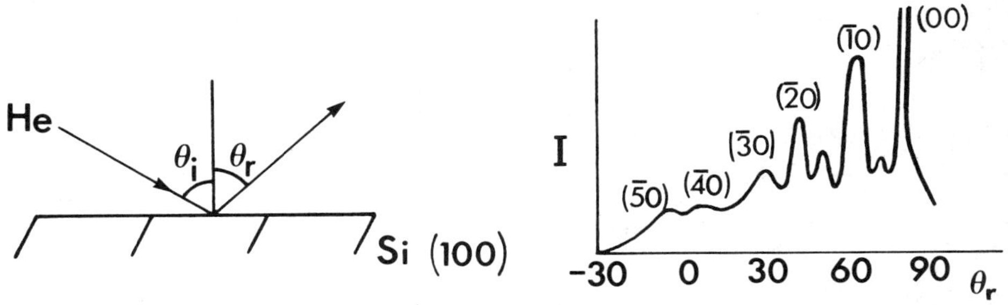

Fig. 4. Schematic representation of He-Si atomic diffraction experiment.

We have found a simple way to analyze the diffraction experiment shown above. We use a conventional X-ray diffraction model where the electron density is represented by a sum of atomic contributions.

That is

$$n_e(\vec{r},\vec{R}_A\cdots) = \Sigma_{A'}\, n_e(\vec{r}-\vec{R}_{A'}) \tag{14}$$

where A' runs over all surface atoms. The force on the scattering atom denoted by A due to this density is

$$\vec{\nabla}_A U(\vec{R}_A,\cdots) = \Sigma_{A'}\int d^3r\ \rho_T(\vec{r}-\vec{R}_{A'})\ \vec{\nabla}_A \left(\frac{-Z_A}{|\vec{r}\ \vec{R}_A|}\right) \tag{15}$$

which because of the convolution theorem allows one to Fourier transform with respect to R_A in a simple matter. We find

$$U(\vec{G}) = Z_A S(\vec{G}) f(\vec{G})/G^2 \tag{16}$$

where $S(\vec{G}) \equiv \Sigma_{A'}\exp(j\vec{G}\cdot\vec{R}_{A'})$ is the X-ray or crystal structure factor, $f(\vec{G}) = \int d^3r\rho_T(\vec{r})e^{j\vec{G}\cdot\vec{r}}$ is the atom form factor and Z/G^2 is the coulomb term. We have assumed spherical atom densities (including the scattering atom). If the scattering atom is allowed to distort as a function of the distance R_A then another term of a complicated form must be added to Eq. (16). We will ignore this term for the present, i.e. we assume the scattering atom exerts no force on itself in Eq. (15).

It is clear from Eq. (16) that in this model atomic beams do probe the electron density just as an X-ray beam does. We, therefore, expect to find the same reciprocal lattice structure as if an X-ray beam were the probe. The reason for this is that in the Born approximation the scattering cross section is essentially $|U(\vec{G})|^2$ and therefore the only difference between $I(\vec{G})$ in Fig. 4 and an X-ray $I(\vec{G})$ is the additional form factor Z_A/G^2.

Another possible use of atomic beam diffraction is to choose a material with surface charge density waves and scatter from this surface oscillatory electron density. Such an experiment has recently been reported[13] in the material Ta_2S. A simplistic application of Eq. (15) using $\rho_T(\vec{r},\vec{R}_A\cdots) = \rho_T^{(0)}(\vec{r},\vec{R}_A\cdots)\ [1 + \rho^{(1)}\cos\vec{\theta}\cdot\vec{r}+\phi]$ to represent the CDW. Clearly, the scattering amplitude is now related to

$$U(\vec{G}) = U_o(\vec{G}) + U_+(\vec{G}+\vec{\theta}) + U_-(\vec{G}-\vec{\theta}) \tag{17}$$

where $U_\pm(\vec{G}) \equiv \rho^{(1)}U_o(\vec{G})e^{\pm i\phi}$ and depends on the amplitude of the SCDW. This model, however, is too simplistic since it assumes that the lattice exactly follows the CDW. We presented it only for heuristic purposes and a detailed analysis will appear later.

C. Atomic Beam Induced Desorption: Knock-Out Reactions

Probably because of the difficulty of observing neutral particles, no experiments have been done wherein a neutral particle is simply knocked off of the surface by impact with a heavy particle and then the knocked-out particle is observed. The analogue of such experiments are well-known in nuclear physics and much information is obtained about not only the binding of the particle but also the elastic cross section for scattering of the bound and impact particles. The diagram below describes the simplest experiment one could imagine.

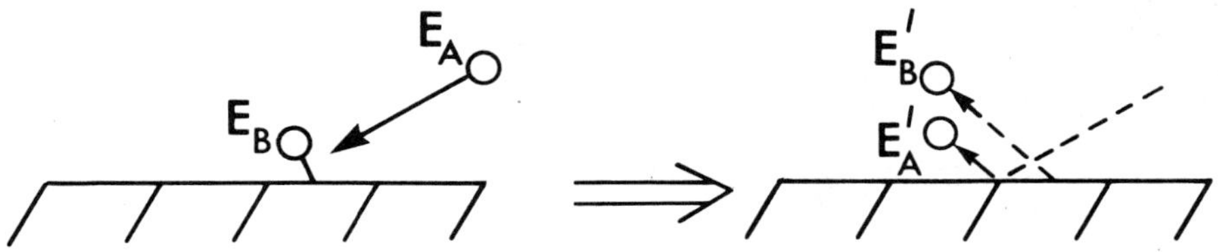

Fig. 5. Simple atomic beam knock-off reaction

In the simplest experiment, one would try to measure the total flux of desorbed particles as a function of temperature and coverage. This data should give one a handle on the distribution of binding energies because the scattering process is so simple compared to other desorption methods. A more ambitious and necessarily more elaborate experiment would be the exact analogue of what is measured in nuclear knock-out reactions namely coincidence measurements of desorbed and scattered particles as a function of angle and energy. Such information would in principle give a thorough view of the state of the bound particle (chemisorption bond) as well as the two-particle elastic scattering cross section. The reason for this is that at least at high incident energy the knock-out cross section is[14]

$$\sigma_{ko} = C\,\sigma_{elastic}\,|\phi_\nu(\vec{q})|^2 \qquad (18)$$

where $|\phi_\nu(\vec{q})|^2$ is the momentum distribution of the bound particle (i.e. $\phi_\nu{}^\nu(\vec{q})$ is just the momentum space vibrational wave function of the adatom). Of course, if one could extract $|\phi_\nu(\vec{q})|^2$ from σ_{ko} then such an experiment, difficult as it may be, would be worth the effort.

It is in the calculation of $\sigma_{elastic}$ that our previous considerations of relating the force to the potential enter. In the Born approximation, of course, one can calculate $\sigma_{elastic}$ using Eq. (16). More generally, Eq. (15) can be used to generate a $U_{\vec{G}_{11}}(Z)$ to be inserted in the scattering integral equation, as was already mentioned in section II.B on atomic diffraction.

D. The Precursor State

There has now accumulated considerable evidence that a precursor state may be involved in both adsorption and desorption processes.[15] It is not known what the precise nature of such a 'state' is, since no models have been adequately tested. One possible model of the precursor state is that it is a short-lived capture of an adatom in a shallow potential well analogous to a physisorbed state. We will show here how the force formulae of previous sections along with the pseudo-atom representation of the metallic electron density can give rise to a potential curve that has multiple binding states the weaker of which could be considered as giving rise to precursor states.

We first note that the force exerted on an atom by another atom a distance R away with a spherical charge distribution ρ is coulombic with an effective charge $Q(R) = 4\pi\int_0^R drr^2\rho(r)$. For many such atoms, then, we have

$$\vec{F}_A = \Sigma_{A'\neq A}\, Q_{A'}(R_{AA'})\, \vec{R}_{AA'}/R^3_{AA'} \tag{19}$$

where $Q_{A'}(R_{AA'}) = \int_0^{R_{AA'}} drr^2\rho_{A'}(r)$ is just the total charge in a sphere of radius $R_{AA'}$. Now, if $\rho_{A'}$ is an ordinary atomic density including the nuclear charge i.e. $\rho_{A'}(r) = n_{A'}(r) - Z_{A'}\delta(\vec{r})$ with $\int d^3r\, n_{A'}(r) = Z_{A'}$ then

$$Q_{A'}(R_{AA'}) = 4\pi\int_0^{R_{AA'}} drr^2 n_{A'}(r) - Z_{A'} \leq 0 \tag{20}$$

the equality holding only for $R_{AA'} \to \infty$. Thus, $\vec{F}_A$ can never be zero for such a system and binding is impossible. In an ordinary molecular system, then one argues that polarization of the atomic charge clouds, overlap, charge transfer and other effects are necessary for binding. In a metal, however, a certain aspect of the adatom-surface binding may already be built in via the metallic pseudopotential and the concomitant pseudoatom electron density. The latter is a spherical charge density that is constructed so as to fit the known pseudopotential and as such is not necessarily everywhere positive and indeed exhibits Friedel oscillations at large distances[16]. Thus, it is possible to have $\vec{F}_A = 0$ for a metal even in this simple model. It would appear to be useful to investigate the nature of the binding using known pseudo-atom densities since equilibrium distances as well as binding energies would be relatively easy to calculate. The qualitative situation is depicted in Fig. 6.

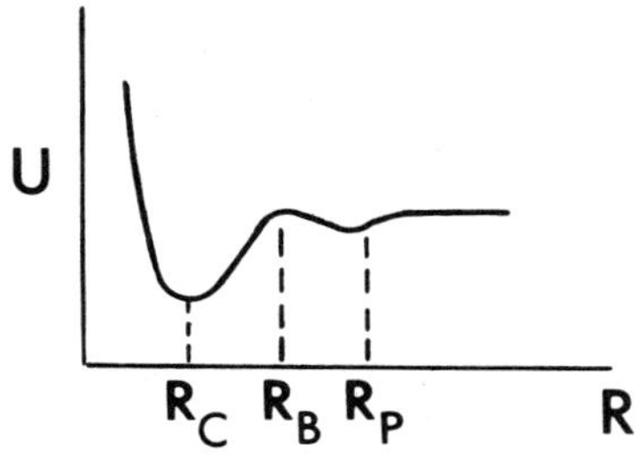

Fig. 6. Adatom-pseudoatom potential curve calculated using Eq. (18). R_c is the chemisorption bond distance, R_p the physisorbed or precursor bond distance - and R_B the barrier distance.

Clearly, the barrier to chemisorption is simply given by

$$E_B = -\int_{R_p}^{R_B} dRF_A \equiv U(R_B) - U(R_p) \tag{21}$$

if only one surface atom were involved.

E. Interadsorbate Interactions

When two adsorbates interact on a surface, it has been customary to distinguish between the direct and indirect part of the interaction.[17] The latter involves the substrate as an intermediary while the direct is simply that which would occur in the gas phase except as modified by the environment but not involving it per se. We will discuss these interactions in the force model using a molecular orbital representation of the electron density.

$$n_e(\vec{r},\vec{R}_A\cdots) = \Sigma_j |\Phi_j(\vec{r},\vec{R}_A\cdots)|^2 \tag{22}$$

$$= \mathrm{Tr}\ \underset{\sim}{P}\ (\vec{R}_A,\cdots)\underset{\sim}{\psi}^{\dagger}\psi(\vec{r},\vec{R}_A\cdots)$$

where the row matrix $\underset{\sim}{\psi} \equiv (\psi_A, \psi_B - - -)$ is an atomic basic and

$$\underset{\sim}{P} = \begin{pmatrix} \underset{\sim}{P}_{AA} & \underset{\sim}{P}_{AB} & \cdots \\ \underset{\sim}{P}_{BA} & \underset{\sim}{P}_{BB} & \cdots \end{pmatrix} \tag{23}$$

which refer to atomic or overlap densities depending on whether A=B or A≠B. Now, the indirect interaction is that which arises from a force on atom A due to an overlap density between another atom B and the metal M as depicted in Fig. 7. That is

$$\vec{F}_A^I = -\int d^3r n_{BM}(\vec{r},\vec{R}_A\cdots)\vec{\nabla}_A \left(\frac{-Z_A}{|\vec{r},\vec{R}_A|}\right) \tag{24}$$

which is just the 'three-center' overlap force where the metal now plays the role of a third atom.

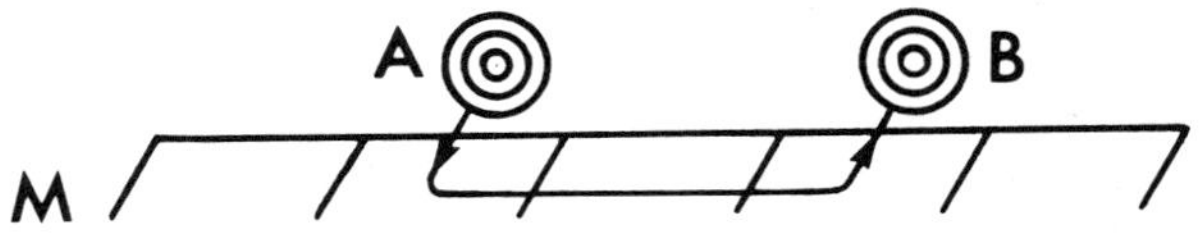

Fig. 7. Indirect or 3-center overlap force between adatoms A and B.

One can use a jellium model for the metal, wherein $\psi_{\underset{\sim}{m}}$ is just a collection of plane waves, and one obtains the usual Friedel oscillatory term for large R. In this case, the overlap density is

$$n_{BM} = \Sigma_{B'} \sum_{\vec{k}} P_{B'k}\, \psi_{B'}(\vec{r}_{B'})\, e^{i\vec{k}\cdot\vec{r}} + c_1 c_1 \qquad (25)$$

where B′ sums over all basis functions on atom B. The usual methods allow one to derive the $\cos(k_F R_{AB})/R_{AB}^3$ term in the potential. A more detailed calculation will appear later.

F. Ion Neutralization Processes

An ion neutralization process occurs, for example, in ion neutralization spectroscopy (INS) wherein an incident ion is neutralized by electrons from a metallic substrate, or in electron-stimulated desorption (ESD) in which the ion is produced at the surface and similarly neutralized. These processes can be viewed as curve crossing events in which an ion is traversing an electronic state $(MA)^+$ of the metal (M) - atom (A) system then crosses (in general non-adiabatically) to a neutral state $(MA)^*$ which may or may not be the ground electronic state. Such curve crossings can be calculated in a variety of ways. We will discuss one such method in what follows. A molecular process that is directly analogous to desorptive ion neutralization is called dissociative recombination. The graphs in Fig. 8 depict the curve crossing event.

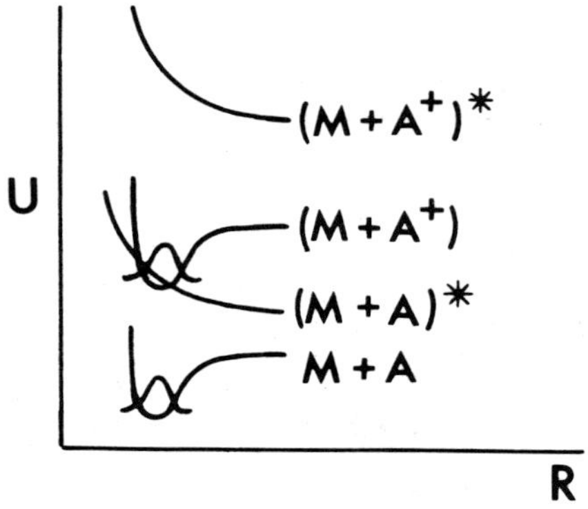

Fig. 8. Various potential curves involved in ion neutralization.

It can be shown[18] that the cross section for neutralization in the Born approximation is

$$\sigma_n = \sigma_o |\langle \phi_{MA+} | V(R) | \phi_{MA*} \rangle|^2 \qquad (26)$$

Where σ_o is a constant, $\phi(R)$ is a vibrational wave function, the potential $V(R) \equiv \langle \Psi_{MA+} \psi_s | H_I | \Psi_{MA*} \rangle$, the function ψ_s

is the substrate electron wave function and H_I is the electronic interaction hamiltonian. In the reflection approximation, the vibrational wave function $\phi_{MA*}(R)$ is just $|U'_{MA*}|^{-\frac{1}{2}}\delta(R-Rc)$ where the crossing distance R_c is determined from the energy transfer condition $U_{MA*}(R_c) - U_{MA+}(R_c) = \varepsilon_s$ the substrate electron energy and the factor involving the force $U'_{MA*}(R_c)$ is necessary because of delta function normalization. This force can be estimated in a frozen MO approximation as follows

$$U'_{MA*} = \int d^3r[|\psi_i|^2 - |\psi_j|^2] \frac{\partial}{\partial R} \left(\frac{-Z_A}{|\vec{r}-\vec{R}|} \right) \tag{27}$$

where ψ_j and ψ_i are the MO's involved in the excitation producing $(MA)^*$ from the ground state MA. The neutralization cross section is

$$\sigma_N = \sigma_o \, |V(R_c)|^2 \, |\phi_{MA+}(R_c)|^2/|U'_{MA*}(R_c)| \ . \tag{28}$$

Even without being able to calculate the transition matrix element $V(R_c)$, one can make some qualitative conclusions based on Eq. (28). Clearly, if the ion vibrational wave function is large at R_c or if the magnitude of the force in the state $(MA)^*$ is very small, then the neutralization cross section will be very large. The reason for the former, of course, is that unless the ion has a large probability of being at R_c, the distance at which energy is conserved on crossing, the neutralization cannot occur. The condition that a small force is required for a large-crossing probability is more difficult to see, but is related to the fact that only transitions that disrupt a particle's momentum as little as possible are strongly allowed. A more quantitative analysis is now in progress.

III. SUMMARY

In this section, we will summarize some of the previous ideas with a view towards ultimately using them in reaction kinetics. It was shown that the interatomic potential U and force on a given atom $\vec{\nabla}_A U$ are given in terms of the one electron density and density matrix via the electrostatic and virial theorems

$$-\vec{F}_A \equiv \vec{\nabla}_A U = \int d^3r \rho_T \vec{\nabla}_A \left(\frac{-Z_A}{|\vec{r}-\vec{R}_A|} \right)$$

$$\Sigma_{A'} \vec{R}_{A'} \cdot \vec{\nabla}_{A'} U + U = -T_e \ . \tag{29}$$

These equations can be used to generate various expressions for the momentum transform $U(\vec{k})$. This function can be used to calculate cross sections in the Born approximation or used directly in the scattering integral equation

$$\psi^{+}(\vec{k}') = \psi_j(\vec{k}) + \int d^3k' G(\vec{k};\vec{k}) U(\vec{k}') \psi^{+}(\vec{k}') \tag{30}$$

These results will ultimately be useful in calculating rate constants in, for example, atomic beam reactive scattering experiments.

Another fundamental quantity in reaction kinetics is the activation energy $E_a \equiv U(x_2) - U(x_1)$. We have discussed two distinct ways of calculating activation energies

$$E_a = - \int_{X_1}^{x_2} dx F_x(x)$$

$$E_a = - (T_e(x_2) - T_e(x_1)) \tag{31}$$

where X is a reaction coordinate. Both of these expressions for E_a depend only on the one electron density matrix and as such should be more easily modeled than the total energy expression. The message in all of this discussion is that Eqs. (29) and (31) provide independent methods for estimating reaction cross sections and activation energies which may be useful when bona fide self-consistent calculations are out of the question. The results reported in this paper are largely heuristic. We hope to supplement this with more quantitative calculations in forthcoming work.

REFERENCES

1. See D.R. Hamman and N.D. Lang, these Proceedings.
2. H. Hellmann, in Einführung in die Quantenchemie (Deuticke, Leipzig, 1937); R.P. Feynman, Phys. Rev. 56, 340 (1939).
3. J.C. Slater, J. Chem. Phys. 1 , 687 (1933); P.O. Löwdin, J. Mole Spectroscopy 3, 46 (1959).
4. H.F. Budd and J. Vannimenus, Phys. Rev. Letts. 31, 1218, (1973); J.E. van Himbergen and R. Silbey, Phys. Rev. 20, 1 (1979).
5. W.L. Clinton, J. Chem. Phys. 33, 1603 (1960) and W.L. Clinton, J. Chem. Phys. 33, 632 (1960).
6. W.L. Clinton, G.A. Henderson and J.V. Pristia, Phys. Rev. 177, 13 (1969).
7. D. Menzel and R. Gomer, J. Chem. Phys. 41, 3311 (1964); P.A. Redhead, Can. J. Phys. 42, 886 (1964).
8. W.L. Clinton, to be published.
9. J.J. Czyzewski, T.E. Madey, and J.T. Yates, Phys. Rev. Lett. 32, 777 (1974).
10. W.L. Clinton, Phys. Rev. Lett. 39, 965 (1977); T.E. Madey, Surf. Sci. 79, 575 (1979), W.L. Clinton, to be published.
11. M.W. Cole and D.R. Frankel, Surf. Sci. 70, 585 (1978).
12. M.J. Cardillo and G.E. Becker, Phys. Rev. Letts. 40, 1148 (1978).

13. O. Boato, P. Cantini, and R. Colella, Phys. Rev. Lett. 42, 1635 (1979).
14. G.F. Chew, Phys. Rev. 80, 196 (1950).
15. See L.D. Schmidt, these Proceedings.
16. M.L. Cohen, V. Heine and D. Weeire, Solid State Physics 24, 271, (1970).
17. T. Einstein, Phys. Rev. B12, 1262 (1975).
18. J.N. Bardsley and M.A. Biondi, Advances in Atomic and Molecular Physics 6, 1 (1970).

THE POTENTIAL OF EXAFS IN ELUCIDATING SURFACE REACTIONS

Edward A. Stern
Physics Department, University of Washington, Seattle, WA 98195

ABSTRACT

The extended X-ray absorption fine structure (EXAFS) can be analyzed to give atomic structure information about each type of atom, separately, for matter in a condensed state, irrespective of whether the sample has long range order or not. The utility of this technique for elucidating surface reactions and kinetics is described. The analysis of EXAFS measurements on the Br_2-graphite system is used to illustrate the possibilities of EXAFS giving details about what occurs in surface reactions.

I. WHAT IS EXAFS

The absorption of X-rays by atoms is characterized by sudden increases at photon energies characteristic of the absorbing atom called absorption edges (Fig. 1). In the photon range of less than 100 keV the absorption of x-rays is predominantly photoelectron excitation where all of the photon's energy is converted directly to electron excitations. The absorption edges are at the energy where new channels of excitation become available corresponding to knocking bound electrons into unoccupied states.[1] The K and L edges correspond to just exciting the n=1 and n=2 shells, respectively, of bound electrons in the atom.

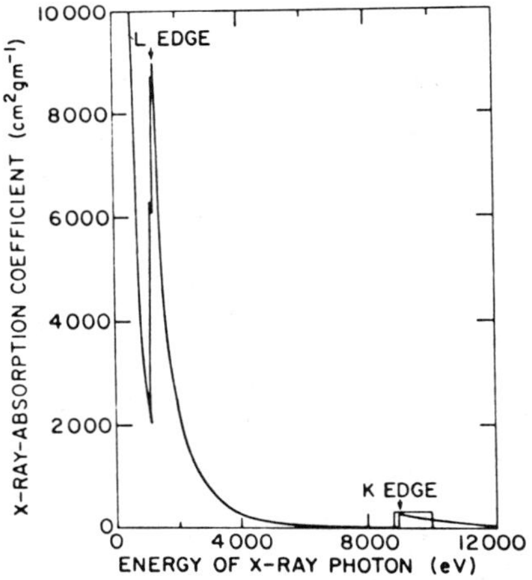

Fig. 1. The absorption coefficient of copper metal as a function of x-ray photon energy.

If the absorption of x-rays at the high energy side of an edge is investigated with higher energy resolution a fine structure is observed as shown in Fig. 2 for the K-edge of Cu. Note that the fine structure extends of the order of of 1000 eV past the edge. For reasons discussed later it is convenient to divide the fine structure into "near edge structure" extending from the edge to about 50 eV above and "extended x-ray absorption fine structures" (EXAFS) covering the range greater than 50 eV past the edge. We will focus on the EXAFS in this paper because it has been shown to contain information on atomic structure.[2,3,4]

The absorption of x-rays in matter usually follows the relation

$$I/I_0 = e^{-\mu x} \quad , \tag{1}$$

ISSN:0094-243X/80/610197-10$1.50

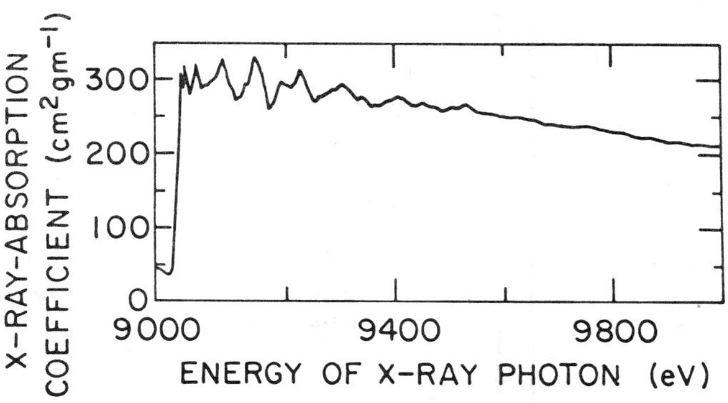

Fig. 2. The K-edge region boxed in Fig. 1 is drawn on an expanded scale showing the fine structure.

where I is the transmitted intensity through a sample of thickness x of an initial beam of intensity I_0. The physical mechanism causing EXAFS can be understood using the "golden rule" expression governing the absorption of X-rays in the dipole approximation.

$$\mu = 4\pi^2 e^2 N_0 \frac{\omega}{c} |\langle f|z|i\rangle|^2 \rho(E_f) \quad , \tag{2}$$

where the initial bound state $|i\rangle$ makes a transition to the photoelectron state $|f\rangle$ as the photon of energy $\hbar\omega$ is absorbed. The density of states corresponding to the photoelectron energy is $\rho(E_f)$, the number of atoms per unit volume is N_0 and the other symbols have their usual meaning. In the EXAFS regime where the photoelectron has excitation energies greater than 50 eV, $\rho(E_f)$ can be approximated by the free electron value which has a monotonic variation. The only remaining factor that can produce the fine structure is the variation in $|f\rangle$.

Figure 3 illustrates the cause of the variation in $|f\rangle$ that leads to the EXAFS. The state $|f\rangle$ is a superposition of an outgoing part and a part scattered by surrounding atoms. The matrix element is appreciable only in the volume where the initial state $|i\rangle$ exists. For the inner shells this region is near the center of the absorbing atom and it is sufficient to discuss the form of $|f\rangle$ only in this region. If the outgoing and scattered parts add

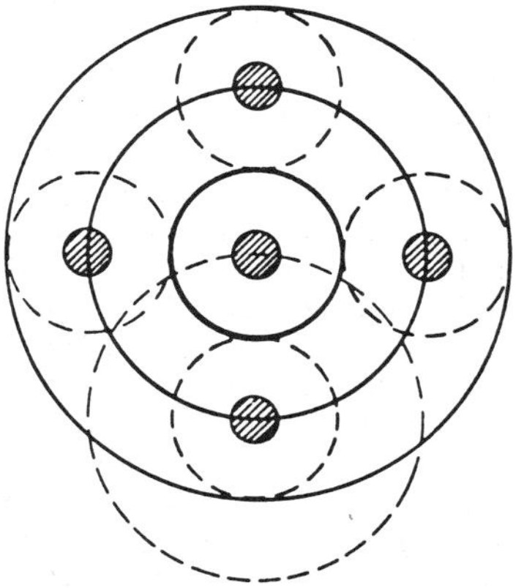

Fig. 3. Illustrating the excited photoelectron state with its backscattering from the surrounding atoms.

constructively at the atomic center, $|f>$ is increased over what it would be for an isolated atom with a corresponding increase in μ. Similarly if the two parts add destructively μ is decreased. The EXAFS monitors the relative phase between the outgoing and backscattered parts of the wave function $|f>$. This in turn depends on the product $2kr_i$ where r_i is the distance between the center atom and the ith backscattering atom, and $k=p/\hbar$, the wave number of the photoelectron of momentum p.

Translating the above picture into mathematical form, the expression for the EXAFS becomes in a randomly oriented sample[3-5]

$$\chi(k) = \frac{\mu-\mu_o}{\mu} = \frac{\sum_i N_i t_i(2k)}{kR_i^2} e^{-2R_i/\gamma} P_i(k)\sin 2[kR_i+\delta_i(k)] \qquad (3)$$

where μ_o is the absorption coefficient for an isolated atom; N_i is the number of atoms in the ith coordination shell of atoms; $t_i(2k)$ is the magnitude of the backscattering from the ith atom which depends on k ; $\delta_i(k)$ is the phase shift introduced as the photoelectron senses the potential of the center atom and backscatters from the ith atom type; and $P_i(k)$ is a factor that takes in account the phase mismatch in the backscattered waves from the atoms in the ith shell because they have a distribution in space about their average value of R_i. If this distribution is gaussian, i.e., proportional to $\exp -\frac{(r_i-R_i)^2}{2\sigma_i^2}$, where σ is the root mean square deviation about the average, then

$$P_i(k) = e^{-2k^2\sigma_i^2} \qquad (4)$$

An important thing to note about (3) is that different coordination shells contribute different frequencies in k-space. Therefore a simple fourier transform of $\chi(k)$ reveals peaks in r-space corresponding to the location of the coordination shells. This is illustrated in Fig. 4, which is a plot of the magnitude of the transform of copper EXAFS. By appropriately analyzing this transformed data it is ideally possible to determine N_i, σ, R_i and the types of surrounding atoms.[5] There are complications in the analysis introduced by $\delta_i(k)$ and t(2k) since they have to be ascertained either experimentally[5,6] or theoretically,[7] but these are surmountable.

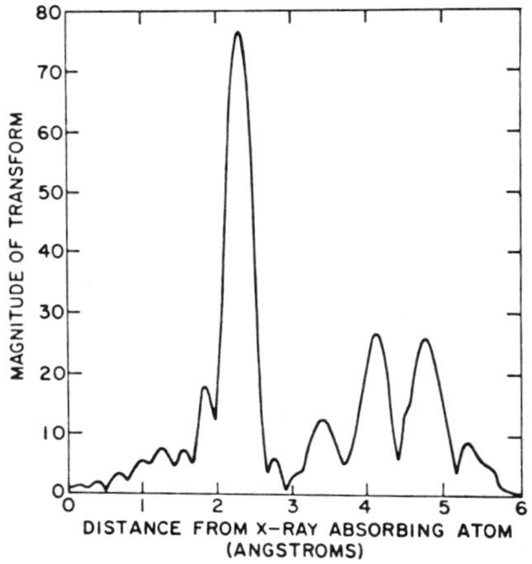

Fig. 4. The magnitude of the fourier transform of the EXAFS of copper. The peaks correspond to the location of coordination shells of atoms.

II. STRENGTHS AND LIMITATIONS

By analyzing the EXAFS spectra it is possible, in appropriate cases, to determine the local atomic structure about the absorbing atom. To ascertain which system would be best suited for study by the EXAFS technique it is necessary to understand the strengths and limitations of the technique.

The strengths of EXAFS are:

1. Species specificity: By tuning the x-ray energy to the absorption edge of a given type of atom, the atomic environment around that atom type can be isolated.

2. Short range order only: Non-crystalline and crystalline samples are treated on exactly the same basis since only the short range order is sensed. Thus structure information on samples without long range order can be obtained with EXAFS. The unknowns $\delta(k)$ and $t(2k)$ can be calibrated experimentally by measuring known crystalline structures.

3. Interpretation simple and direct: No involved quantum mechanical calculations are necessary to interpret EXAFS. The interpretation is direct and comes simply from analyzing the spectra by means such as fourier transforming. In an ideal case it is possible to obtain $\rho_{AB}(r)$, the two particle distribution function between the A and B atoms. The type of neighbor can be identified by $t(2k)$ and $\delta(k)$ which are characteristic of each atom. This is illustrated for $t(2k)$ in Fig. 5. Note that both low and high Z atoms have the same magnitude of $t(2k)$ but their k-dependence differs. Thus low Z atoms are not obscured by higher Z atoms as in the case for x-ray scattering, which is greater from higher Z atoms.

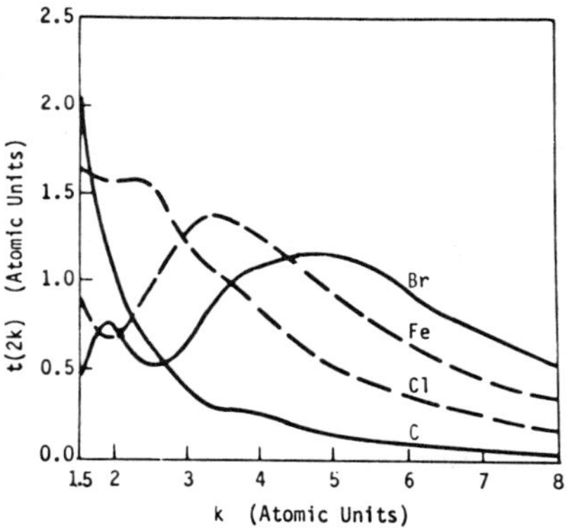

Fig. 5. The backscattering amplitude of various atoms as a function of the photoelectron wave number.

4. Angular information sometimes available: For oriented samples with less than 3-fold symmetry, angular information of bond directions can be obtained with polarized radiation. Otherwise, only radial information can be ascertained.

5. Near edge structure: The near edge structure and shifts in its energies contain information on chemical bonding, valence state, electronic charge distribution, coordination symmetry, etc. Unfortunately, to fully mine this information requires detailed quantum mechanical calculations, but it is possible in many cases to obtain parts of this information reasonably easily.

The limitations of EXAFS are:

1. No long range order information: Even in crystalline systems EXAFS can probe only the short range structure. Diffraction techniques are not superseded by EXAFS.

2. Multiple scattering effects: The simplicity in interpreting EXAFS mentioned in (3) above stems from the fact that a single back-scattering from each surrounding atom is adequate to calculate |f>. However, this approximation is only rigorously true for the first coordination shell[3,4] and multiple scattering effects become more important the farther the shell becomes. This limits the usefulness of EXAFS to only the first several coordination shells.

3. Limited angular information: As mentioned in (4) above, the number of cases of angular information on atom locations is limited.

4. High sensitivity to disorder: High sensitivity to disorder limits the maximum disorder detectable before the signal becomes lost. This limits measurable σ to less than 0.1 Å. On the other hand the sensitivity of EXAFS to disorder leads to the advantage of being able to more accurately measure the disorder when it is less than 0.1 Å.

5. Chemical effects: Both $\delta(k)$ and $t(2k)$ must be calibrated by comparison to known structures as mentioned in (2) of strengths. The basic assumption in such a calibration is that the quantities so determined can be accurately transferred to the unknown. This transferability is not completely precise with changing chemical environment, limiting the accuracy of the determined quantities, e.g., R_j can be obtained to an accuracy of about 0.01 Å.

III. APPLICATION TO SURFACE REACTIONS

EXAFS' characteristics open up new possibilities in studying surface reactions that are not amenable to more standard techniques. In this section some of these new possibilities are listed.

The species specificity property permits one to distinguish the surface reactant from the substrate and also to focus on the environment of contaminants. Each atom type in the reacting molecule can have its local environment measured in the ideal case. One is not limited in these measurements to surface layers with long range order.

For each atom type on the surface, EXAFS ideally can determine bond lengths, types and numbers of neighbors, and the distribution of the nearest neighbors in space. From the thermal dependence of the vibration amplitude one can monitor the frequency of local modes and ascertain how the surface affects their frequencies. When the molecule has less than three-fold symmetry its molecular orientation relative to the surface can be determined. Finally, the sum of the information listed above should determine the adsorption site of the reactant.

The above properties open up the possibility of determining the reaction species at various steps if the reaction can be frozen in time at these steps. We will illustrate such a case in Section V, namely for the Br_2-graphite system.

Finally, because EXAFS measurements on large surface area systems can be done in x-ray transmission, it is possible to use EXAFS to study technologically important surfaces under actual operating conditions. This follows because x-rays can penetrate reasonable thicknesses of the atmosphere around operating technologically important surfaces.

IV. EXPERIMENTAL CONSIDERATIONS

Since normally the surface atoms are a very small percent of the total atoms in a bulk sample, the experimental measurement of EXAFS must be chosen to enhance the surface. The simplest way to enhance the contribution of the surface atoms is to use a sample with a large surface area. If the surface area is large enough, the fraction of adsorbed atoms can be appreciable. In that case a simple x-ray transmission measurement can be used to study the EXAFS of the surface atoms. Since most technologically important surfaces are themselves large surface areas, this approach has much practical importance.

If one wants to measure EXAFS on a single crystal surface, as much of the fundamental studies are presently doing, then some EXAFS detection method must be used which enhances the surface. Electron detection is such a method. When the x-ray is absorbed the excited atom decays by emitting Auger electrons and fluorescent x-rays. The Auger electrons can only travel a few Angstroms before losing energy and thus those that escape the sample originate in surface atoms.[9,10] The Auger electrons have energies characteristic of their parent atom and their intensity is proportional to the number of excited atoms. Thus they can be used to monitor EXAFS with an enhanced surface sensitivity.

Low Z atoms decay with a high probability of Auger electron emission while high Z atoms decay mainly by emitting fluorescent x-rays. Thus, to detect high Z surface atoms it may be useful to detect fluorescent x-rays. Although fluorescent x-rays do not enhance the surface sensitivity as do Auger electrons they can be used if the surface atoms are of a different type than the rest of the sample by taking advantage of the species specificity of EXAFS. In that case surface studies are not limited to high vacuum conditions, in contrast to electron detection studies.

Another electron detection scheme is secondary yield where the electrons emitted from the surface are detected without energy discriminating for a particular Auger electron.[11,12] In this case the majority of electrons detected have come indirectly from the excited atom, losing energy along the way. They therefore come from a greater effective depth, about 50 Å, than do Auger electrons and are not as surface specific.

One possibility permitted by EXAFS is to span the gap between single crystal fundamental surface studies and technologically important surfaces with large and not well characterized surface areas. By use of different detection schemes from Auger electrons to x-ray fluorescence to x-ray transmission, EXAFS can span this gap.

V. THE Br_2-GRAPHITE SYSTEM

The interaction of Br_2 with graphite is a particularly ideal system illustrating some of the possibilities of elucidating chemical reactions by EXAFS. The interaction of Br_2 with graphite[13] can be controlled at room temperature by varying the partial pressure of Br_2. As shown in Fig. 6, the uptake of bromine into the interior of the

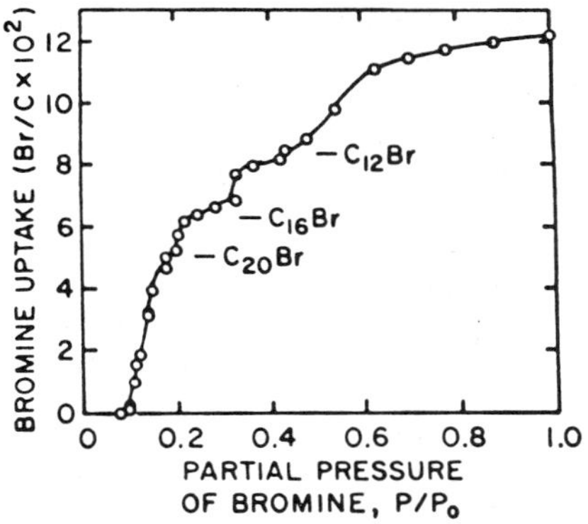

Fig. 6. The intercalation of Br_2 into graphite as a function of the Br_2 pressure at room temperature.

graphite, known as intercalation where the bromine forces apart the c-planes of graphite, starts at a minimum pressure about 0.1 of the saturation pressure p_0. Below this minimum pressure the bromine adsorbs on the surface of the graphite, and the coverage can be controlled by the pressure. Thus, it is possible to stop the reaction at any stage by fixing the bromine pressure.

This property was used to study the Br_2-graphite system from sub-monolayer coverages of 0.2, 0.6, and 0.9 of a monolayer, where a monolayer is defined as one Br_2 per 21 $\mathring{A}^2$, to the early stages of intercalation.[14,15] The graphite was in the form of Grafoil which has a large surface area of about 20 m^2 per gram,[16] thus permitting transmission measurements of EXAFS. The Grafoil has the additional convenient feature of being in the form of sheets along which the microscopic surfaces of graphite have a preferred orientation. About 0.6 of the surface area is oriented along the macroscopic surface and essentially all of the microscopic surface is the basal plane of the graphite, i.e., the (0001) surface. Thus Grafoil combines the best of two worlds. It has the simplicity of exposing a well defined single crystalline surface, whose orientation is known, with the added advantage of a large surface area and the attendant experimental simplification of detecting the EXAFS.

In what follows I briefly outline some of the properties of the Br_2-graphite system determined by EXAFS. The references can be consulted for more details.

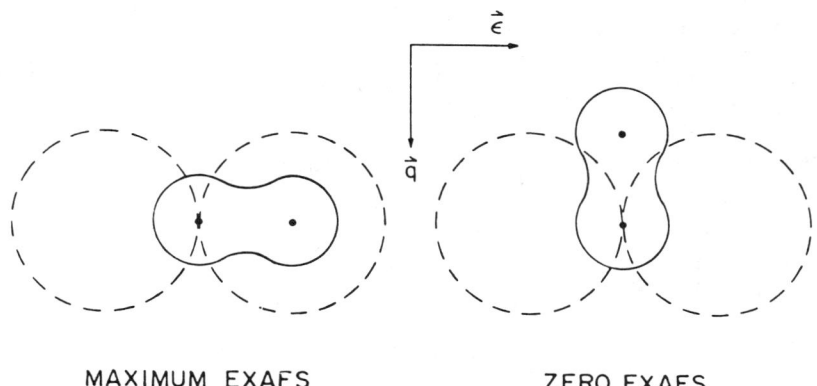

Fig. 7. Illustrating the mechanism of the angular variation of EXAFS from the Br_2 molecule.

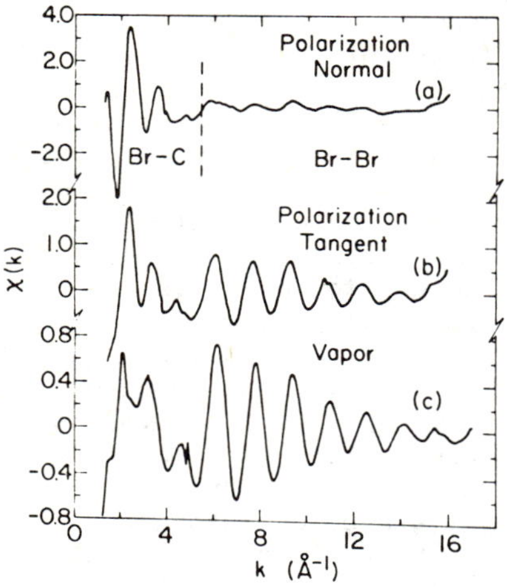

Fig. 8. The measured EXAFS of the Br_2 adsorbed on graphite as a function of the polarization of the x-rays relative to the graphite surface. In (a) the polarization is normal to the surface. In (b) it is parallel to the surface. The EXAFS for Br_2 vapor is shown in (c). The EXAFS in (a) and (b) has contributions from the surrounding C and Br atoms as indicated.

First, results are presented for the adsorbed Br_2 on graphite.[14] The directness of the interpretation of EXAFS is best illustrated in the determination of the orientation of the Br_2 molecule. As illustrated in Fig.7, the K-shell EXAFS contribution from the accompanying Br atom in a molecule is maximum when the x-ray polarization is along the molecular axis. This is so because the photoelectron state has p-symmetry with its axis along the polarization direction. The p-state has maximum amplitude along the polarization direction and zero amplitude perpendicular. The EXAFS amplitude is proportional to the p-state amplitude at the scattering atom.

Fig. 8 shows the variation with polarization of the measured EXAFS for 0.9 monolayers of Br_2 adsorbed on Grafoil. Note that there are two distinct EXAFS contributions with opposite orientation dependence. The EXAFS contribution which decreases rapidly with k is due to carbon backscattering while the one extending to higher k is due to the Br backscattering in the molecule (see Fig. 5). It is immediately obvious from the fact that the maximum EXAFS from Br backscattering occurs when the polarization is parallel to the surface that the Br_2 molecules are lying flat on the surface. The clear separation between Br and C neighbors is also obvious.

By using their different k-dependences it is possible to separately determine the C and Br neighbors to each Br atom. Such an analysis leads to the following conclusions.[14] Bromine adsorbs as a molecule oriented flat on the surface for all measured coverages. The adsorbed molecule is stretched 0.03 Å relative to the vapor to 2.31 ± .01 Å and is situated with each atom in the molecule centered as well as possible over adjacent hexagonal holes of the graphite basal plane, which are 2.46 Å apart, as illustrated in Fig. 9.

The analysis also yields the root mean square distribution of Br-Br distances in the molecule as a function of coverage and temperature. The wagging of the molecular axis relative to the surface can also be ascertained.

The early stages of the intercalation of Br_2 into graphite are not possible to study by conventional diffraction techniques because the bromine has no long range order and is a small percentage of the total sample. However, EXAFS studies have successfully elucidated some features of the bromine transition from adsorbed to intercalated.[15]

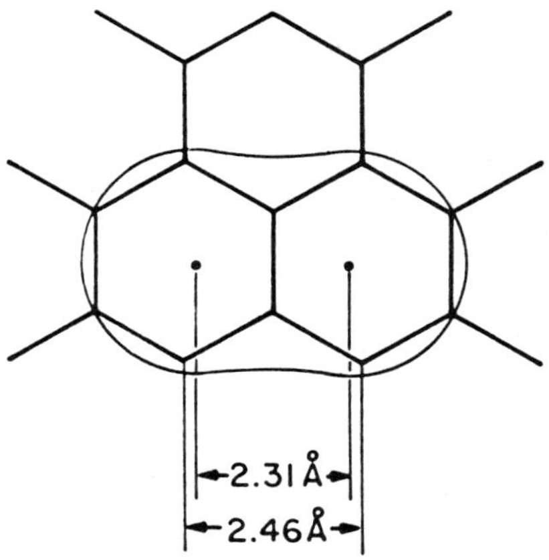

Fig. 9. Illustrating the location of adsorbed Br_2 molecules on the graphite surface. The carbon atoms are at the hexagon corners while the Br_2 molecule is represented by the dumbbell shaped object.

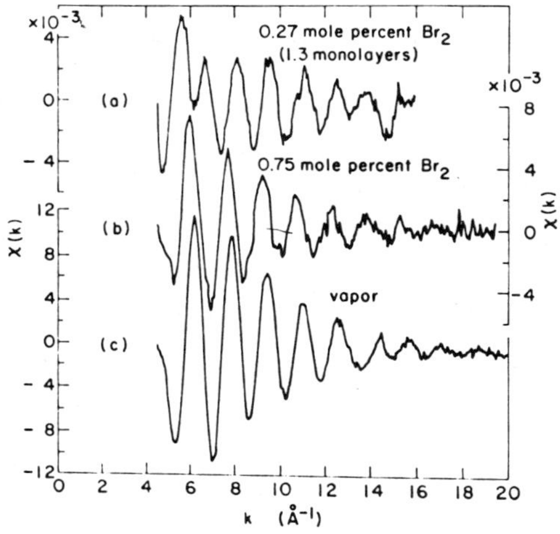

Fig. 10. The $\chi(k)$ for Br_2 intercalated into graphite for various molecular percentages of Br_2.

The most striking is the identification of a new species occurring at the initiation of intercalation. This is illustrated in Fig. 10 for the $\chi(k)$ plot of 0.27 mole % of Br_2, which would correspond to 1.3 monolayers if all of the Br_2 would adsorb, which they do not. By comparing with the 0.75 mole % of Br_2 and the Br_2 vapor curves it is noted that $\chi(k)$ has a decreased amplitude around k = 6 Å^{-1}. This decreased amplitude is caused by a beating of two different frequencies corresponding to two different Br-Br distances. In addition to the adsorbed Br-Br distance of 2.31 Å there is also present a distance of 2.53 ± 0.03 Å. The amount of the new species corresponding to this large distance quickly saturates as the amount of Br_2 increases. The further increase in added Br_2 is taken up by the intercalated species which has a Br-Br spacing of 2.34 ± 0.02 Å. The new species appears to serve the function of a "can opener" to pry apart the graphite planes and prepare them for subsequent intercalation.

The information obtained or obtainable by EXAFS on the Br_2-graphite system pertinent to reaction kinetics can be summarized as follows:

1. Temperature dependence of vibration
   a. Adsorbed Br_2 stretching mode
   b. Wagging
   c. Br-C vibration
2. Adsorption site
   a. Bond distances
   b. Orientation of Br_2 molecule
3. Changes with coverage
   a. Temperature dependence of vibrations
   b. Site of adsorption
4. Steps in the reactions - various species identified
   a. Adsorbed
   b. "Can opener"

c. Intercalated
d. Residual in defect sites

VI. SUMMARY

Because of the unique characteristics of EXAFS as a structure determining analytic tool it opens up the possibility of studying reaction kinetics and dynamics. What is EXAFS and its basic physics are described. The strengths and limitations of EXAFS are listed and some experimental considerations for looking at surfaces are summarized. Some of these features of EXAFS are illustrated by applications to the Br_2-graphite system.

REFERENCES

1. R.B. Leighton, Principles of Modern Physics (McGraw-Hill Co., Inc., New York, 1959) p.421-430.
2. D.E. Sayers, E.A. Stern, and F.W. Lytle, Phys. Rev. Lett. 27, 1204 (1971).
3. E.A. Stern, Phys. Rev. B10, 3027 (1974).
4. P.A. Lee and J.B. Pendry, Phys. Rev. B11, 2795 (1975).
5. E.A. Stern, D.E. Sayers, and F.W. Lytle, Phys. Rev. B11, 4836 (1975).
6. F.W. Lytle, D.E. Sayers, and E.A. Stern, Phys. Rev. B11, 4825 (1975).
7. P.A. Lee and G. Beni, Phys. Rev. B15, 2862 (1977).
8. E.A. Stern, J. Vac. Sci. Technol. 14, 461 (1977).
9. P.A. Lee, Phys. Rev. B13, 5261 (1976); P.H. Citrin, P. Eisenberger, and R.C. Hewitt, Phys. Rev. Lett. 41, 309 (1978).
10. U. Landman and D.L. Adams, Proc. Nat. Acad. Sci. USA, 73, 2550 (1976).
11. G. Martens, P. Rabe, N. Schwentner, and A. Werner, J. Phys. C11, 3125 (1978).
12. J. Stohr, L. Johansson, I. Lindau, and P. Pianetta, Phys. Rev. B20, 664 (1979).
13. T. Sasa, Y. Takahashi, and T. Mukaibo, Carbon 9, 407 (1971).
14. S.M. Heald and E.A. Stern, Phys. Rev. B17, 4069 (1978).
15. S.M. Heald and E.A. Stern, (to be published in J. of Synthetic Metals).
16. J.G. Dash, Films in Solid Surfaces (Academic, New York, 1975).

SURFACE DIFFUSION OF ADSORBATES AND RELATED MATTERS

R. Gomer
The Department of Chemistry and The James Franck Institute
The University of Chicago, Chicago, Illinois 60637

ABSTRACT

A method for determining diffusion coefficients of adsorbates on microscopic single crystal planes from the time auto-correlation function of field emission current fluctuations is described. Some diffusion results and related observations are presented and discussed.

INTRODUCTION

Knowledge of the mobility of adsorbates is central to understanding the mechanism of surface reactions and is also interesting in its own right. Activation energies of diffusion on different single crystal planes give information on the potential corrugation seen by adsorbates, and the prefactor of the diffusion coefficient contains information on the dynamics of diffusion. Since diffusion is quite sensitive to surface imperfections, steps, and edges, and since vacuum requirements limit the time of measurement, a microscopic method, capable of being applied to single crystal surface is highly desirable. Ideally, it should work at effectively constant coverage but be useable over a wide range of coverages. It turns out that the field emission current fluctuation method[1] to be described here meets these requirements for a wide class of adsorbates and in addition provides information on their equation of state. It also raises some as yet unanswered but very interesting questions. The chief drawback of the method, which might as well be stated at the outset, is that measurements must be made while field emission is occurring so that fields of $2\text{-}5\times10^7$ volts/ cm must be applied. This is not a serious limitation for adsorbates of small polarizability or dipole moment, like oxygen, hydrogen or even inert gases, but would be serious for alkali metals, if absolute values of activation energy were the prime objective. In view of the extremely detailed and fine-grained studies of metal atom diffusion, which can and have been carried out with the field ion microscope,[2] why is it necessary to devise another method? The answer is two-fold. First, most if not all non-metallic adsorbates are invisible in the field ion microscope, so that atoms like O, H, or Xe cannot be observed. Second, the field ion microscope is best suited to the observation of a very small number of ad-atoms, that is, it works best at effectively zero coverage, although pairs and even triplets of ad-atoms can be studied. On the other hand, the fluctuation method works best when several hundred atoms are observed.

SALIENT FEATURES OF THE FIELD EMISSION MICROSCOPE

Since the field emission microscope is central to the method, a very brief

ISSN:0094-243X/80/610207-13$1.50

summary of its relevant features is in order. More detailed reviews have been given elsewhere.[3] The microscope consists of an emitter in the form of a metal tip of 100- 1000 Å radius, surrounded by an anode in the form of a fluorescent screen. The tip is prepared by etching a fine wire to a needle-like shape, suitably annealed in high vacuum. Since the field F at the emitter apex is of order

$$F \cong V/5r \qquad (1)$$

where V is the applied voltage and r radius, very modest voltages suffice to produce fields of 2 -6 x 10^7 volts/cm, required for field emission. The latter consists of electron tunneling through the field deformed potential barrier at the metal surface. The field emitted current density is given by the Fowler-Nordheim equation

$$i = (B/F^2)\ \exp{-6.8\times10^7 \phi^{3/2}/F} \qquad (2)$$

where F is the applied field in volts/cm, ϕ the emitter work function in eV, and B a field independent quantity.

Electrons emerging from the tunneling barrier are accelerated along the lines of force; the latter diverge almost radially from the emitter apex, since the latter is a conductor. (Fig. 1) Thus, to a first approximation, the linear magnification of the microscope is given by the ratio of tip to screen distance/emitter radius, i.e. 10^5-10^6. Resolution is limited by the transverse electron velocity distribution to 20 -25 Å. Since metal electrons obey Fermi-Dirac statistics, field emission and resolution are almost temperature independent from 0 - 1000°K. Assuming that the emitter is smooth, the image on the screen is thus an emission map of the apex region, where the field is the highest. Because of its small size, the apex is almost invariably part of a single crystal, even if the emitter was etched from a polycrystalline wire; its shape after anneal is hemi-speroidal with flat facets corresponding to low energy planes blending smoothly into regions of varying Miller indices. Since different regions have different work functions, the emission anisotropy has the crystal symmetry and allows crystallographic indexing of the emitter. (Fig. 2).

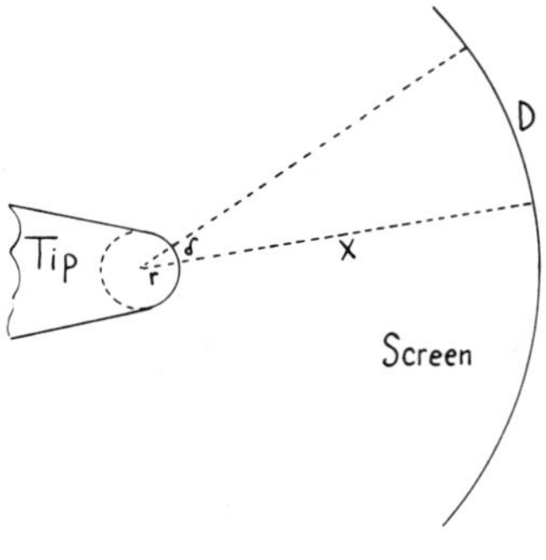

Fig. 1. Mechanism of image formation and enlargement in the field emission microscope. Linear magnification $D/\delta \cong x/r$.

We have already seen from Eq. 2 that emission is highly sensitive to work function. Since there is invariably some charge transfer from or to adsorbates, the latter has dipole moments, P and thus causes a work function change

$$\Delta \phi = 4\pi P (\bar{n}/A) \qquad (3)$$

where $\bar{n}$ is the average number of adsorbate particles in a region of area A. Thus field emission is very sensitive to even small changes in adsorbate coverage.

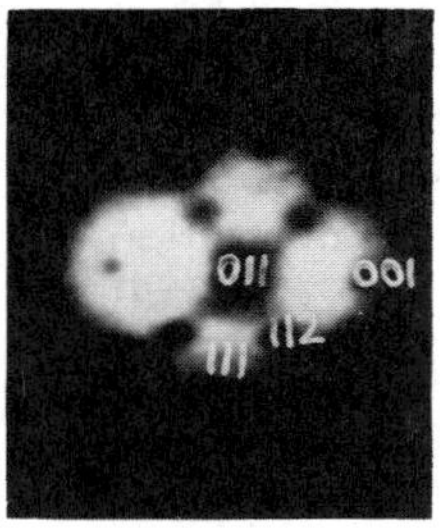

Fig. 2. Field emission pattern from a clean tungsten emitter. Some crystallographic directions are labelled.

This sensitivity, coupled with high linear magnification and moderate resolution makes the field emission microscope a very useful tool for studying surface diffusion. In the 1950's, the author and his students immersed a sealed off field emission microscope in liquid He, and allowed gas from a source located at one side of the emitter to impinge on it. At 4.2 K the vapor pressure of all gases except He is negligibly low, so that ultra-high vacuum is established. Further, the sticking coefficients on the walls are practically unity, so that only those portions of the emitter "seeing" the gas source receive a deposit. It is then possible to heat the emitter and to follow the migration of the gas deposit visually, since this leads to emission changes. (Fig. 3) A great deal of useful information can be obtained in this way,[3] but obviously coverage gradients are involved and even if diffusion is observed on a single crystal plane, there is no assurance that the observed diffusion is not limited by supply of adsorbate elsewhere. However, the field emission microscope can also be used in a different way to study diffusion at effectively constant coverage on single crystal planes, and I shall now describe this approach.

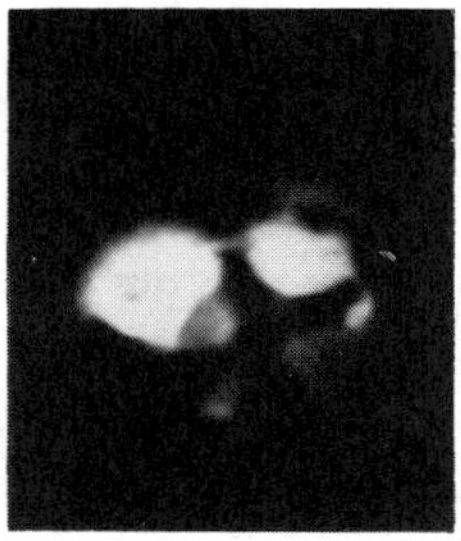

Fig. 3. Diffusion of chemisorbed hydrogen on a tungsten emitter. Note the sharp boundaries and the anisotropic nature of the diffusion.

PRINCIPLE OF THE FLUCTUATION METHOD

The high magnification and adequate resolution of the field emission microscope make it possible to measure emission from only a small portion of well developed single crystal planes, by putting a small probehole into the screen, and suitably amplifying the current passing through the probehole. By means of electrostatic deflection plates the emission pattern can be steered so that the desired region of the emitter is located over the probehole. Regions of 50-100 Å effective radius can be examined in this way, even allowing for the fact that the finite resolution of the microscope increases the size of the region examined. If the emitter is uniformly covered with adsorbate and heated to a temperature where the latter becomes mobile, adsorbate concentration fluctuations will occur in the probed region. Since the latter is very small, these fluctuations will be quite appreciable. In fact, for non-interacting point particles the mean square number fluctuation, $\langle \Delta n \rangle^2$ is

$$\langle \Delta n \rangle^2 = \bar{n} \tag{4}$$

the mean number of particles in the region examined. These fluctuations will build up and decay with a mean relaxation time τ_o related to the adsorbate's surface diffusion coefficient D through

$$\tau_o = r_o^2/4D \tag{5}$$

where r_o is the effective radius of the region examined. Thus we should expect the time autocorrelation function $f_n(t)$ of the density fluctuations δn

$$f_n(t) = \langle \delta n(0)\, \delta n(t) \rangle \tag{6}$$

to decay approximately in a time τ_o. Since each adsorbate particle carries its dipole moment with it and thus affects the work function, which in turn affects emission through Eq. 2, we should expect a close relation between the density fluctuations and the field emission current fluctuations. In fact, if the contributions to the potential from dipoles outside the probed region are neglected, δn is proportional to $\delta \ell n i$ for small fluctuations, since expansion of Eq. 2 about the mean work function $\bar{\phi}$ gives

$$\delta \ell n i = \delta \ell n B - [\,\tfrac{3}{2} \times 6.8 \times 10^7\, (\bar{\phi})^{1/2}/F\,]\, \delta \phi \tag{7}$$

$$= C\, \delta n$$

where

$$C = d\ell n B/dn + [\,\tfrac{3}{2} \times 6.8 \times 10^7 (\bar{\phi})^{1/2}/F\,]\, \frac{d\phi}{dn} \tag{8}$$

and the first term takes into account that B is usually a function of adsorbate coverage.

It is possible to derive an expression for the correlation function of the density fluctuations, if it is assumed that diffusion of individual ad-particles is uncorrelated. In that case $f_n(t)$ is

$$f_n(t) = (\bar{n}/A)\ \frac{1}{4\pi Dt} \int_A e^{-|\vec{x} - \vec{x}'|^2/4Dt}\, d\vec{x}'\, d\vec{x} \tag{9}$$

where $\vec{x}$ and $\vec{x}'$ are position vectors. Introduction of the non-dimensional variable $x/r_o = \rho$ immediately expresses f_n as a function of t/τ_o, τ_o being the relaxation time defined in Eq. 5,

$$f_n(t/\tau_o) = (\bar{n}/A)\ (\tau_o/t)\ \frac{1}{4\pi} \int_A e^{-|\vec{\rho} - \vec{\rho}'|^2 \tau_o/t}\ d\vec{\rho}\, d\vec{\rho}'$$

$$= (\bar{n}/A)\ \frac{1}{A}\ g_1(t/\tau_o) \tag{10}$$

where $g_1(t/\tau_o)$ is a dimensionless quantity. The field emission current correlation function is conveniently defined as

$$f_i(t) = \langle \delta\ln i\,(0)\ \delta\ln i\,(t)\rangle_2$$

$$\cong \langle \delta i\,(0)\ \delta i(t)\,\rangle/\bar{i} \tag{11}$$

where the second line follows by expansion for small fluctuations. Thus the measured quantity $f_i(t)$ is related to $f_n(t)$ by

$$f_i(t) = C^2\ f_n(t) = C^2\ (\bar{n}/A)^2\ g_1(t/\tau_o) \tag{12}$$

Expression 9 is valid if dipoles outside the region A can be neglected. It is possible to take the effect of dipoles outside the probed region into account. The resulting expressions are more complicated than Eq. 12 but the form of $f_i(t)$ is not very different; in essence the effective radius of the probed region is increased.[1]

Eq. 10 can be evaluated in closed form only for a rectangular probed region. The result for a square region is

$$f_n(t) = (n/A)\ [\mathrm{erf}\ (\tau_o/t)^{1/2}\ +\ t/\pi\tau_o\,(\,e^{-\tau_o/t} - 1)\ ]^2 \tag{13}$$

It is noteworthy, that f_n is much more complicated than a simple exponential. In particular for $t/\tau_o << 1$ it decays as t^{-1}. This result also carries over to the general case, as expected. Specifically for a circular aperture

$$g(t/\tau_o) \rightarrow \tau_o/t \quad \text{for} \quad t/\tau_o << 1 \tag{14}$$

The t^{-1} behavior of the correlation function has nothing to do with two-dimensional divergences: our correlation function is an area integral over a van

Hove function and explicitly assumes the existence of a diffusion coefficient.

Eq. 9 suggests that the correlation function can be looked at from a slightly different point of view. The quantity

$$P(\vec{x}|\vec{x}', t) = \frac{1}{4\pi Dt} \exp -|\vec{x}-\vec{x}'|^2/4Dt \tag{15}$$

corresponding to the probability that a particle in $d\vec{x}$ at $\vec{x}$ at $t = 0$ will be in $d\vec{x}'$ at $\vec{x}'$ at t becomes $\delta(\vec{x}-\vec{x}')$ as $t \rightarrow 0$. We may thus think of the correlation function as follows: A delta-function fluctuation is placed at $\vec{x}$ at $t = 0$ and allowed to evolve in time. The strength of the fluctuation left at t is obtained by integrating with respect to $\vec{x}'$ over the probed area A. Since we have no a priori reason for picking the location $\vec{x}$, we must also average over $\vec{x}$, i.e. integrate with respect to $\vec{x}$ over A. This procedure is equivalent to assuming a step excess concentration profile over A at $t = 0$, letting this evolve in time according to the diffusion equation

$$D \nabla^2 c = \partial c/\partial t \tag{16}$$

(of which $P(\vec{x}|\vec{x}', t)$ is a solution) and then integrating over the strength of the concentration fluctuation left in A at time t. Thus for a probed region of area A contained in an infinite plane

$$f_n(t) = \bar{n} \int_A c(\vec{x}, t)\, d\vec{x} \tag{17}$$

with $c(\vec{x}, 0) = 1$ in A and 0 elsewhere. The correct magnitude of the fluctuation is insured by the factor $\bar{n} = \langle \Delta n \rangle^2$. If the area of the plane containing A, A_{tot}, is finite Eq. 17 can be shown[4] to require only the subtraction of $\bar{n}(A/A_{tot})$, corresponding to the fact that for $t = \infty$ Eq. 17 predicts this amount to remain in A. It is straightforward to show that Eq. 17 leads to the same result as Eq. 9. The usefulness of Eq. 17 is two-fold. First, it permits the calculation of $f_n(t)$ for finite planes under various boundary conditions when $P(x|x', t)$ no longer takes the simple form 15. Second, and perhaps of more fundamental importance, it suggests that whenever the diffusion equation, Eq. 16, is valid the correlation function can be considered to be the time evolution, as given by $g_1(t)$, of the mean square fluctuation, even when the latter has a more complicated form than $\bar{n}$. Thus the form of the correlation function used here is probably correct even for some types of interactions between ad-particles, as long as an effective diffusion coefficient can be defined.

MEAN SQUARE FLUCTUATIONS

We have just alluded to the fact that $\langle \Delta n \rangle^2$ is not in general equal to $\bar{n}$. It can be shown[5] that it is in fact

$$\langle \Delta n \rangle^2 = (\bar{n}/A)\, K\, kT \qquad (18)$$

where K is the two-dimensional compressibility of the adsorbed phase. While it is difficult to obtain $\langle \Delta n \rangle^2$ accurately from the measured quantity $f_i(0)$ because of uncertainties in A and C, its temperature variation gives that of $\langle \Delta n \rangle^2$ i.e. of K, and thus is related to the equation of state of the adsorbate. Since the thrust of this volume concerns kinetic rather than equilibrium phenomena, no further discussion of $\langle \Delta n \rangle^2$ will be given here. Some results are given elsewhere.[4, 6, 7]

EXPERIMENTAL ASPECTS

Experimental procedures and apparatus have been described in some detail elsewhere.[4] We give here only a very brief sketch and mention some important points. The apparatus consists of a field emission tube with provisions for heating the emitter assembly, steering the beam, amplifying the probehole current, and depositing gas on the emitter. For the particular tube shown in Fig. 4, designed to be used in a liquid H_2 cryostat, the gas source consists of a small, electrically heatable Pt platform on which gas can be condensed and reevaporated. After suitable amplification the a.c. component of the probehole current passes into an autocorrelator, which samples current at preset intervals Δt (which can range from 10^{-5} to 0.1 seconds), and carries out the appropriate multiplications $\delta i(0)\delta i(\Delta t), \ldots \delta i(0)\delta i(n\Delta t)$, n = 1 to 400, repeats the process many times, and adds and stores the results.

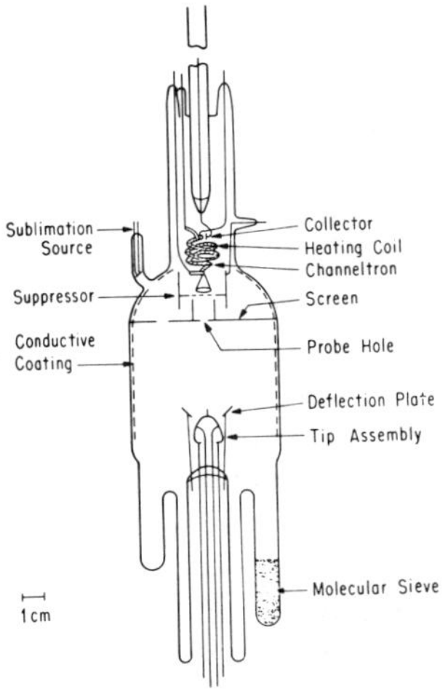

Fig. 4. Field emission tube for fluctuation measurements.

In practice the measurements are not wholly trivial, since very small currents must be amplified, and the band-width of the amplification system imposes limits on both τ_o and the time range t/τ_o. The minimum value of Δt is limited by the high frequency cut-off of the amplification system and thus restricts experiments to a temperature range where τ_o is sufficiently large to make $\tau_o >> \Delta t_{min}$; in practice $\tau_o = 0.1 - 10^{-4}$ sec can be used. Since the correlation function is the Fourier transform of the power spectrum, the low frequency cut-off leads to oscillations with that frequency for $t/\tau_0 >> 1$, thus limiting the observable time domain of the correlation function.

τ_0 values are obtained by comparing experimental with theoretical correlation functions. (Fig. 5) The temperature dependence of τ_0 then gives the activation energy of diffusion in the usual way.

SOME DIFFUSION RESULTS

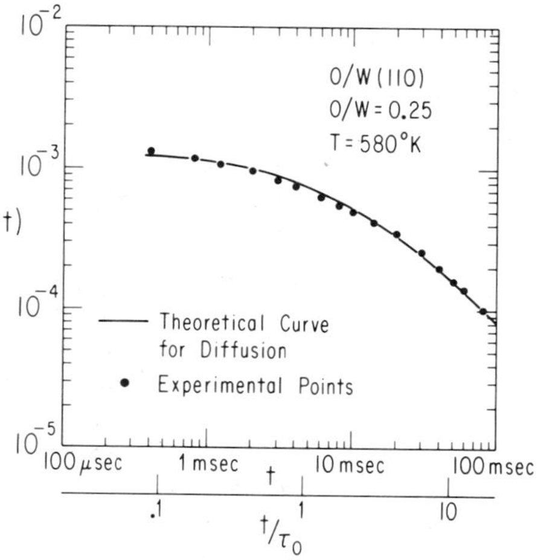

Fig. 5. Experimental and theoretical correlation functions, superimposed by horizontal and vertical displacements.

Experiments carried out to date by this method have been limited to oxygen,[4] CO,[6] and Xe[7] on the tungsten (110) plane, which is the closest packed plane for bcc metals. The diffusion results for oxygen are summarized in Figs. 6 and 7. The diffusion coefficient is here defined as

$$D = D_0 \, e^{-E_{dif}/kT} \quad (19)$$

Two things stand out. First, both E_{dif} and D_0 are highly coverage dependent, increasing with increasing coverage. Second, over most of the coverage range accessible, D_0 is very much smaller than the value $a^2\nu \simeq 10^{-3}$–10^{-4} cm^2/sec predicted by simple kinetic models. The coverage dependence of E_{dif} is not difficult to understand in a qualitative way, since there is evidence from LEED for strong interactions between adsorbed O atoms. At low coverage E_{dif} is presumably independent on adsorbate-adsorbate interactions, and gives an indication of the potential corrugation of the substrate. The desorption energy, E_{des}, of O atoms from the (110) plane is not accurately known, but probably lies between 92 and 126 Kcal/mole.[8] Thus at low coverage $E_{dif}/E_{des} = 0.11$ to 0.15, depending on the value of E_{des}. Discussion of D_0 will be deferred until some other results have been presented.

In the case of Xe, coverages of $\theta = 0.3$, 0.5, and 0.9 monolayers were investigated. One monolayer in this case corresponds[9] to 5.5×10^{14} Xe atoms/cm^2. For comparison the W atom density on the (110) plane is 1.42×10^{15} atoms/cm^2. For $\theta = 0.3$ and 0.9 E_{dif} was ~ 1.4 Kcal, corresponding to $E_{dif}/E_{des} = 0.14-0.25$, depending on the value of E_{des} which is not known accurately at this time, although the lower value and hence the higher ratio seem more likely.[10] It is interesting that even for the very weak chemisorption (or strong physisorption) case represented by Xe on W(110) the potential corrugation seen by the adsorbate is so strong. This finding is particularly interesting in view of the fact, that it has been known for some time that He diffraction from clean metal surfaces is extremely weak, because of the lack of sufficient potential corrugation. On the other hand, it is also known now from the work of L. Wharton and his students,[11] that heavy inert gas atoms like Xe do not scatter elastically at all from metal surfaces, so that there maybe no real contradiction. For $\theta = 0.5$ it was not possible to obtain E_{dif}, because of anomalous temperature behavior of the correlation functions. There are several possible explanations, (including of course inadequate measurements, since the fluctuation signal for Xe is particularly small, and measurements particularly difficult), but we will not discuss these here, except to mention that the assumption of independent particle or hole diffusion is particularly bad at $\theta = 0.5$. For $0 = 0.3$ and 0.9, D_o values are extremely low, $< 10^{-8} cm^2/sec$.

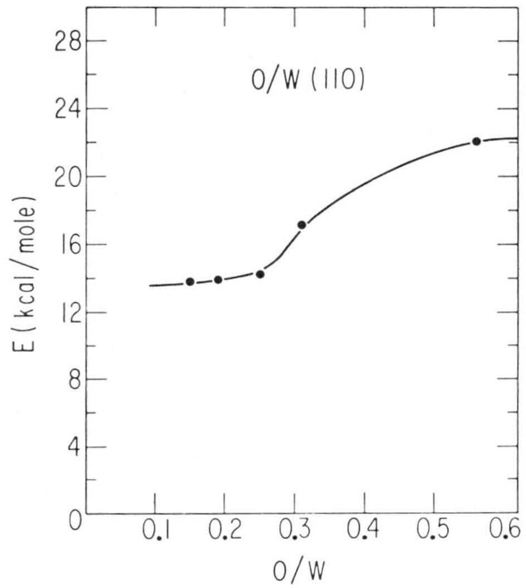

Fig. 6. Activation energy of surface diffusion for O atoms on the tungsten (110) plane as function of oxygen coverage, expressed as ratio of O/W.

All the D_o values quoted here are based on the actual radius of the probed region, with no allowance made for the effective increase in r_o arising from finite resolution. Since the calculated values of D and hence D_o are proportional to r_o^2, the true values of D_o could be increased by factors of up to 4; this still leaves them in most cases much smaller than $a^2\nu$. It is possible that the fluctuation method always gives an effective diffusion coefficient which incorporates correlation effects, so that there is no particular reason to expect D_o to approximate $a^2\nu$. There may also be another reason for the smallness of D_o (and for its increase with E_{dif} in the case of oxygen diffusion). For surface

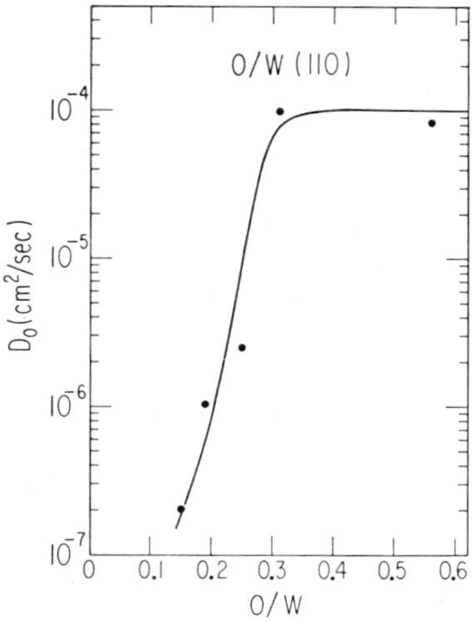

Fig. 7. Prefactor D_o as function of O atom concentration for oxygen diffusion on the tungsten (110) plane.

diffusion there is no such thing as a single barrier energy. Even if the substrate atom positions were frozen at some specified configuration, many physical paths are available to the adsorbate and correspond to different activation energies and entropies. If the substrate is "unfrozen," the possibilities are of course increased. Thus, it seems quite possible that trade-offs between activation energy and entropy will occur, and that the system will choose a path (in phase space), which maximizes rate by minimizing activation energy at the expense of decreasing entropy. Physically, this would correspond, for example, to adsorbate movement occurring only for a very favorable, but not very probable configuration of the relevant substrate atoms. The change in D_o with E_{dif} in the oxygen case may also be explicable along these lines. As coverage increases, E_{dif} goes up because of O - O interactions. At the same time, these may block low energy - low entropy paths, so that the net effect is an increase in both E_{dif} and D_o. Over modest temperature intervals a single activation energy and prefactor would still be seen; if a sufficiently wide interval were accessible to observation, one should expect to find curvature in the Arrhenius plots of log D vs. 1/T. There is a hint of this in the oxygen results, but for reasons to be explained presently, we have not been able to demonstrate such an effect unequivocally. On the other hand, Butz and Wagner [12] who investigated O diffusion on W(110) by a macroscopic method at much higher temperatures found a "normal" prefactor and a slightly higher activation energy for O/W = 0.5 than obtained by us. In view of the considerable difficulties associated with measurements on macroscopic planes, one should not push the comparison too far, but the effect is in the right direction. It is interesting that the experiments of Ehrlich and his coworkers on single metal atom diffusion[2] by means of field ion microscopy in general yield much higher prefactors, 10^{-5} to $10^{-3} cm^2/sec$, than those obtained by the fluctuation method. It would obviously be interesting to attempt fluctuation measurements in the same systems that have or can be studied by field ion microscopy, and hopefully this will be done. Possibly, such measurements will answer the question whether metal diffusion is inherently more "normal" as far as prefactor goes than that of highly non-metallic adsorbates, or whether coverage effects account for the difference.

PREDIFFUSION

In almost all cases examined so far, fluctuations leading to an exponentially decaying correlation function were observed at temperatures below those where the diffusional form is seen.[4, 6] It is not difficult to show[1] that an exponentially decaying $f_i(t)$ will result if an adsorbate particle can flip from one state, say A, into another, say B, with mean relaxation time $\tau_{AB} = 1/k_{AB}$ and back with a relaxation time $\tau_{BA} = 1/k_{BA}$, if emission in the vicinity of the ad-particle is different in states A and B respectively. As might be expected, the observed decay time τ of the correlation function is given by

$$1/\tau = 1/\tau_{AB} + 1/\tau_{BA} \qquad (20)$$

i.e. in terms of first order rate constants

$$1/\tau \equiv \bar{k} = k_{AB} + k_{BA} \qquad (21)$$

It is tempting to associate the exponential decay regime with prediffusion -- flip-flop among nearly equivalent sites, without true diffusion. It can be seen by inspection of Fig. 8 that this is indeed possible on the (110) plane. Purely configurational changes at constant coverage, i.e. rearrangements of the ad-atoms in A without change in their number could in principle also lead to small emission changes, although the average work function remains unchanged in such rearrangements.

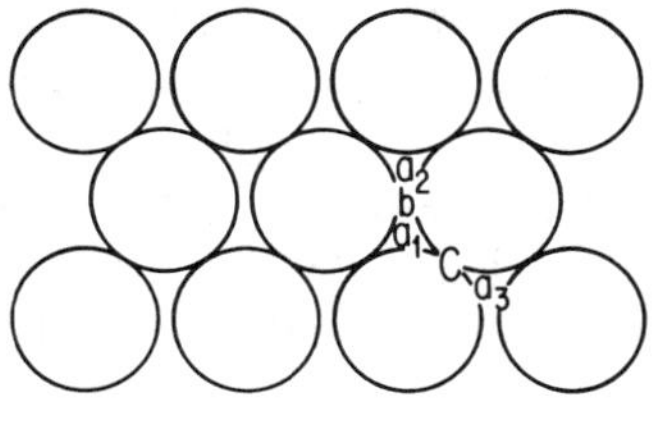

Fig. 8. Schematic diagram of the (110) plane of a bcc metal, indicating possible sites involved in flip-flop.

For oxygen two exponential decay regimes have in fact been seen,[4] suggesting two kinds of flip-flop, or possibly flip-flop and spatial rearrangement. It also happens that a good diffusional correlation function mimics two decaying exponentials when plotted as log $f_i(t)$ vs. t. Thus it is not always possible to distinguish between prediffusion and diffusion. If the prediffusive regime, in those cases where two exponentials are seen, is interpreted as diffusion, considerably decreased values of E_{dif} and D_o would result. In the absence of more data the matter must rest here at present.

LONG TIME BEHAVIOR OF THE DIFFUSIONAL CORRELATION FUNCTION

The last subject to be discussed concerns observations on the long time behavior of the correlation function. We have already seen that simple theory predicts 1/t decay. This is in fact observed, but only for moderately low values of the diffusion coefficient, $D \leq 10^{-11} cm^2/sec$.[4, 6] For higher values, i.e. higher temperatures, decay is slower and approaches a $t^{-1/2}$ law. In the absence of a general theory of correlation functions, not neglecting adsorbate-adsorbate interactions, one can only speculate about the meaning of this intriguing result. We have argued previously[4] that the effect is probably not caused by the finite size of the (110) plane, but I will not go into details of this argument here. Suffice it to say that the most likely effect of finite size would be to increase rather than decrease decay rate at long times.

The following hypothesis is at least interesting -- its correctness is of course another matter. Let us assume that at moderately low temperatures, where $D \leq 10^{-11} cm^2/sec$, on the average only a single particle is in motion in the probed region at any one time. The diffusing particle feels the other ad-particles, but in a static way -- they contribute very much like substrate atoms to the net potential felt by the diffusing particle. At higher temperatures, for $D > 10^{-11} cm^2/sec$, the probability increases that several or most particles are in motion simultaneously. Under these circumstances, interactions between the ad-particles become dynamical, and quite different kinds of correlations can come into play. Specifically, multi-particle density fluctuations can now occur, and may have a longer decay time than single particle fluctuations. If this were the case, the behavior at short times would still be dominated by single particle fluctuations, and thus the normal correlation function would be seen there. However, at long times the correlation function would be governed by the slowly decaying component, i.e. the multi-particle fluctuations, and these could lead to the observed deviations from t^{-1} decay. Obviously, this is highly speculative and a great deal of theoretical work remains to be done.

CONCLUSION

I have attempted to give a brief sketch of the main features of the fluctuation method of measuring surface diffusion. It should be clear from the preceding that the method holds considerable promise for a much wider range of systems than have so far been studied by it. It should also be clear that a number of quite puzzling theoretical questions have been raised by the results obtained to date, and that much remains to be done, both experimentally and theoretically.

REFERENCES

1. R. Gomer, Surf. Sci. 38, 373, (1973).
2. G. Ehrlich, Surf. Sci. 63, 422, (1977).
3. R. Gomer, Field Emission and Field Ionization, Harvard Univ. Press. Cambridge, Mass. (1961).
4. J. -R. Chen and R. Gomer, Surf. Sci. 79, 413, (1979).
5. B. Bell, R. Gomer, and H. Reiss, Surf. Sci. 55, 494, (1976).
6. J. -R. Chen and R. Gomer, Surf. Sci. 81, 589, (1979).
7. J. -R. Chen and R. Gomer, Surf. Sci. submitted.
8. K. Kohrt and R. Gomer, J. Chem. Phys. 52, 3283 (1970).
9. C. Wang and R. Gomer, Surf. Sci. 84, 329, (1979).
10. C. Wang and R. Gomer, Surf. Sci., in press.
11. J. E. Hurst, C. A. Becker, J. P. Cowin, K. C. Janda, L. Wharton and D. J. Auerbach, Phys. Rev. Lett., submitted.
12. R. Butz and H. Wagner, Surf. Sci. 63, 448, (1977).

TRANSPORT AND REACTION ON SURFACES: A STOCHASTIC APPROACH*

Michael F. Shlesinger and Uzi Landman
School of Physics, Georgia Institute of Technology
Atlanta, Georgia 30332

I. INTRODUCTION

In recent years surface physics has received a resurgent interest because of its importance in fields such as heterogeneous catalysis, thin film and crystal growth, interfacing electronic devices, and photovoltaic cells. Essential to this renewed interest in surface physics has been the advent of new and improved experimental techniques which enable detailed investigation of structure (geometrical, electronic and vibronic) and reaction processes under well controlled conditions.

It is convenient to describe surface reactions in terms of "elementary processes" such as adsorption, desorption, migration, dissociation and association. The fundamental understanding of each of the above steps comprises rather broad fields of study. In this paper we will concentrate on the migration or diffusion step and on some aspects of certain diffusion controlled reactions on surfaces. Basic knowledge of these processes which would enable the optimization of the parameters of catalytic materials, requires an understanding on a microscopic level. Recent developments and refinements of measurement techniques, such as Field Ion Microscopy (FIM) and Field Emission Microscopy (FEM) serve as the impetus for the development of theoretical microscopic models and refined methods of data analysis. In Section II we will review recent developments in the analysis of FIM diffusion experiments. These new methods when applied to the analysis of experimental data provide a spectroscopy of the migration mechanism, i.e. allow the detailed determination of kinetic parameters (frequency factors and activation energies) corresponding to elementary steps of the diffusion process. The recent direct observations using FIM of the diffusion of adatom clusters where the migration of the cluster proceeds via alternating spatial configurations (internal states of the cluster), provide data which when properly analyzed yields the above mentioned spectroscopy.

Our basic approach employs stochastic techniques. While single particle diffusion is described commonly in terms of a random walk process, a generalization of the formalism is needed in order to describe complex diffusion mechanisms, and diffusion in defective systems. Propagating systems may be endowed with internal states which may represent spatial configurations, mobile and immobile states, spin states, different band components in a Wannier site localized representation or even a temporal correlation memory.

*Supported by DOE Grant No. EG-77-S-05-5489

ISSN:0094-243X/80/610221-28$1.50

Mapping the internal states onto a random walk lattice and using matrix Green's function techniques enables us to analyze such complex systems. These methods and their generalization to periodically defective lattices by introducing supercell Green's function are described in Section II.

In Section III we present a different approach to studying diffusion on defective lattices which has computational advantages (particularly for small defect concentrations) over the method described in Section II. The method is based on a defect renormalized Green's function.

Kinetics of unimolecular and bimolecular diffusion controlled reactions on surfaces, catalyzed by active sites is the subject of Section IV. The reaction rate which appears in the master equation description of the reaction is expressed as the conditional probability first passage time density which is related to the more well known first passage time density. The reaction rate is related to the probability distributions governing migration on the surface which reflects the geometrical structure and potential surface. It is found that the reaction rate is in general time dependent, approaching a constant value at long times, and that it depends on the atomic arrangement at the surface. The structural dependence of the rates can become quite pronounced for a low concentration of active sites.

Finally in Section V we propose an explanation to a puzzle in heterogeneous catalysis. The bimolecular reaction of the disproportionation of propylene on the surface of a catalyst is seen to have a reaction rate maxima as a function of temperature. However, thermodynamic inequalities which must be satisfied for a rate maxima to occur are violated by the experimental data. We show how the thermodynamic inequalities may be satisfied if multistate diffusion mechanisms, such as described in Section II and III, are assumed.

The unifying feature to our treatment of the above studies of diffusion and reaction on surfaces is a stochastic description via random walk lattice Green's function propagators.

II. PARTICLE AND CLUSTER MOTION ON IDEAL AND DEFECTIVE SURFACES: GREEN'S FUNCTION WITH INTERNAL STATES METHOD

In order to analyze diffusion controlled reactions, as we do in Sections IV and V, we must first discuss the process of diffusion on a surface. The diffusive motion of adatoms and clusters of adatoms on surfaces has been dramatically revealed by Field Ion Microscope (FIM)[1-5] studies. Field Ion Microscopy which was conceived and developed by E. W. Müller in the early 1950's was used first for the investigation of adatom migration on surfaces by Ehrlich and Hudda[6] in 1966. Later studies revealed that adatoms on metal surfaces can become correlated to move as a single cluster.[7-16]. One example we will analyze in detail is the motion of rhenium dimers on a W(211) surface[7].

The FIM is, under certain conditions, able to give images from which one can determine the distance traveled by an adatom in a time t at a temperature T. For example, the motion of a single tungsten adatom on a W(211) surface is seen to occur[6] (away from boundaries) as a one dimensional random walk with symmetric nearest neighbor hopping. Standard random walk theory gives for the mean squared displacement, $\sigma^2(t)$, after a time t,

$$\sigma^2(t) = \lambda L^2 t \tag{2.1}$$

where λ is the hopping transition rate, L is the lattice spacing, and $\sigma^2(t)$ can be obtained from FIM pictures. The transition rate λ is seen to be in the Arrhenius form

$$\lambda = \nu \exp(-E/kT) \tag{2.2}$$

since a semilog semilogrithmic plot of $\sigma^2(t)L^2t$ vs. $1/kT$ yields a straight line of slope - E and ordinate intercept log ν. Thus, an analysis of the FIM pictures can yield the activation energy, E, for diffusion, as well as, the frequency factor ν. We will now discuss how the maximum amount of information can be extracted from FIM data when the motion of a cluster occurs and several transition rates are involved. This will lead us to the study of random walks with several internal states.

The nature of the motion of a cluster on a surface depends on the substrate compasition and morphology as well as on the type and number of atoms in the cluster. For example rhenium dimers[7] are seen to undergo one dimensional motion on W(211) by alternating between straight and staggered configurations as shown in Fig. 1. If only one staggered position is allowed the center of mass motion of the dimer can be mapped onto a perfect lattice with two states per unit cell (Fig. 1b), and the motion is characterized by the transition rates, a,α,b, and β. If there is no bias caused by say, an external electric field then α = a and b = β. We express the rates in activated form,

$$\begin{aligned} a &= \nu_a \exp[(-E_a + V)/kT] , \\ \alpha &= \nu_a \exp[(-E_a - V)/kT] , \end{aligned} \tag{2.3}$$

As seen, a and α are not independent quantities even in the presence of a biasing electric potential V. Thus the dimer motion in Fig. 1b is characterized by only two transition rates. If a third more extended (non-dissociated) state is allowed then the center of mass motion can be mapped onto a lattice with three states per unit cell (two of which overlap) as shown in Fig. 1c. The motion is then characterized by the four transition rates a,b,c, and d. If the dimer can move in two dimensions then the center of mass motion can be mapped onto a two-dimensional lattice, as shown in Fig. 2., characterized by four transition rates, i.e, four activation energies and four frequency factors.

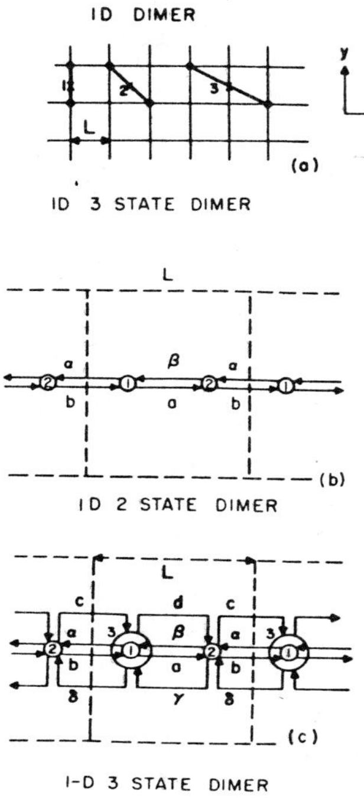

FIG. 1. One-dimensional dimer migration. (a) Three possible spatial configurations of a dimer (filled circles connected by heavy line) moving along the x direction (the allowed equivalent mirror-image configurations are not included): if only states 1 and 2 are allowed, a 2-state dimer; if all states are allowed, a 3-state dimer. The location of the dimer centroid is marked X. (b) Random-walk lattice describing the motion of the centroid of a 2-state dimer in (a). The unit cell is denoted by numbered circles. Lettered arrows indicate transitions to and from states. Note that transition rates connecting states can be different for transitions to the left or right (i.e., a = α, b = β). (c) Random-walk lattice for the 3-state dimer shown in (a). Note that the centroid location is the same for states 1 and 3; however, they are distinguished by different transition rates.

FIG. 2. Two-dimensional dimer migration: (a) spatial configurations; (b) random-walk lattice.

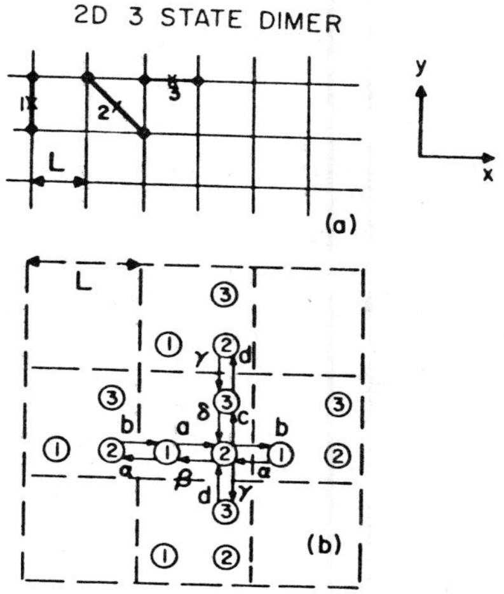

Mathematical Formalism of Random Walks with Internal States

We will now develop the mathematical formulation of random walks on these lattices with internal states and show how to relate the unknown transition rates in terms of the known FIM observables such as diffusion distances and equilibrium occupation probabilities of the different internal states. Our generating (Green's) function analysis will be based on the semi-Markiv continuous time random walk of Montroll and Weiss[17] and its generalizations.[18-23]. In addition this approach has been shown to be equivalent to a generalized master approach.[24]

In the course of our study many probablistic quantities will be introduced in order to calculate the values of FIM observables. Let us first introduce $R_{ij}(\vec{\ell},t|\vec{\ell}_o)$ which is the probability density for reaching site $\vec{\ell}$ in internal state i, $(\vec{\ell},i)$, exactly at time t given that $(\vec{\ell}_o,j)$ was attained at t=o. This quantity satisfies the following recursion relation[20], and identifies R as a Green's function propagator

$$R_{ij}(\vec{\ell},t)|\vec{\ell}_o) = \sum_{\vec{\ell}'}\sum_{m}\int_0^t \Psi_{im}(\vec{\ell}-\vec{\ell}',\tau)R_{mj}(\vec{\ell},t-\tau|\vec{\ell}_o)dt + \delta_{\vec{\ell},\vec{\ell}_o}\,\delta_{ij}\,\delta(t) \qquad (2.4)$$

where $\Psi_{im}(\vec{\ell},t)$ is the probability density that at time t a single jump occurs from (o,m) to $(\vec{\ell},i)$ given that the state (o,m) was attained at t=o. If there are s internal states then $\underline{\underline{R}}$ and $\underline{\underline{\Psi}}$ are sxs matrices and in matrix notation Eq. (2.4) becomes

$$\underline{\underline{R}}(\vec{\ell},u|\vec{\ell}_o) - \sum_{\vec{\ell}'}\underline{\underline{\Psi}}(\vec{\ell}-\vec{\ell}',u)\underline{\underline{R}}(\vec{\ell}',u|\vec{\ell}_o) = \delta_{\vec{\ell},\vec{\ell}_o}\,\underline{\underline{1}} \qquad (2.5)$$

where we have Laplace transformed over time (t→u). To proceed further one must examine the waiting time density matrix $\underline{\underline{\Psi}}$. We write

$$\Psi_{ij}(\vec{\ell},t) = p_{ij}(\vec{\ell})\psi_j(t) \qquad (2.6)$$

where $\psi_j(t)$ is the probability density that a transition occurs at time t from an internal state j which was attained at time t=o. To keep a matrix notation we treat ψ_j as the jj element of a diagonal matrix. The probability that this jump goes into an internal state i is given by $p_{ij}(\vec{\ell})$. Using Eq. (2.5) in Eq. (2.6) and Fourier transforming $(\vec{\ell}\to\vec{k})$ over all lattices sites $\vec{\ell}$ we arrive at in matrix notation

$$\underline{\underline{R}}(\vec{k},u|\vec{\ell}_o) = [\underline{\underline{1}} - \underline{\underline{p}}(\vec{k})\underline{\underline{\psi}}(u)]^{-1}\exp(+i\vec{k}\cdot\vec{\ell}_o) \qquad (2.7)$$

where the Fourier transform and its inverse are defined in Appendix A. In our notation, functions of u have always been Laplace transformed over time, and functions of $\vec{k}$ have been Fourier transformed over the lattice space.

The probability for being at $(\vec{\ell},i)$ at time t, $P_{ij}(\vec{\ell},t|\vec{\ell}_0)$ when the stochastic process began at $(\vec{\ell}_0,j)$ is related to $R_{ij}(\vec{\ell},t|\vec{\ell}_0)$ by

$$P_{ij}(\vec{\ell},t|\vec{\ell}_0) = \int_0^t R_{ij}(\vec{\ell},t-\tau|\vec{\ell}_0)\,[\int_\tau^\infty \psi_i(\tau')d\tau']d\tau$$
$$\equiv \int_0^t R_{ij}(\vec{\ell},t-\tau|\vec{\ell}_0)\,\Phi_i\,(\tau)d\tau \qquad (2.8)$$

where the factor Φ takes into account that the system may have reached $(\vec{\ell},i)$ at on earlier time $t-\tau$, and no transition out of $(\vec{\ell},i)$ occurs in the remaining time τ.

All the quantities one wishes to calculate are derivable from $P_{ij}(\vec{\ell},t|\vec{\ell}_0)$ which in turn only depends on $\Psi_{ij}(\vec{\ell},t)$ as can be seem from Eqs (2.7) and (2.8). We choose $\underline{\underline{\Psi}}$ to be normalizable, i.e,

$$\sum_i \sum_{\vec{\ell}} p_{ij}(\vec{\ell}) = 1, \quad \int_0^\infty \psi_i(\tau)d\tau = 1 \qquad (2.9)$$

We now show how to calculate positional moments and equilibrium occupation probabilities from a knowledge of $\underline{\underline{\Psi}}$.

The positional moments of the probability distribution are given by

$$<\ell_r^n(t)>= \sum_{\vec{\ell}} \sum_{i,j} (\vec{\ell}_r)^n\, P_{ij}(\vec{\ell},t|o)f_i \qquad (2.10)$$

where f_i is the probability that an internal state j is occupied initially. Since from Eq. (2.8)

$$\underline{\underline{P}}(\vec{k},t) = \mathscr{L}^{-1}\left(u^{-1}[\underline{\underline{1}}-\underline{\underline{\psi}}(u)] \sum_{\vec{\ell}} e^{i\vec{k}\cdot\vec{\ell}}\,\underline{\underline{R}}(\vec{\ell},u)\right) \qquad (2.11)$$

where $\mathscr{L}^{-1}$ is the inverse transform and r=x,y, or z, we see that Eq. (2.10) can be rewritten as

$$<\ell_r^n(t)> = \lim_{\vec{k}\to o} \mathscr{L}^{-1}\left\{(-i)^n \sum_{i,j} \partial^n R_{ij}(\vec{k},u|o)/\partial k_r^n \times u^{-1}[1-\psi_i(u)]\, f_i\right\} \qquad (2.12)$$

where $\underline{\underline{R}}(\vec{k},u|o)$ is given in terms of $\underline{\underline{\Psi}}$ in Eq. (2.7).

Note that the matrix $\underline{\underline{R}}$ is the inverse of the matrix $[\underline{\underline{1}} - \underline{\underline{p}}(\vec{k})\underline{\underline{\psi}}(u)]$ (see Eq. 2.7). Performing the matrix inversion we write $\underline{\underline{R}}$ as

$$\underline{\underline{R}}(\vec{k},u) = \underline{\underline{M}}(\vec{k},u)\Delta^{-1}(\vec{k},u) \tag{2.13}$$

where Δ is the determinant of the cofactors in $\underline{\underline{M}}$. All the physical quantities we will be interested are in the $t\to\infty$ limit and will only involve $\underline{\underline{R}}$ and its derivatives in the limit of both $\vec{k}$ and u going to zero. In this limit[26] the elements of $\underline{\underline{M}}$ will approach constants, while Δ will diverge as u^{-1}. Thus $\partial^2\underline{\underline{R}}/\partial k^2$ in Eq. (2.12) (which enters the calculation of the variance $\sigma^2(t)$ will diverge as u^{-2} causing the mean squared displacement of the random walker to grow linearly with time as was given in Eq. (2.1). This is the standard diffusion limit ($t\to\infty$) result. It can be shown[26] that in the diffusion limit Eq. (2.12) for $\sigma^2(t)$ reduces to

$$\begin{aligned}\sigma^2(t) &= \langle \ell^2(t)\rangle - \langle \ell(t)\rangle^2 \\ &= \lim_{u\to o}\ \lim_{k\to o}\ \frac{u^2}{\Delta^2}\ \frac{\partial^2\Delta}{\partial k^2}\ t\end{aligned} \tag{2.14}$$

Thus one only needs to calculate Δ and one does not have to perform the tedious matrix inversion to calculate M.

Another quantity of interest which can be obtained from FIM data on cluster motion by simply counting the number of micrographs in which the cluster is found in the various spatial configurations is the equilibrium probability of occupying an internal state j, P_j, eq. This quantity is defined as

$$P_j,\text{eq.} = \lim_{t\to\infty} \sum_i \sum_{\vec{\ell}} P_{ji}(\vec{\ell},t|o)f_i \ . \tag{2.15}$$

The RHS can be written in Laplace space as

$$P_j,\text{eq.} = \lim_{t\to\infty} \sum_i \sum_{\vec{\ell}} u\, P_{ji}(\vec{\ell},u|o)f_i \ .$$

Since $\lim_{\vec{k}\to o} P_{ji}(\vec{k},u|o) = \sum_{\vec{\ell}} P_{ji}(\vec{\ell},u|o)$, we can express P_j, eg. as

$$P_j,\ \text{eq.} = \lim_{u\to o}\ \lim_{\vec{k}\to o} \sum_i R_{ji}(\vec{k},u|o)[1-\psi_j(u)]f_i$$

Except in extreme cases where the mean time to make a transition between states is infinite, R_{ji} will not depend on the initial state i so

$$P_j, eq = \lim_{u \to 0} \lim_{\vec{k} \to 0} R_{ji}(\vec{k}, u|o)[1-\psi_j(u)] \qquad (2.16)$$

Dimer Diffusion in 1D

The set of transition rates {a} connecting the different internal states of a cluster are assumed to be in an activated form, $a = \nu_a \exp(-E_a/kT)$.

To find all the individual activation energies and frequency factors characterizing the diffusion we need to consider a random walk with internal states.

Consider first the two state dimer in Fig. 1. The effect of a bias can be incorporated by choosing $a \neq \alpha$, $b \neq \beta$. The total rate of leaving state 1 is $A=a+\alpha$, and the probability that the transition is to the right is a/A, and the probability that the transition is to the left is α/A. We choose $a=\alpha$, $b=\beta$. The waiting time density matrix is then given by

$$\underline{\underline{\Psi}}(\ell,t) = \begin{pmatrix} 0 & \tfrac{1}{2}B \exp(-Bt)\ (\delta_{\ell,0} + \delta_{\ell,1}) \\ \tfrac{1}{2}A \exp(-At)\ (\delta_{\ell,0} + \delta_{\ell,-1}) & 0 \end{pmatrix} \qquad (2.17)$$

where if the transition is within the unit cell ℓ does not change value, and it changes by $\pm L$ depending on whether the transition moves the dimer centroid to the unit cell on the right or the left. We will measure lengths in units of the unit cell size L. The matrix $\underline{\underline{R}}$ is given by, from Eq. (2.7),

$$\underline{\underline{R}}(k,u) = \underline{\underline{M}}/\Delta = \left[1 - \tfrac{1}{2}\frac{AB}{(A+u)\ (B+u)}\ (1+\cos\ kL)\right]^{-1} x \begin{pmatrix} 1 & \tfrac{1}{2}\frac{B}{B+u}\ (1+e^{ikL}) \\ \tfrac{1}{2}\frac{A}{A+u}\ (1+e^{-ikL}) & 1 \end{pmatrix} \qquad (2.18)$$

In the diffusion limit $(t \to \infty)$ using Eq. (2.14) we find

$$\sigma^2(t) = \tfrac{1}{2} L^2 \frac{AB}{A+B}\ t = L^2 \frac{ab}{a+b}\ t \qquad (2.19)$$

From the knowledge of $\underline{\underline{\Psi}}$, detailed balance relations, Eq. (2.16) can be calculated to give

$$\frac{P_{1,\,eq}}{P_{2,\,eq}} = \frac{b}{a} \equiv R_{12}(T) \qquad (2.20)$$

Eqs. (2.19) and (2.20) allow us to solve for the individual rates, i.e.,

$$\nu_a \exp(-E_a/kT) = a = L^{-2}\, t^{-1}\, \sigma^2(t)\, [1+R_{12}(T)] \qquad (2.21a)$$

$$\nu_b \exp(-E_b/kT) = b = L^{-2}\, t^{-1}\, \sigma^2(t)\, [1+R_{12}^{-1}(T)] \qquad (2.21b)$$

Experimentally $\sigma^2(t)$ and $P_{1,eq.}(T) = 1-P_{2,eq.}(T)$ are measurable and L^2t is known. Thus, a semilogarithmic plot of the RHS of Eq. (2.21a) vs. 1/kT would yield a straight line of slope $-E_a$, and ordinate intercept log ν_a. Similarly, the same plot for Eq. (2.21b) would yield E_b and ν_b. Note, that for this case merely plotting log $\sigma^2(t)$ vs. 1/kT (where $\sigma^2(t)$ is the variance of the dimer centroid position), does not allow the determination of E_a, ν_a, E_b and ν_b. Such a plot would in fact yield a "curved" Arrhenius line as can be seen by substituting activated forms for a and b in Eq. (2.19) for $\sigma^2(t)$. However, in a <u>limited temperature range</u> a plot of log $\sigma^2(t)$ vs. 1/kT may appear to be a straight line, but its slope and intercept will not characterize the individual transition rates of the dimer motion. We emphasize that <u>full use of all the FIM data</u> such as both $\sigma^2(t)$ and detailed balance relations $R_{12}(T)$ must be employed to calculate the individual dimer transition rates. Reed and Ehrlich[10] have also obtained Eq. (2.19) for the positional variance using Kolmogorov birth and death equations for the study of dimer motion. In a later paper[7] pertaining to the motion of rhenium dimers on W(211) Graham, Stolt, and Ehrlich find a = 17.5±.4kcal/mole, while b = 18.2±.3 kcal/mole, where as by just plotting log $\sigma^2(t)$ for the dimer centroid vs. 1/kT yields a "straight line" with slope 18.0±.3 kcal/mole. This demonstrates the spectroscopic kinetic information available from such studies. We note that our matrix continuous-time random walk approach can be applied to a system with any number of states per unit cell, even if different states of the cluster have the same center of mass. This point was an obstacle to extending the approach of Reed and Ehrlich.[10]

<u>Three State Dimer in 1D</u>

We now consider the last case in Fig. 1 where the dimer is allowed to extend into a third (non-dissociative) state which we term the extended state. Note that the center of mass of the dimer is the same for states 1 and 3 and thus states 1 and 3 coincide spatially but are distinguished by the different transition rates connecting them to state 2. Two new transition rates, c and d, are introduced to give us a total of four independent transition rates. We now inquire, "Is there sufficient FIM data to determine all four rates?" We can derive one equation for $\sigma^2(t)$, and two other independent detailed balance equations for $R_{12}(T)$ and $R_{23}(T)$. To obtain a fourth equation the diffusion experiment must be done under the influence of an electrical bias so the first spatial moment $<\ell(t)>$ is non-vanishing. Such experiments have been performed by Tsong and Walko.[25] Thus the amount of information one wishes to extract from the data will determine what type of experiment should be undertaken. The solution for each of the four individual transition rates in terms of the FIM observables ($\sigma^2(t)$, $<\ell(t)>$, $R_{12}(T)$, $R_{23}(T)$) has been given in reference 21, along with other examples, including diffusion on periodically defective lattices. The defects can be partially reflecting or absorbing or may just be described by different transition rates. The method used[21] to include periodic defects was to construct identical super unit cells so that each of them contained the same defect structure. For example, in 1D consider the motion of a single adatom on a lattice with every n-th site being defective (e.g. described by a different transition rate from a normal site). We would construct identical unit cells (see Fig. 3) and a corresponding nxn matrix $\underline{\underline{\Psi}}$ and use the internal state method of analysis described above. However, this can become difficult for large n and thus we present in the next section an alternative renormalization method.

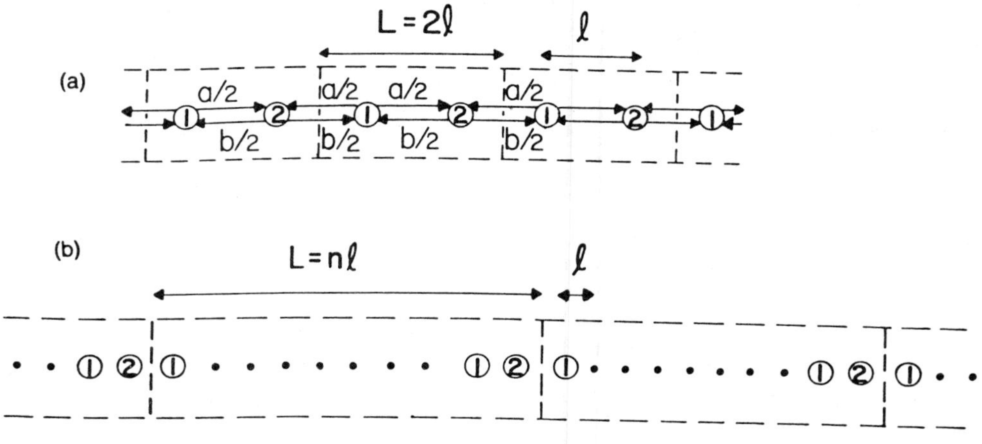

FIG. 3 (a) A 1D random-walk lattice with two alternating states is shown. The distance between sites 1 and 2 is ℓ, and the unit cell (dashed lines) has length 2ℓ. The total rates of leaving states 1 and 2 are a and b, respectively. The probabilities of going to the left or right from states 1 or 2 are ½. The lattice may represent the centroid positions of a dimer performing a 1D "channeled" motion on a crystalline surface [e.g., W. dimer on a W(211) suface, or a 1D ordered, alternating two-component system. The random motion is solved by two methods. First, the lattice is treated as having two states per unit cell and 2 x 2 Green's-function propagator is derived. An alternative approach treats state 2 as a periodically occurring defect, and a defect-renormalized scalar Green's-function propagator is derived. (b) The defects (state 2) are now spaced a distance $n\ell$ apart. In the text, it is shown that the defect-renormalized-propagator solution is much simpler than the $n \times n$ matrix internal-state approach.

III. MOTION ON DEFECTIVE SURFACES: RENORMALIZED GREEN'S FUNCTION METHOD

In this section we develop an alternative to the internal state supercell approach to random walks an defective lattices.[22,26,27] The fact that we only treat periodically placed defect does not render the model inapplicable to physical situations. It pertains to a.c. conductivity in ordered overlayer assemblies[28], transport in alloy or multicomponent systems, and ordered overlayer adsorption systems.[29] The results are shown to be similar to the average T-matrix approximation for diffusion in the presence of a low concentration of randomly placed defects.[30,31] The coherent potential approximation may also be used for randomly placed low defect concentrations.[32]

We return to the original Green's function equation (2.4) and treat the effects of defects as inhomogeneities rather than as extra internal states. Consider the case in 1D periodically placed defects spaced n lattice sites apart (see Fig. 3). The normal sites are characterized by

$$\Psi(\ell,t) = \psi_o(t)\ p(\ell) = A \exp(-At)\tfrac{1}{2}(\delta_{\ell,1} + \delta_{\ell,-1}) \qquad (3.1)$$

and defect sites differ only by changing the transition rate from A to D, i.e.,

$$\Psi_d(\ell,t) = \psi_d(t)p(\ell) = D \exp(-Dt)\tfrac{1}{2}(\delta_{\ell,1} + \delta_{\ell,1}) \qquad (3.2)$$

Only nearest neighbor jumping is allowed. Eq. (2.4) becomes (when Laplace transformed over time) for a walker beginning at ℓ=o,

$$\begin{aligned} R(\ell,u) - \sum_{\ell'} p(\ell-\ell\ )\ \psi_o(u)\ R(\ell',u) - \delta_{\ell,o} \\ = \sum_{j=-\infty}^{\infty} p(\ell-nj)\left\{\psi_d(u)-\psi_o(u)\right\}\ R(nj,u) \end{aligned} \qquad (3.3)$$

where the LHS represents propagation on an ideal lattice and the RHS is non-zero only for terms which involve transitions from a defect.

We wish to solve for R(k,u). The solution to the ideal lattice problem in Fourier space (k) and Laplace space (u) is given by

$$R_o(k,u) = [1-p(k)\psi_o(u)]^{-1}$$
$$= [1-\cos(k)\psi_o(u)]^{-1} \qquad (3.4)$$

Fourier transforming Eq. (3.3) we obtain

$$[R_o(k,u)]^{-1} R(k,u) = 1 + p(k) [\psi_d(u)-\psi_o(u)] \sum_{j=-\infty}^{\infty} e^{iknj} R(nj,u) \qquad (3.5)$$

Using the calculation of partial discrete Fourier transforms from Appendix A (Eqs. (A-3) and A-6)) we find the sum on the RHS of Eq. (4.5) can be written as

$$\frac{1}{n} \sum_{m=0}^{n-1} R(k + m \frac{2\pi}{n},u) \qquad (3.6)$$

where the first argument of R is calculated modulo 2π and $0 \leq k \leq 2\pi$. next write the defect term D(k,u) as

$$D(k,u) = p(k) [\psi_d(u) - \psi_o(u)] . \qquad (3.7)$$

For small u (long times) note that

$$D(k,u) \sim u \cos k [A^{-1} - D^{-1}] + 0(u^2) \qquad (3.8)$$

Using the notation

$$L(k,u) = [R_o(k,u)]^{-1} R(k,u) - 1 \qquad (3.9)$$

Eq (4.5) becomes

$$L(k,u) = n^{-1} D(k,u) \sum_{m=1}^{n-1} R(k + m \frac{2\pi}{n}, u) . \qquad (3.10)$$

Let us first consider k values in the first "Brillouin zone" denoted by K, such that $0 \leq K \leq \frac{2\pi}{n}$. Since

$$L(K + m \frac{2\pi}{n}, u) = n^{-1} D(K + m \frac{2\pi}{n},u) \sum_{j=0}^{n-1} R(K + \frac{2\pi}{n}, u) \qquad (3.11)$$

the L(k,u) function for all values of k, $(0 \leq k \leq 2\pi)$ and thus R(k,u) can be related to L(K,u) evaluated just in the first Brillouin zone by

$$L(K + m\frac{2\pi}{n},u) = D(K + m\frac{2\pi}{n},u)\ D^{-1}(K,u)\ L(K,u) \tag{3.12}$$

for $\quad m = 0,1,\ldots, n-1$

and where D(k,u) is easily evaluated for any k.

We now write $R(k = K + m\frac{2\pi}{n}, u)$ from Eq. (3.9) in terms of known quantities D(k,m) and $R_0(k,u)$ evaluated just in the first Brillouin zone using Eq. (3.12)

$$\begin{aligned} R(K + m\frac{2\pi}{n}, u) &= R_0(K + m\frac{2\pi}{n}, u)\ [L(K + m\frac{2\pi}{n},u) + 1] \\ &= R_0(K + m\frac{2\pi}{n}, u)\ [D(K + m\frac{2\pi}{n}, u)\ D^{-1}(K,u)L(K,u) + 1] \end{aligned} \tag{3.13}$$

Summing both sides of Eq. (3.13) from m = o,..., n-1 we find

$$\begin{aligned} \sum_{m=0}^{n-1} R(K + m\frac{2\pi}{n}, u) &= R(K,u) \\ &+ \sum_{m=1}^{n-1} R_0(K + m\frac{2\pi}{n}, u)[D(K + m\frac{2\pi}{n}, u)D^{-1}(K,u) \\ &\times \{[R_0(K,u)]^{-1} R(K,u)-1\} + 1\ ] \end{aligned} \tag{3.14}$$

where the m=o term has been separated from the sum on the RHS. Using Eq. (3.11) with m=o the LHS of Eq. (3.14) can also be expressed in terms of R(K,u). Thus one can solve for R(K,u) in the first Brillouin zone and relate $R(K + m\frac{2\pi}{n}, u)$ to R(K,u) using Eq. (3.12). The result is

$$R(K,u) = W^{-1}(K,u)\ V(K,u) \tag{3.15a}$$

where

$$\begin{aligned} W(K,u) &= [(R_0(K,u))^{-1} - n^{-1}\ D(K,u)] \\ &- n^{-1}\ (R_0(K,u))^{-1} \sum_{m=1}^{n-1} R_0(K + m\frac{2\pi}{n}, u)D(K + m\frac{2\pi}{n}, u) \end{aligned} \tag{3.15b}$$

and

$$V(K,u) = 1 + n^{-1} \sum_{m=1}^{n-1} \left(R_0(K + m\frac{2\pi}{n}, u) \left[D(K,u) - D(K + m\frac{2\pi}{n}, u)\right] \right) \tag{3.15c}$$

Note that $R_0(k,u) = [1-\psi(u) \cos k]^{-1}$ diverges in the $u\to 0$ and $k\to 0$ limits only as u^{-1}, and that $D(k,u)\to u$ for small u (see Eq. (3.8)). The propagator $R_0(K + m \frac{2\pi}{n}, u)$ for $m \neq 0$ does not diverge. Thus all the terms m Eq. (3.15) do not contribute in the diffusion limit. As $u\to 0$, $k\to 0$ the reduced expression becomes in the diffusion limit.

$$R(K,u) = \{[R_0(K,u)]^{-1} - \frac{1}{n} D(K,u)\}^{-1} \tag{3.16}$$

We see that the effect of the defects is to renormalize the Green's function propagator $R_0(k,u)$ to $R(K,u)$ as in Eq. (3.15). In the long time limit ($u\to 0$) the renormalization is a simple "self-energy" term

$$n^{-1} D(K,u) = n^{-1} \cos k\, [u(D-A)]/[(A + u)\ (D + u)] \quad ,$$

In a previous publication we omitted the sum in Eq. (A.5) and kept only the m=o term, which led us directly to Eq. (3.16) rather than Eq. (3.15). We thank Dr. Harvey Scher and Dr. C. H. Wu for pointing this out to us. In a preprint they have considered random walks on lattices with periodic defects along with a temperature dependent defect to defect hops instead of just nearest neighbor hops. This is shown to lead to different channels of transport as a function of temperature.

Using the renormalized $R(K,u)$ we can calculate the variance as

$$\begin{aligned}\sigma^2(t) &= \int_0^t \sum_\ell \ell^2\, R(\ell,t-\tau)\Phi(\ell,\tau)d\tau \\ &= \sum_\ell \ell^2 \int_0^t R(\ell,t-\tau)\ \{\ \Phi_A(\tau) + [\Phi_B(\tau) - \Phi_A(\tau)]\delta_{\ell,nj}\ d\tau \\ &= -\ell^2 \mathscr{L}^{-1}\ \{\partial^2 R(K,u)/\partial K^2\big|_{K=0}\ [\Phi_A(u) + n^{-1}(\Phi_B(u) - \Phi_A(u))]\} \\ &= \frac{n\ AB}{A + (n-1)B}\ \ell^2 t \end{aligned} \tag{3.17}$$

where $\Phi_{A,B}(u) = u^{-1}[1 - \psi_{A,B}(u)]$ has been used (see Eq. (2.8). For $n\to\infty$ or $A = B$, (no defects), we obtain the perfect lattice result, Eq. (2.1), $\sigma_0^2(t) = A\ell^2 t$. If the transition rate A is known then one can solve for the defective transition rate B to find

$$\frac{A\sigma^2(t)}{An\ell^2 t - (n-1)\sigma^2(t)} = \nu_B \exp(-E_B/kT) = B \tag{3.18}$$

Thus a semilogarithmic plot of the experimentally known quantities on the LHS vs. 1/kT will yield the activation energy and frequency factor characterizing the rate B. Note again that $\sigma^2(t)$ itself is not of the Arrhenius forms, and a plot of log σ^2 vs. 1/kT will not allow a determination of A and B.

By treating the defects as inhomogeneities in our Green's function Eq. (3.3) we were able to calculate $R(K,u)$ as a scalar quantity.

In contrast, with the supercell method of Section II, we would need to use nxn matrices and their inverses. Unless special symmetries exist the calculations with nxn matrices can become prohibitive, thus showing the advantage in those cases of renormalized propagator method.

It is interesting to observe that to first order in the concentration of defects $C=n^{-1}$ the expansion of Eq. (4.17) yields

$$\sigma^2(t) = \sigma_o^2(t)\ \{1 - C[(\nu_B/\nu_A)e^{-(E_A-E_B)/kT} - 1]\} + O(C^2) \tag{3.19}$$

Which is similar to the effective diffusion constant in a system containing randomly placed defects as given by the average -T-matrix approximation.[30,31]

The renormalization procedure used for calculating the variance for the example in Eq. (3.17) is actually more general and applies to cases in higher dimensions where the structure factor $p(\vec{k})$ is different at defective sites and when the diffusant cartains internal states (configuration, energetic, etc.) and a set of defects can be periodically repeated. When internal states are present the renormalization equations become matrix equations. The solution in Eq. (3.15) is already in the correct order for the proper matrix multiplication. Only the definition of the "self-energy" needs to be generalized to

$$\underline{\underline{D}}(\vec{k},u) = \sum_d \left[\underline{\underline{p}}_d(\vec{k})\underline{\underline{\psi}}_d(u) - \underline{\underline{p}}_o(\vec{k})\underline{\underline{\psi}}_o(u)\right] \tag{3.20}$$

where the sum is over the set of periodically repeating defects. When $p_d(\vec{k}) \neq p_o(\vec{k})$ care must be exercized arriving at asymptotic results since now $D(\vec{k}=o,\ u\to o) \sim$ constant while if $p_d(\vec{k})=p_o(\vec{k})$ then $D(\vec{k}=o,\ u\to o)\sim u$.

Before we proceed to discuss certain aspects of diffusion controlled reactions we comment on an interesting result concerning the effect of defects (inhomogeneities) on the lattice structure dependence of the diffusion constant. In reference (26) we have studied particle and cluster motion on defective surfaces. In one example, we considered square (sq.) hexagonal (hex), and triangular (tri) lattices with equal bond lengths and equal defect concentrations. In addition we let the effect of the defects to extend to nearest-neighbor sites. Normal, defect, and defect nearest neighbor sites were ascribed transition rates A,D, and B respectively. It was found for single particle motion that the variances in position (and thus the diffusion constants) order as $\sigma^2_{tri}(t) \leq \sigma^2_{sq}(t) \leq \sigma^2_{hex}(t)$ when $B \leq A$, i.e, when the migration of the particle is slowed in the vicinity of the defect. It should be noted that this structural (lattice connectivity) dependence occurs only for diffusion on defective lattices in which the range of influence of the defects extendes to at least nearest neighbors.

Studies of particle diffusion on surfaces containing defects are not abundant. Recently an FIM investigation of diffusion on a 3% R_e substituted W(100) surface was performed[10], since defects are of importance in catalytic reactions on surfaces, either as promoters or inhibiturs of migration, and in eight of the above results it is suggested that detailed studies of diffusion on such systems be carried out.

IV. DIFFUSION CONTROLLED REACTIONS ON SURFACES

In this section we examine closely the meaning of "reaction rate" as it appears in probabilistic master equation descriptions of chemical kinetics. We show that the reaction rate is a conditional probability density, and describe its evaluation for certain model systems using the formalisms outlined in the previous sections. An explicit example of a diffusion controlled bimolecular reaction is discussed in Section V.

Unimolecular Reactions: What is a Reaction Rate?

The evolution of many physical systems can be viewed as unimolecular, bimolecular, or pseudo-unimolecular (if from two reactive species one species is vastly more abundant than the other) reactions. Unimolecular decay reactions involve the irreversible loss of independent reactants. McQuarrie[34] has reviewed the master equation approach to unimolecular and bimolecular reactions. This stochastic approach allows the calculation of fluctuations, which deterministic equations do not. McQuarrie's solutions are in terms of rate constants, but he does not discuss how to calculate these rate constants from first principles. In this section, we show that the rate constants are conditional first passage (coincidence) probability densities and we relate them to the probability that the lifetime of a reactant is greater than a time t.

The master equation governing unimolecular decay is

$$dP(N,t)/dt = K(t)[P(N+1,t) - NP(N,t)] \,, \tag{4.1}$$

where N is a random variable representing the number of reactants which have not yet reacted (decayed). The solution of Eq. (4.1) for the mean is

$$\langle N(t) \rangle = N_o \exp[-\int_0^t K(\tau)d\tau] \tag{4.2a}$$

where N_o is the initial number of reactants, and N(t) satisfies the equation

$$d\langle N(t)\rangle/dt = -K(t)\langle N(t)\rangle \,. \tag{4.2b}$$

The major task involved with Eq. (4.1) is to calculate the reaction rate K(t) which contains all the physics of the reaction under study.

The quantity K(t) is the conditional probability of a reaction occurring, of a particular reactant, in the interval (t,t + dt), given that no reaction occurred in the interval (0,t). We assume the stochastic process began at t=0. The quantity K(t) dt says that the stochastic process began at t=0 and that at time t one has the information that the particular reactant of interest has not yet decayed, and then asks with what probability will the decay occur at time t+dt. Thus, K(t) is a prediction at time t of what will happen at time t+dt, given the information that the reactant did not decay in (o,t). The conditional first passage density K(t) can be related to the unconditional first passage density F(t). The quantity F(t)dt says that the stochastic process began at time t=0 and then asks for the probability that the decay takes place in the interval (t,t + dt). So F(t) is a prediction at time t=0 of what will happen at time t+dt.

To calculate K(t), we first define P(L>t) to be the probability that the lifetime of a particle is greater than t, i.e., that the decay takes place in the interval (t,∞):

$$P(L>t) = \int_t^\infty F(\tau)\, d\tau \quad . \tag{4.3}$$

Now, P(L>t + dt) can be written terms of a conditional probability

$$P(L>t + dt) = P[L>t + dt \,|\, \text{no decay in } (o,t)]P(L>t), \tag{4.4}$$

where the last factor on the RHS is the probability that no decay has occurred in the interval (o,t). Dividing both sides of Eq. (4.4) by P(L>t) and expanding P(L>t + dt) in a Taylor series about t, we find

$$P[L>t + dt \,|\, \text{no decay in } (o,t)]$$

$$= 1 + dt(d/dt) \log P(L>t) + O(dt)^2 \tag{4.5}$$

Now,

$$K(t)dt \equiv P[L\leq t + dt \,|\, \text{no decay in } (o,t)]$$

$$= 1 - P[L>t + dt \,|\, \text{no decay in } (o,t)] \tag{4.6a}$$

or

$$K(t) = -(d/dt) \log P(L>t)$$

$$= F(t)/ \int_t^\infty F(\tau)d\tau. \tag{4.6b}$$

Using Eq. (4.6b) in (4.2a) yield the intuitive result

$$< N(t) > /N_o = \exp[- \int_o^t K(\tau)d\tau] = \int_t^\infty F(\tau)d\tau \tag{4.7}$$

and shows the connection between the unconditional and conditional first passage distributions.

We see that if $P(L>t) \sim \exp(-\lambda t)$ as $t \to \infty$, then $K(t) \to \lambda$, a constant. This is the reason why constant reaction rates can often be used in Eq. (4.1). Note that

$$K(t) = - (d/dt)\log(\langle N(t) \rangle / N_o), \tag{4.8}$$

which is similar in structure to Eq. (4.6b). In Eq. (4.8), one uses experimental results to find K(t), while in Eq. (4.6), one uses assigned microscopic parameters to predict K(t). Together, Eqs. (4.6) and 4.8) help allow for a determination of the microscopic parameters governing K(t).

Calculation of a Diffusion Controlled Unimolecular Reaction Rate

Consider now the situation where Eq. (4.1) represents N_o reactants per active site at t=o, and where there is one active site per V lattice sites. A reaction occurs at the instant when a reactant reaches an active site on our periodic lattice. We divide our system into identical unit cells each initially with N_o reactants and a active site at the origin of a unit cell with Vsites. Periodic boundary conditions are used in each unit cell, so the study of one cell will yield the kinetics of the concentration of the reactants $C(t) = N(t)/V$. We could endow our reactants with internal states (energetic, configurational, spin, etc.) and the lattice with various types of defects (promoters or inhibitors of diffusion) using our matrix renormalized propagator as discussed in Section III. However we will choose for simplicity a single particle on a perfect 2D square lattice governed by nearest neighbor jumps and $\psi(t) = A\exp(-At)$. This model is then similar to Montroll's[37] and Lindenberg[32], Heminger, and Pearlstein's studies of exciton trapping. The equations which are now presented to calculate K(t) (Eq. 4.6) should not be confused with the unimolecular reaction Eq. (4.1).

Given $\Psi(\vec{\ell},t) = \psi(t)p(\vec{\ell})$ we now calculate the unconditional first passage time F(t) for a reactant initial at a non-active site. The probability density $f(o,t|\vec{\ell}_o)$ for reaching the active site $\vec{\ell}$=o at time t for the reactant starting at $\vec{\ell}_o$, enters the following equation

$$\sum_{\vec{\ell}_o \neq o} R(o,t|\vec{\ell}_o)g(\vec{\ell}_o) = \sum_{\ell_o \neq o} \int_o^t f(o,t-\tau|\vec{\ell}_o)g(\vec{\ell}_o)R(o,\tau|o)d\tau \tag{4.9}$$

where we have averaged over the initial probability of occupying $\vec{\ell}_o$, and the RHS takes into account that for the reactant to reach the origen at time t it could have reached there at an earlier time $t - \tau$ and returned to the origin (any number of times) in the remaining time

τ. Here we are treating the origin as a normal site and calculating the first passage into $\vec{\ell}=0$. The propagator R was first discussed in Eq. (2.4). Eq (4.9) can be solved by Fourier and Laplace transforms to yield

$$F(t) = \frac{1}{V-1} \sum_{\vec{\ell}_0 \neq 0} f(0,t|\vec{\ell}_0) = \frac{1}{V-1} \mathcal{L}^{-1}\left[\frac{R(\vec{k}=0,u|0)}{R(\vec{\ell}=0,u|0} - 1\right] \tag{4.10}$$

The denominator on the RHS is the random walk Green's function for return to the origin and is given by, in 2D,

$$R(\vec{\ell}=0,u|0) = \frac{1}{V} \sum_{k_1=1}^{V_1} \sum_{k_2=1}^{V_2} [1 - p(\vec{k})\psi(u)]^{-1} \tag{4.11}$$

where $V=V_1V_2$ and $k_i=2\pi h_i/V_i$, $i=1,2$. While the Green's function has a simple form in Fourier and Laplace space (see Eq. (2.9) it is not known, in general, in closed form in real space other than as in Eq. (4.11). Note however, that $R(\vec{\ell}=0,u|0)$ diverges for small u.[17] This is because $\lim_{u\to 0} R^{-1}(\vec{\ell}=0,u|0)$ represents the probability that a random walker never returns to the origin after any number of steps, and this is zero for a finite lattice. Montroll[37] has analyzed the behavior of Eq. (4.11) in the small u limit. Using Montroll's results we obtain in 2D.

$$F(u) \sim 1 - S\,V[1 - \psi(u)] + O[1 - \psi(u)]^{3/2} \tag{4.12}$$

where $S=S_1 \log V + S_2 + S_3/V + S_4/V^2 + \ldots..$ and the values of $S_1,\ldots,S_4$ are of the order of 0.1 to 1.0 and differ for hexagon, square, and triangular lattices. In our example $1-\psi(u) \sim u\langle t\rangle$ where $\langle t\rangle = \int_0^\infty t\psi(t)dt$ is the mean time between jumps. In the long time limit F(t) and $\psi(t)$ will have the same form, but different parameters such that

$$\int_0^\infty t\,F(t)dt = S\,V \int_0^\infty t\,\psi(t)dt \tag{4.13}$$

so approximately SV steps are taken before the reaction occurs. Thus we arrive at the equation for the concentration of reactants C(t) in the long time limit

$$\frac{C(t)}{C(0)} = \exp\left(-\frac{A}{SV}\,t\right) \tag{4.14}$$

For shorter times the reaction rate (K(t) = A/SV) will be time dependent. The structure of the substrate enters the rate through the quantity S. Using Montroll's values for $S_1,S_2,\ldots$,it can be shown that the rates order as $K_{tri} > K_{sq} > K_{hex}$, which is the same ordering as the coordination numbers. The structural effect is more pronounced for large V (low active site concentrations). For V=10, K_{sq} and K_{tri} are 23% and 29% larger, respectively, than K_{hex}.

Diffusion Controlled Bimolecular Reactions

Bimolecular reactions on a surface can be treated in a similar fashion as unimolecular reactions, but they are inherently more difficult.[36] First, the bimolecular master equations (Eqs. 4.15) and (4.16) are more complicated than the simple unimolecular one. Secondly, the calculation of K(t) involves conditional first pair coincidences rather than first passage times. Thirdly, a single reactant upon reaching a reactive site will leave if the second reactant (which is necessary for the reaction) does not arrive in sufficient time. It is likely that the transition rate for a single particle to leave the reactive site is different than for it to leave a non-reactive site. In this case single particles will migrate on a defective lattice in the bimolecular reaction due to the nature of the reactive site, but not in the unimolecular reaction where they immediately react at the reactive (defective) site.

The bimolecular reaction master equation for identical particles $A+A \rightarrow 2A$ is[34]

$$\frac{dP(N,t)}{dt} = K(t) \left[\binom{N+2}{2} P(N+2\ ,\ t) - \binom{N}{2} P(N,t) \right] \qquad (4.15)$$

where N is the number of reactants which have not decayed at time t. We assume N is initially even, and changed by two after a reaction.

For two different species $A+B \rightarrow C$ the bimolecular master equation is given by[34]

$$\frac{dP(N,t)}{dt} = K(t) \left[(N+1)P(Z_o + N + 1) - N(Z_o + N)\ P(N,t) \right] \qquad (4.16)$$

where N is the number of type A reactants, and $M = Z_o + N$ is the number of B reactants. Recombination reactions such as annealing are of this type.

Eqs. (4.15) and (4.16) can be solved by generating function techniques[34] to yield respectively,

$$\langle N(t) \rangle = \sum_{N=2}^{N_o} A_N \exp\left[-2^{-1} N(N-1) \int_0^t K(\tau)d\tau\right] \qquad (4.17)$$

where

$$A_N = \frac{1-2^N}{2^N} \left(\frac{\Gamma(N_o+1)\Gamma\left(\frac{N_o - N - 1}{2}\right)}{\Gamma(N_o - N + 1)\Gamma\left(\frac{N_o + N + 1}{2}\right)} \right)$$

and N is even, and

$$\langle N(t)\rangle = \sum_{N=0}^{N_o} \frac{(2N+Z_o)\Gamma(N_o+1)\Gamma(N_o+Z_o+1)}{\Gamma(N_o-N+1)\Gamma(N_o+Z_o+N+1} \exp\left(-N(N+Z_o)\int_0^t K(\tau)d\tau\right) \quad (4.18)$$

Here K(t)dt is the conditional probability distribution that a reaction takes place between two particular reactants in the interval (t,t+dt) given that the reaction did not occur before in (o,t). As before we will be interested in diffusion controlled reactions at reactive sites which we take to be the origins of the defect superlattice cells of volume V with periodic boundary conditions. In analogy to Eq. (4.9) to find the probability density for a first coincidence at the reactive site $\vec{\ell}=o$, at time t, $f(o,t|\vec{\ell}_1,\vec{\ell}_2)$ of two reactants which were at $\vec{\ell}_1$ and $\vec{\ell}_2$ at t=o, we first need to calculate the probability density of any coincidence $C(o,t|\vec{\ell}_1,\vec{\ell}_2)$, of two particles which initially were situated at sites $\vec{\ell}_1$, and $\vec{\ell}_2$ and coincide at the origin of their unit cell at time t.

$$\begin{aligned} C(t) &= \sum_{\vec{\ell}_2}\sum_{\vec{\ell}_1}{}' C(o,t|\vec{\ell}_1,\vec{\ell}_2)g(\vec{\ell}_1)g(\vec{\ell}_2) \\ &= \sum_{\vec{\ell}_2}\sum_{\vec{\ell}_1}\int_0^t f(o,t-\tau|\vec{\ell}_1,\vec{\ell}_2)g(\vec{\ell}_1)g(\vec{\ell}_2)C(o,\tau|o\ o)d\tau \end{aligned} \quad (4.19)$$

where we have averaged over all initial positions of the two particular reactants, except for both being at the origin initially. The above equation can be solved for the first coincidence density f to give

$$\begin{aligned} F(t) &\equiv \sum_{\vec{\ell}_1}\sum_{\vec{\ell}_2}{}' f(o,t|\vec{\ell}_1,\vec{\ell}_2)g(\vec{\ell}_1)g(\vec{\ell}_2) \\ &= \mathcal{L}^{-1}[C(u)/C(o,u|o\ o)] \end{aligned} \quad (4.20)$$

To proceed further we need to specify the allowable states of the reactants and the probability distributions governing transitions. We will treat the simplest case where all reactants are of the same type and have the waiting time density $\psi(t)=A\exp(-At)$ for hopping, and only nearest nieghbor jumps occur. We also consider that the release rate from the reactive site is unchanged from that of the normal site when only one reactant is there at the origin. A reaction will occur at time t if one reactant already resides on the reactive site (having arived there earlier) and a second reactant arrives there exactly at time t, or vice versa. Remembering that R is the probability density for just arriving at a site, and $\underline{P}$ is the probability for being at a site, we have

$$C(t) = \frac{2}{V(V-1)}\sum_{\vec{\ell}_1}\sum_{\vec{\ell}_2}{}' P(o,t)|\vec{\ell}_1)R(o,t|\vec{\ell}_2) \quad (4.21)$$

Thus for large V we have

$$F(t) = \mathscr{L}^{-1}\left(\frac{P(\vec{k}=o,u|o)R(\vec{k}=o,u|o)}{P(\vec{\ell}=o,u|o)R(\vec{\ell}=o,u|o)}\right) \qquad (4.22)$$

Again using Montroll's[37] asymptotic ($u\to o$) results for the terms in the denominator we find for the rate constant in Eq. (4.17)

$$\lim_{t\to\infty} K(t) = A/2S \qquad (4.23)$$

where S is given in Eq. (4.12). Note that this rate is one half of the unimolecular reaction rate, but both rates enter completely different equations (Eq. (4.17) and Eq. (4.18)) for the mean number of reactants). For reaction between two species whose transition rates are characterized by rates A and B we find for large t the K to be used in Eq. (5.22) is

$$K(t) = (A+B)/(S[2 + A/B + B/A]) \qquad (4.24)$$

V. CONDITIONS FOR A RATE MAXIMIZING TEMPERATURE IN LANGMUIR-HINSHELWOOD REACTIONS

In this section we illustrate the manner in which multistate diffusion mechenisms might provid an interpretetion of certain experiments in which the rate of a catalytic reaction can be maximized as a function of temperature.

A number of heterogeneous catalytic reactions exhibit a maximum rate of reaction as a function of temperature. This may be due to the reaction mechanism or to changes in the catalytic structure as a function of temperature or a combination of both. We will consider the experiment of Moffat and Clark[38] who in 1969 found a rate temperature maximum for the olefin disproportion ation reaction of propylene into ethylene on the cobaltmolybdate-alumina catalyst C_o-M_o-$A\ell_2O_3$. They also found that the reaction obeys the bimolecular Langmuir-Hinshelwood law. This means that two adsorbed propylenes react, perhaps at an active site. Other reaction mechanisms are possible. For example, Begley and Wilson[39] found an Eley-Rideal mechanism (an adsorbed species reacting with a gas phase species) for olefin disproportionation on a tungsten-selica catalyst.

The Langmuir-Hinshelwood law is derived as follows. Let Θ be the fractional surface coverage by the reactants. Then

$$\frac{d\Theta}{dt} = Pk_1(1-\Theta) - k_2\Theta \ , \qquad (5.1)$$

where k, is the adsorption rate and k_2 is the desorption rate and P is the pressure. At equilibrium

$$\Theta = KP\ /(1 + KP), \tag{5.2}$$

where $K = k_1/k_2$. The bimolecular rate of reaction is proportional to Θ^2 and to the diffusion controlled rate of reaction on the surface denoted by k. Thus the Langmuir-Hinshelwood rate of reaction r is

$$r = k\left[\frac{KP}{1+KP}\right]^2 . \tag{5.3}$$

Note that Eq. (5.3) can be rewritten as

$$\frac{1}{\sqrt{r}} = \frac{1}{\sqrt{R}} \quad (1 + \frac{1}{KP}) \tag{5.4}$$

Thus a plot of $r^{-\frac{1}{2}}$ vs. P^{-1} should yield a straight line, as is seen in the propylene disproportionation reaction studied by Moffat and Clark[38].

Usually, Arrhenius forms are assumed for both k and K[38-40], i.e.

$$k = A\ \exp(-E/RT)$$

$$K = \exp(\Delta S/R)\ \exp(-\Delta H/RT) \tag{5.5}$$

where E is the activation energy for diffusion, ΔS is the differential entrophy of adsorption, ΔH is the heat of adsorption, and A is the frequency factor. Since k and K can be determined from the $r^{-\frac{1}{2}}$ vs. P^{-1} plot, one can further calculate the activation energy E by

$$E = - \frac{\partial\ \ell n\ k}{\partial(1/RT)} \tag{5.6}$$

and the heat of absorption by

$$-\ \Delta H = \frac{\partial\ \ell n\ K}{\partial(1/RT)}$$

Moffat and Clark[38] found the values E= 8.2 K cal/mole and $|\Delta H|$=2.8 K cal/mole as well as a rate maximizing temperature T_m ∿420°F. If Eq. (5.5) is substituted into the rate equation (5.3) and the derivative with respect to temperature of the resulting expression is set equal to zero, a condition for the appearance of a rate maximum as a function of temperature is obtained. When the adsorption step of the reaction is exothermic, $\Delta H<0$, and the rate maximum condition is[38]

$$|\Delta H| > E(\beta + 1)\ /2 \tag{5.8}$$

where $\beta = P\ \exp(\Delta S/R)$ and is usually much smaller than unity. This condition Eq. (5.8) is not satisfied by the measurements of Moffat and

Clark and yet a rate maximum still occurs.

To explain this apparent paradox we[41] propose mechanisms where the Langmuir-Hinshelwood form of Eq. (5.3) is preserved, but the rate maximum condition Eq. (5.8) is modified to a form which can be consistent with the experimental results. Common to the mechanisms which we propose is the property that k, which is related to the diffusive motion on the surface, is characterized by more than one rate constant. For example the motion of the nine atoms comprising the propylene $CH_3CH{=}CH_2$ may involve transitions between different configurations of this cluster each with its own transition rate. Out of all the possible transitions those with the highest activation energies will be the rate limiting steps. Considering the two most important rate limiting transitions with rate A and B, the diffusion constant in Eq. (5.3) will take the form, as in Eq. (2.19)

$$k = \frac{AB}{A+B} \tag{5.9}$$

We write both A and B in the Arrhenius form

$$A = \nu_A \exp(-E_A/RT)$$

$$B = \nu_B \exp(-E_B/RT) \tag{5.10}$$

so k will not be of the Arrhenius form. If however a straight line results from an experimental plot of log k vs. $(RT)^{-1}$ this does not necessarily imply that $k=\nu \exp(-E/RT)$ and that k cannot be written as in Eqs. (5.9) and (5.10). In practice, a plot of log k from Eqs. (5.9) and (2.10) vs. $(RT)^{-1}$ will yield a straight line of slope $-E_A$, in an appropriate temperature range, if

$$(1) \quad E_A \approx E_B \tag{5.11a}$$

or

$$(2) \quad E_A \gg E_B + RT\,\ln(\nu_A/\nu_B) \tag{5.11b}$$

Using the k in Eqs. (5.9) and (5.10) the following condition is found for the occurance of a rate maximizing temperature T_m in Eq. (5.3)

$$\frac{1 + P\exp[\Delta S/R - |\Delta H|/RT_m]}{2|\Delta H|} = \frac{A + B}{E_B A + E_A B} \tag{5.12}$$

Let us denote the RHS by Γ. A linear analysis shows that a small decrease in pressure δP will decrease T_m by the amount δT_m, where

$$\delta T_m = \delta P \exp[\Delta S/R + |\Delta H|/RT_m]RT_m^{\,2}/|\Delta H| \times$$

$$\left\{2|\Delta H|\Gamma - 1 + 2AB[\Gamma(E_A - E_B)/(A + B)]^2\right\}^{-1} \tag{5.13}$$

This is in accord with the experimental result that a decrease in pressure lowers the maximizing temperature.[38] Neglecting P exp($\Delta S/R$) compared to unity we arrive at the following inequality for the existence of a rate maximum

$$\nu_A(2|\Delta H| - E_B) > \nu_B(E_A - 2|\Delta H|) \tag{5.14}$$

where we have assumed, without loss of generality, that $E_A > E_B$. In contrast to Eq. (5.8) this inequality can be satisfied by the measurements of Moffat and Clark[38], E_A = 8.2 Kcal/mole and $|\Delta H|$ = 2.8 k cal/mole, if

$$0 < E_B < 5.6 - 2.6 \frac{\nu_B}{\nu_A} \text{ K cal/mole} \tag{5.15}$$

For k to be adequately described by an Arrhenius plot, Eq. (5.11b) must also be satisfied. A similar inequality can be derived by assuming the surface contains two types of sites with two different release rates[41].

Not all reactions have reversible rate maxima. For example, Maatman et. al[40]. studied the cracking of isopropylbenzene [$C_6H_5CH(CH_3)_2$] into benzene [C_6H_6] and propylene [C_3H_6] on a silica-alumina catalyst. They derived the condition that the heat of desorption from active sites must be greater than the activation energy for diffusion for a rate maxima to exist. This condition is not met and no rate maxima is seen.

It would be interesting to find the conditions for the existence of a rate maximizing temperature for the production of various hydrocarbons in a Fischer-Tropsch reaction where progressively heavier hydrocarbons are produced from a catalyzed gaseous CO and H_2 mixture.

$$nCO + (2n + 1)\, H_2 \rightarrow C_nH_{2n+2} + nH_2O \tag{5.16}$$

APPENDIX A

PARTIAL DISCRETE FOURIER TRANSFORMS

We define the discrete Fourier transform (Eq. A.1) and its inverse (Eq. A.2) on an infinite lattice of dimension d, as

$$f(\vec{k}) = \sum_{\ell_1=-\infty}^{\infty} \cdots \sum_{\ell_d=-\infty}^{\infty} f(\vec{\ell})\, e^{i\vec{k}\cdot\vec{\ell}} , \tag{A.1}$$

$$f(\vec{\ell}) = (2\pi)^{-d} \int_0^{2\pi} \cdots \int_0^{2\pi} f(\vec{k})\, e^{-i\vec{k}\cdot\vec{\ell}}\, d\vec{k} . \tag{A.2}$$

Consider first a 1D lattice and define the partial Fourier transform

$$f_{\gamma,n}(k) = \sum_{\ell=-\infty}^{\infty} f(\gamma\ell + n)\, e^{ik(\gamma\ell + n)} . \tag{A.3}$$

Using Eq. (A.2) we obtain

$$f_{\gamma,n}(\vec{k}) = \sum_{\ell=-\infty}^{\infty} (2\pi)^{-1} \int_0^{2\pi} f(k')\, e^{i(k-k')(\gamma\ell + n)}dk$$

$$= \frac{1}{\gamma} \int_0^{2\pi} f(k')\, \delta(k-k' + \frac{2\pi m}{\gamma})\, dk' \tag{A.4}$$

$$= \frac{1}{\gamma} \sum_m f(k + \frac{2\pi m}{\gamma}) \tag{A.5}$$

where the summation is over all integer values of m (positive and negative) such that $k + \frac{2\pi m}{\gamma} \varepsilon (0,2\pi)$. Eqs. (A.3) and (A.5) can be easily generalized to higher dimensions, say d, to yield

$$(\gamma_1 \ldots \gamma_d)^{-1} \sum_{m_1} \ldots \sum_{m_d} f(k_1 + \frac{2\pi m}{\gamma_1}, \ldots, k_d + \frac{2\pi m_d}{\gamma_d}) \tag{A.6}$$

For a finite lattice of d dimensions (the number of lattice points being $N_1 x \ldots x\ N_d$ the discrete Fourier transforms are defined as

$$f(\vec{k}) = \sum_{\ell_1=1}^{N_1} \cdots \sum_{\ell_d=1}^{N_d} f(\vec{\ell})\, e^{i\vec{k}\cdot\vec{\ell}} \tag{A.7}$$

and

$$f(\vec{\ell}) = (N_1 \ldots N_d)^{-1} \sum_{S_1=1}^{N_1} \cdots \sum_{S_d=1}^{N_d} f(\vec{k})\, e^{-i\vec{k}\cdot\vec{\ell}} \tag{A.8}$$

where $k_i = 2\pi S_i/N_i$, $1 = 1,\ldots,d$; and $s_i = 1, \ldots, N_i$.

For the partial discrete transform

$$f_\gamma(\vec{k}) = \sum_{\ell_1=1}^{N_1} \cdots \sum_{\ell_d=1}^{N_d} f(\gamma_1\ell_1,\ldots,\gamma_d\ell_d)\, e^{i\vec{k}\cdot(\gamma_1\ell_1,\ldots,\gamma_d\ell_d)} \quad (A.9)$$

we obtain the same result as in Eq. (A.6), except k can only take on the $N_1 x \ldots x\ N_d$ values in Eq. (A.8)

REFERENCES

1. E. W. Müller and T. T. Tsong, Field Ion Microscopy (Elsevier, N.Y., 1969).
2. D. W. Bassett, in Surface and Defect Properties of solids (The Chemical Society, London, 1973) Vol. 2, P. 34.
3. G. Ehrlich, Crit. Rev. Solid. State Sci. 4, 205 (1974).
4. W. R. Graham and G. Ehrlich, Thin Solid Films, 25, 85 (1975).
5. G. E. Rhead, Surf. Sci. 47, 207 (1975).
6. G. Ehrlich and F. G. Hudda, J. Chem. Phys. 44, 1039 (1966).
7. K. Stolt, W. R. Graham, and G. Ehrlich, J. Chem. Phys. 65, 3206 (1976).
8. T. T. Tsong, Phys. Rev. 136, 417 (1972).
9. W. R. Graham and G. Ehrlich, Phys. Rev. Lett. 31, 1407 (1973); 32, 1309 (1974); J. Phys. F 4, L212 (1974).
10. D. A. Reed and G. Erhlich, J. Chem. Phys. 64, 4616 (1976).
11. S. Nishigaki and S. Nakamura, Jap. J. Appl. Phys. 14, 769 (1975).
12. T. T. Tsong, P. Cowan, and G. Kellogg, Thin Solid Films. 25, 97 (1975).
13. D. W. Bassett, J. Phys. C., 9, 2491 (1976).
14. D. W. Bassett and M. J. Parsley, Nature 221, 1046 (1969).
15. D. W. Bassett, Surf. Sci. 21, 181 (1970).
16. M. Audiffren, P. Traimond, J. Bardon, and M. Drechsler, Surf. Sci. 75, 751 (1978).
17. E. W. Montroll and G. H. Weiss, J. Math. Phys. 6, 67 (1965).
18. H. Scher and M. Lax, Phys. Rev. B7, 4491, 4502 (1973).
19. E. W. Montroll and H. Scher, J. Stat. Phys. 9, 101 (1973). M. F. Shlesinger, J. Stat. Phys. 10, 421 (1974).
20. U. Landman. E. W. Montroll, and M. F. Shlesinger, Phys. Rev. Lett. 38, 285 (1977), Proc. Nat. Acad. Sci (USA) 74, 430 (1977).
21. U. Landman and M. F. Shlesinger, Phys. Rev. B16, 3389 (1977).
22. U. Landman and M. F. Shlesinger, Annals Israel Phys. Soc. 2, 682 (1978).
23. M. F. Shlesinger and U. Landman, in Applied Stochastic Processes, ed. G. Adomian, Academic Press (to be published).
24. V. M. Kenkre, E. W. Montroll, and M. F. Shlesinger, J. Stat. Phys. 9, 101 (1973).
25. T. T. Tsong and R. T. Walko, Phys. Stat. Sol. (a), 12, 111 (1972).

26. U. Landman and M. F. Shlesinger, Phys. Rev. B19, 6207, 6220 (1979).
27. U. Landman and M. F. Shlesinger, Solid State Comm. 27, 939, (1978).
28. M. Sugi and S. Iizima, Phys. Rev. B15, 574 (1977).
29. H. P. Bonzel in Surface Physics of Materials, J. M. Blakely, ed. (Academic Press, N.Y. 1975).
30. K. Schrader, Z. Phys. B25, 91 (1976).
31. K. W. Kehr and D. Richter, Solid State Comm. 20, 477 (1976).
32. K. Lakatos-Lindenberg, R. P. Hemenger, R. M. Pearlstein, J. Chem. Phys. 56, 4852 (1972).
33. P. L. Cowan and T. T. Tsong, Surf. Sci. 67, 158 (1977)
34. D. A. McQuarrie, in Meuthen's Monographs on Applied Probability and Statistics, J. Gani, ed. (Meuthen, London 1967).
35. M. F. Shlesinger, J. Chem. Phys. 70, 4813 (1979); and J. Non-Cryst. Solids (to appear).
36. U. Landman and M. F. Shlesinger, Phys. Rev. Lett. 41, 1174 (1978).
37. E. W. Montroll, J. Math. Phys. 10, 753 (1969).
38. A. T. Moffat and A. Clarke, J. Catal 17, 264 (1970).
39. J. W. Begley and R. T. Wilson, J. Catal. 9, 375 (1967).
40. R. W. Maatman, D. L. Leenstra, A. Leenstra, R. L. Blankespoor, and D. N. Rubingh, J. Catal. 7, 1 (1967).
41. M. F. Shlesinger and U. Landman, J. Catal, (to be published).

INELASTIC ELECTRON SCATTERING: SURFACE VIBRATIONAL SPECTROSCOPY

E.W. Plummer and W. Ho
Department of Physics, University of Pennsylvania
Philadelphia, Pennsylvania 19104

S. Andersson
Department of Physics, Chalmers University of Technology
Fack, S-402 20 Goteborg 5, Sweden

ABSTRACT

The excitation mechanism for inelastic electron scattering from vibrational modes of adsorbed atoms and molecules is discussed, and the capabilities of inelastic electron scattering are compared to infrared absorption. Our present understanding of the scattering mechanism is discussed. Several experimental examples are presented to illustrate theoretical predictions of the angular and energy dependence of its differential cross section. Experimental examples will be shown which do not agree with our understanding of the scattering mechanism. Finally, we will outline theoretical and experiment problems that must be addressed in the future.

ISSN:0094-243X/80/610249-26$1.50

I. INTRODUCTION

The first observation of inelastic scattering of electrons by excitation of a vibrational mode of an adsorbed molecule or atom was made by Probst and Piper in 1967[1]. This technique has developed into a powerful tool for investigating the bonding configuration of adsorbates[2,3]. Much of the development work has been done by Ibach's group[2]; the next paper in this volume by Sexton clearly illustrates the capabilities of the technique. In this paper, we will examine in detail, the scattering mechanism for electron excitation of adsorbates. Our philosophy is that the better we understand the scattering mechanism, the more useful the technique will become. Conceptually, the inelastic electron scattering experiment is quite simple. A collimated monoenergetic (energy E_o) electron beam is scattered from a crystal surface. The electrons scattered into a narrow angular acceptance are energy analyzed. Characteristic vibrational losses of energy $\hbar\omega$ are observed in such a loss spectrum. Figure 1 shows a loss spectrum from an ordered (c(2 x 2) structure) CO overlayer on a Cu (100) crystal[4]. The incident beam energy is 1.3eV and the angle of incidence from the normal 47.7°. This spectrum is collected in the specular direction, as is most common in inelastic loss spectroscopy[2,3]. The elastically scattered electrons appear as the major peak in Fig. 1, at zero energy loss. On the intensity scale of the elastically scattered peak, no other structure is observable in the spectrum. When the low energy tail of the elastic signal is magnified by 100 two characteristic losses are observed, one at a loss energy of 260 meV, the other at 43 meV. These two losses are due to excitation of the C-O and the Cu-CO stretching vibrations, respectively. These modes are shown schematically in Fig. 1.

The objective of this paper is to deliniate what can be learned about the surface-adsorbate system by investigating the energy and angular dependence of the diffenential scattering cross section.

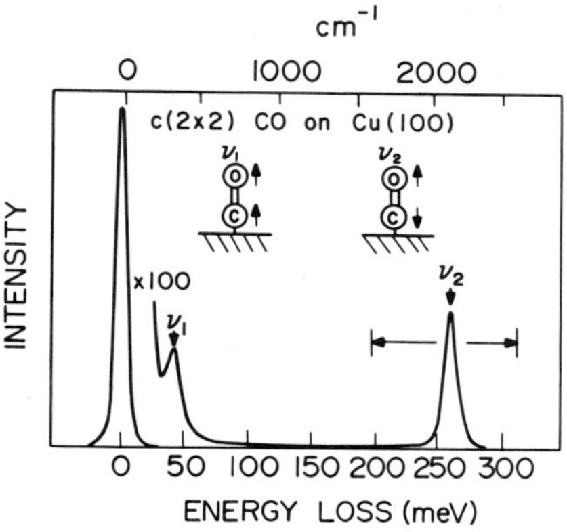

Fig. 1. Electron loss spectrum of c(2 x 2)CO on Cu(100)[4]. The arrow spanning the 2 mode, indicates the effective range of conventional infrared spectroscopy.

For example, we would like to know which vibrational modes will have a large inelastic cross section at specific impact energies E_o and collection angles θ_s. Are the selection rules for inelastic electron scattering equivalent to infrared selection rules? How can we maximize the cross section from a specific mode?

The paper is organized in the following manner. Section II will compare briefly infrared spectroscopy and inelastic electron scattering. Section III is the main section of this paper, where we present the existing theoretical models for inelastic electron scattering from a surface adsorbate and compare the predictions to experimental data. The final section (IV) looks into the future, attempting to anticipate experimental and theoretical questions.

II. INFRARED SPECTROSCOPY INELASTIC ELECTRON SCATTERING

There are several different spectroscopic techniques currently being used to determine vibrational spectra of adsorbed species: Infrared absorption and reflection spectroscopy, high resolution electron loss spectroscopy, inelastic electron tunneling spectroscopy, inelastic neutron scattering and Raman scpectroscopy. Of these methods, infrared reflection spectroscopy, Raman spectroscopy, and inelastic electron scattering have been utilized to determine the vibrational spectra of atoms and molecules adsorbed on single crystal metal surfaces. In general infrared spectroscopy has been applied to the study of the adsorption of CO on metal surfaces while inelastic electron scattering has been applied to a wide range of atomic and molecular adsorbates[2,3].

A quick comparison of infrared spectroscopy and inelastic electron scattering for the case of CO adsorbed on Cu points out the assets and liabilities of each technique. The horizontal arrows in Fig. 1 spans the range of a conventional infrared spectrometer, 1500 to 2500 cm^{-1}. The range can be increased by changing sources, windows, detectors or monochromators[5]. Inelastic electron scattering has a much wider range in one instrument than infrared[2]. Fig. 1 clearly illustrates this point, both the Cu-CO stretch at 43 meV (347 cm^{-1}) and the C-O stretch at 260 meV (2100 cm^{-1}) can be seen in one spectrum. Fig. 2 shows a distinct advantage of infrared spectroscopy, resolution. In this figure the infrared spectra of Hollins and Pritchard[6] for CO on Cu(111) is compared to Andersson's electron scattering data for CO on Cu(100). The resolution in the infrared experiment was ~1 meV while the electron scattering resolution was ~10 meV. A 1 meV shift in the C-O stretching frequence as the CO coverage was increased could easily be detected in the infrared experiment. It is also possible to follow the "coupling shift" in the C-O stretching frequency as a function of the concentration of the two isotopes $C^{12}O^{16}$/ $C^{13}O^{16}$ [5,6]. These small changes are outside of the detectibility of electron scattering. Inproving of the resolution in the infrared experiment is basically a question of signal to noise, so we will undoubtly soon see much higher resolution spectra than

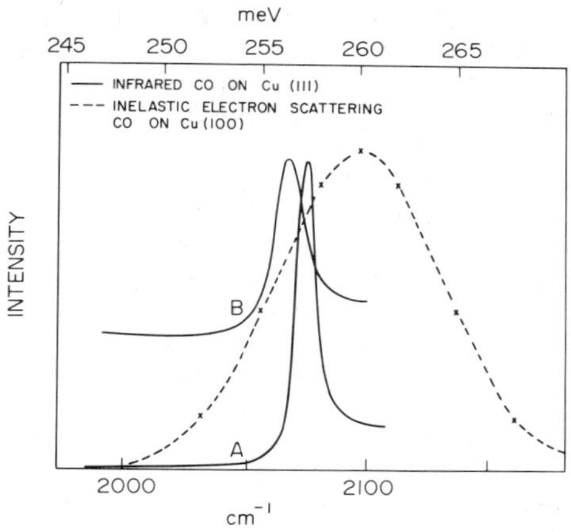

Fig. 2. Comparison of inelastic electron loss spectra from CO of Cu(100)[4] dashed line and two different coverages of CO on Cu(111) seen by infrared spectroscopy[6] solid lines. Curve A is ~1/3 monolayer and curve B is ~1 monolayer.

those seen in Fig.2. On the other hand the resolution limits in an electron spectrometer are not just signal. We may anticipate an improvement in resolution over that shown in Fig. 2, but hardly an order of magnitude. Therefore, infrared spectroscopy will always have an advantage in high resolution studies. The obvious advantage of inelastic electron scattering is that any "allowed" vibrational mode can be observed in one spectrum. The only limitation is that very low energy modes are hidden in the tail of the elastically scattered beam. In general an inelastic loss experiment must be conducted in good vacuum, say better than 10^{6}torr. At higher pressures problems may arise because of cathode poisining, adsorption on spectrometer surface and scattering of the electrons by the gas. The infrared spectrometer is usually outside of the sample chamber, which means that such a spectrometer can be operated at high pressures.

Infrared reflection spectroscopy sees only those vibrational modes with a component of the dynamic dipole moment along the surface normal. This is a consequence of the fact that the reflectivity is high at these frequencies and the field outside the surface has no component parallel to the surface. This fact has lead to the "infrared selection rule". In contrast to this, there are in principle no selection rules for inelastic electron scattering. This would mean that an electron loss spectrum could be much richer in structure than an infrared spectrum. This could be both an advantage and a disadvantage. You could see all of the fundamental modes of the surface complex, but since the resolution is not very good, this may result in a broad band instead of discrete loss lines. In the best possible operational mode, you would like to be able to tune in or out specific modes by changing the energy or collection angle in the inelastic electron scattering experiment. How you do this is the subject of Section III.

It has been proposed that the "infrared selection rule" is applicable to inelastic electron scattering, if you collect in the specular direction[2,3,7-13]. If this was all you could see with inelastic electron scattering, then eventually infrared spectroscopy would eliminate the need for electron loss experiments. New infrared laser sources are being developed, ultra high vacuum monochromators or interferometers are obtainable. One can conceive of a completely UHV system; source, sample, monochromator (transform type) and detector. Fortunately there is information obtainable from inelastic electron scattering that is not accesible by infrared.

III. ELECTRON-VIBRATIONAL SCATTERING

There are numerous theoretical papers in the literature dealing with the excitation of a vibrational mode of an adsorbed atom or molecule by a low energy electron beam[7-15]. Our objective is to present a general outline of how such a calculation can be done and then investigate the physical predictions of the various electron-vibrational coupling potentials used in the theory. Fig. 3 shows schemically four different scattering paths for a point scatterer (X) above a semi-infinite surface. In all four cases, the incident electron is characterized by the wave vector k_o and the inelastic outgoing electron k_1. Conservation of energy requires that

$$\frac{h^2}{2M}(k_o{}^2 - k_1{}^2) = \hbar\omega$$

Where $\hbar\omega$ is the vibrational energy. In process #1 the incident

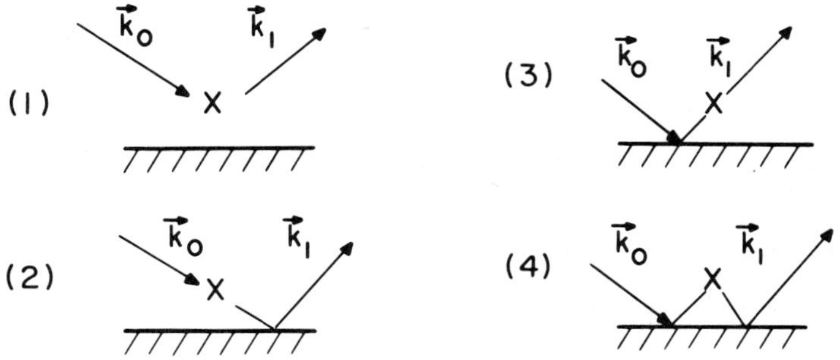

Fig. 3. Schematic drawing of four different scattering paths from a point scattered (X) above a semi-infinite surface.

electron is inelastically scattered through a large angle directly by the scatterer. In process #2, the incident electron is inelastically scattered into the forward direction (small angle scattering) by the scatterer and is subsequently elastically reflected from the substrate. In process #3, the incident electron is elastically scattered before it is inelastically scattered in the forward direction. Process #4, is more like the multiple (elastic) scattering which takes place in LEED. In general, all of these scattering processes can be divided into two categories: (1) forward (small angle) inelastic scattering followed or preceeded by elastic backscattering from the substrate (processes #2 and 3). (2) Large angle inelastic scattering (processes #1 and 4). In this simple separation, we are in essence looking at the long range vs. short range contributions of the electron-vibrational potential. The long range part leads to forward scattering while short range scattering leads to large angle inelastic scattering. Another way to represent these processes mathematically is to look at the momentum transfer parallel to the surface $q_{||}$ if we define

$$\vec{q}_{||} = \vec{k}_{o||} - \vec{k}_{1||}$$

Then if $q_{||}$ is small, we are looking close to the specularly reflected direction and the long range portion of the scattering potential should in general be appropriate for describing $\frac{d^2\sigma}{d\Omega d\omega}$. On the other hand, if $q_{||}$ is large, we must incooperate the short range portion of the scattering potential. In the following theoretical description, we will outline the consequences of the two limiting cases; long range dipole scattering and short range "impact" scattering.

The electron-vibration interaction can be treated in the adiabatic approximation where we divide the Hamiltonian into terms H_e for the electrons with fixed nuclear coordinates, H_ν for the nuclear coordinates and $H_{e\nu}$ which couples the electronic and nuclear coordinates. The wave function for the system is then a product of two wave functions.

$$\psi(r,R)\Phi(R) \qquad (3)$$

Where the electronic wave function ψ is a function of the electronic coordinates r for each fixed position of the nuclear coordinates R. The transition from an initial state i to a final state f is claculated in first order perturbation theory

$$M_{1f} = \langle\phi_f(R)\psi_f(r,R)|H_{e\nu}|\psi_i(r,R)\phi_i(R)\rangle. \qquad (4)$$

In general, $H_{e\nu}$ is a product of two terms, the potential acting on the electron due to the change in nuclear position and the potential acting on the nucleus by the electron[16]. This means that first, we should solve

$$<\psi_f|H_{e\nu}|\psi_i> = f_{ij}(R) \quad (5)$$

For all positions R[15]. This electron scattering term is then integrated over the nuclear coordinates to give M_{ij}. This procedure can be simplified if approximations are made for $H_{e\nu}$.

(a) LONG RANGE DIPOLE SCATTERING

The obvious place to start evaluating Eqn. 4 is to assume that the interaction between the electrons and the vibrational mode is through the long range dipole field of the vibrating molecule[7-13].

$$H_{e\nu} = -e\,\vec{\mu}(R)\cdot\frac{\vec{r}}{r^3} \quad (6)$$

Only the dynamic dipole moment is important, so we expand $\mu(R)$ into

$$\mu(R) = \mu_s + s\frac{\partial\mu}{\partial s}\vec{\mu}_o \quad (7)$$

The notation we are using is that presented by Lenac et al[13]. s is a length in units of the r.m.s. displacement of a harmonic oscillator in its ground state. $\vec{\mu}_o$ is a unit vector in the direction of the dipole and μ_s is the static dipole. For this long range potential, we can write the electronic wave functions as independent of R.

$$\psi_k(r,R) = \psi_k^{\pm}(r) = e^{i\vec{k}_{||}\cdot\vec{r}}\left(e^{\mp ik_\perp z} + A^{\frac{1}{2}}e^{\mp 2i\delta}e^{\pm ik_\perp z}\right) \quad (8)$$

Where $k_\perp$ is the component of k perpendicular to the surface and $A^{\frac{1}{2}}(\delta)$ is the amplitude (phase) of the refected wave[13]. Eqns. 6, 7, and 8 simplify the solution of Eqn. 4. Since the electronic and nuclear coordinates are separable, we must solve two matrix elements

$$<\phi_f|\vec{\mu}|\phi_i> \quad (9a)$$

and

$$<\psi_{k_o}|\frac{\vec{r}}{r^3}|\psi_{k_1}> \quad (9b)$$

If we assume that ϕ_f is the wave function for a harmonic oscillator then

$$|<\phi_n|s\frac{\partial\mu}{\partial s}|\phi_o|^2 = \frac{h^2}{2M\omega_o}\left|\frac{\partial\mu}{\partial s}\right|^2 \quad (10)$$

with n=1. M is the reduced mass and $\frac{\partial \mu}{\partial S}$ is the effective dynamic charge of the dipole

$$q = \frac{\partial \mu}{\partial S} \tag{11}$$

The differential inelastic cross section can now be written as[13]

$$\frac{d^2\sigma}{d\Omega dE^1} = \frac{E^1}{E_o}^{\frac{1}{2}} \frac{1}{E_o \hbar\omega_o} q^2 |f(\vec{k}_o,\vec{k}_1)|^2 \tag{12}$$

with

$$E_o = E_1 + \hbar\omega_o$$

and

$$f(\vec{k}_o,\vec{k}_1) = \frac{k_o}{2} \langle \psi_{k_1}^- | \frac{\vec{\mu}_o \cdot \vec{r}}{r^3} | \psi_{k_o}^+ \rangle \tag{13}$$

$f(k_o,k_1)$ can easily be evaluated once R, δ and q are chosen. The differential cross section can be integrated over the collection solid angle of the analyzer to obtain a differential cross section for any mode $\hbar\omega_o$ as a function of impact energy E_o or collection angle θ_i. Eqn. 12 should be multiplied by the density of the non-interacting oscillating dipoles.

If one wants to be quantitative in evaluating Eqn. 12 then a little more care is required. For example, if you treat the perturbation as the field of the dipole acting on the electron you must include the induced image dipole[13]. If you do the calculations by treating the electrons' electric field acting upon the dipole as the perturbation, then you must include the image of the electron[10a]. The change in the electron trajectory due to the image potential should be included in both treatments[10b]. In either representation, there are two major predictions of the dipole theory.

(1) If the reflectivity A is not very small, the inelastic signal is peaked near the specular direction (forward scattering). This means processes #2 and #3 of Fig. 3 dominate; loss followed reflection or reflection followed by loss. This result leads to the following predictions which can be checked experimentally.

(a) If $A(E_o) = A(E_o - \hbar\omega)$ then the ratio of the inelastic signal to the elastic signal is independent of A.

$$\frac{I(\text{inelastic})}{I(\text{elastic})} = H(E) \tag{14}$$

Where H(E) can be calculated from equation 12. The only parameter is the effective dynamic charge q(Eqn. 11). H(E) is a slowly varying function of E.

(b) Since there is a finite energy loss involved in the process the inelastic scattering is not peaked exactly in the forward direction[10,13]. In general, two lobes would appear, one on either side of the specular direction. Which lobe is biggest depends upon A and the position of the dipole from the surface[13]. When the dipole is very close to the surface, the lobe closest to the surface normal will be larger[10]. The angle of displacement of the lobe from the specular direction is a function of $\hbar\omega/2E_o$[13]. So the lower the incident energy E_o or the larger the loss energy $\hbar\omega_o$ the farther the loss signal will peak from the specular direction.

When Eqn. 12 is integrated over Ω the loss signal increases like $E_o^{-3/2}$ as the incident energy is decreased. For small collection angles the shift of the loss signal away from specular at low excitation energy will dominate the $E_o^{-3/2}$ term so that the ratio of the loss signal to the elastic signal will always decrease as $E_o \rightarrow 0$[13,17].

(2) Collection in the specular direction results in the "dipole selection rule"[2,13]. Only those modes with a dynamic dipole moment perpendicular to the surface can be excited. There are several ways to explain this quasi-selection rule, depending upon how one attacks the matrix element in Eqn. 4. As explained previously, one may either treat the dynamic field of the dipole on the electron or visa versa. In the first case[13] the image dipole of a dipole parallel to the surface cancels out the dipole field when the dipole is in the surface[13]. The perpendicular component of the dipole and its image add. If one considers the field of the electron acting on the dipole, then the image charge completely cancels out the parallel component of the electric field at the surface[10]. This is not an absolute selection rule since it will be broken, though weakly, if the parallel dipole is placed outside the suface image plane.

We now present several experimental checks of the predictions of dipole scattering. Fig. 4 shows the experimental and theroetical variation in intensity of the fundamental C-O stretching mode for the Cu(100) c(2 x 2)CO system (Fig. 1) as a function of the collection angle. $\Delta\theta_s$ is a measure of the angle of collection with repsect to the specular direction with a negative angle indicating a collection towards the surface normal. The experimental data shown in Fig. 4 was taken in a system with a fixed monochromator analyzer geometry; 95.4^o between incident and collect beam. The angular profiles were measured by rotating the crystal through an angle α. This means that $\Delta\theta_s = 2\alpha$, because the specularly reflected beam is moving by an angle 2α with respect to the analyzer, as the crystal is rotated. The crystal rotation was such that the collection plane was the (100) plane of the crystal. If the reflectivity does not change rapidly with incidence angle this experiment is almost equivalent to rotating the analyzer while keeping the crystal fixed.

The solid curves in Fig. 4 denote the elastic intensity distribution measured as the area of the elastic peak in the spectrum (see Fig. 1) while the crosses (X) and the open circles

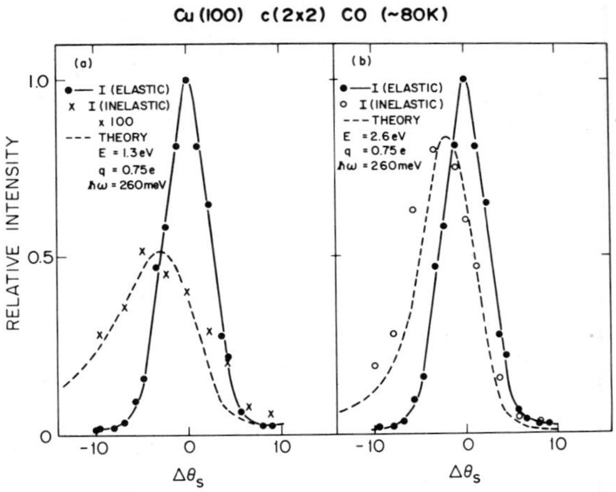

Fig. 4. Angular dependence of elastic beam and inelastic loss (260 meV) for two different incident energies for c(2x2)CO on Cu(111). The dashed curves are theoretical calculations.

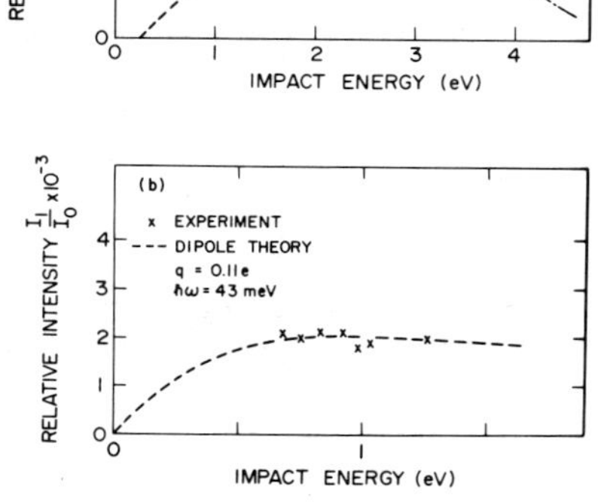

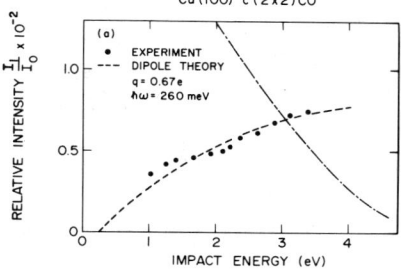

Fig. 5. Relative loss intensity vs. incident impact energy (eV) for c(2x2)CO on Cu(100) (see Fig. 1). Curve (a) for the CO stretch at 260 meV and curve (b) for Cu-CO stretch at 43 meV. The dot-dahsed curve in Fig. (a) is for the OH stretch on Ni(111) at a loss energy of 460 meV[17].

(o) denote the corresponding intensity distribution for the 260 meV C-O stretch loss peak measured at 1.3 eV (a) and 2.6 eV (b) incident electron energy, respectively. The peak in the elastic distribution defines the scattering situation when the specular direction coincides with the analyzer orientation. The experimental data reveal that the inelastic cross section is sharply peaked in a direction which does not coincide with the specular. The weight and width of the peak depends on the primary electron energy. These are essentially also the characteristic features of the corresponding dipole scattering model calculations by Persson[10b] (dashed curves in Fig. 4). These curves were calculated for q = 0.75 e, n_j = 7.67 x 10^{18} molecules per in^2, 1.3 eV and 2.6 eV primary electron energy respectively and for a range of crystal orientations around the specular orientation. The cross section (Eqn. 2) was integrated over a cone of half-angle 3°. For simplicity we assumed that the analyzer transmission coefficient is unity over this angular range and that the elastic reflectivity of the surface remains constant over the range of an angles explored.

A more extended analysis of the energy dependence of the cross section is shown in Fig. 5. The filled circles (●) represent the experimental relative intensity, I_1/I_o, measured in the specular direction. The dashed curve is calculated[10b] for q = 0.67 e and an angle of incidence of 47.7^o. All other conditions are the same as above. The general trend in the energy dependence agrees fairly well. The deviation between the q values above is a measure of the accuracy of the experiment, proper specular adjustment being critical for the data in Fig. 5. These q values are quite close to the value ~0.6 e found from infrared studies of the Cu(100)-CO system (see ref. 7). The dot-dashed curve in Fig. 5a is the O-H stretch of OH adsorbed on NiO(111)[17]. It is included to point out at this stage, that not every adsorbate produces a loss signal in agreement with dipole theory.

The only remaining prediction to be tested is the "dipole selection rule". To test this rule, we need to know the orientation of the molecule. Angle resolved photoemission experiments indicate that CO bonds with its axis perpendicular to the Cu surface[18]. Then we would expect to see the C-O and Cu-CO stretching modes, but the bending or sissor modes should be "invisible" in the specular direction. The data in Fig. 1 is consistent with this picture, but the bending modes could be very low in energy and not resolved. We will show subsequently two examples of H adsorption which clearly illustrates that there are dipole selection rule.

The "dipole selection rule" is the essential ingredient in any arguement about the bonding configuration of a molecule based upon inelastic electron scattering spectra. It is essential to know when this rule is applicable. Since we want to use the rule to determine the bonding, we can't use the bonding configuration to verify the rules. Two simple checks are surely necessary if not sufficient.

1. The loss signal should peak near the specular direction as shown in Fig. 4. This behavior should be checked at several energies. If the signal doesn't peak in the specular direction, then dipole scattering doesn't dominate the process and the argument for the "dipole selection rule" is inappropriate.
2. The inelastic signal divided by the elastic signal should be a smoothly varying function like that of Fig. 5. This is only true when there are no rapid changes in the reflectivity A with energy on a scale of the loss energy.

In our experience, if the loss signal peaks near the specular direction (an order of magnitude or more difference in signal), then the excitation mechanism in the specular direction is dipolar. This means the mode obeys the "dipole selection rule".

II. SHORT RANGE "IMPACT" SCATTERING

Here we will be concerned with the form of $H_{e\nu}$ when the electron is close enough to see the ion core. Fig. 6 shows a more detailed picture of this scattering. The incoming electron may interact with the adsorbed atom or molecule directly as shown by the dashed line (process #1 of Fig. 3) or it may multiple scatter from the substrate before, after or before and after inelastically scattering from the adsorbate. This problem must be solved by finding $\psi_k(r,R)$ for each position of the nuclear coordinates R. The scattering factor $f_{if}(R)$ must be evaluated (Eqn. 5), and the expectation value found for transition between two vibrational states (Eqn. 4). Davenport, Ho and Schrieffer carried out this procedure for molecules with fixed orientation in space[15]. There are several general predictions from this type of theory which are independent of numerical details.

(1) The inelastic signal is spread out over wide angles. In general, the total cross section for inelastic loss via short range scattering is larger than for dipole scattering

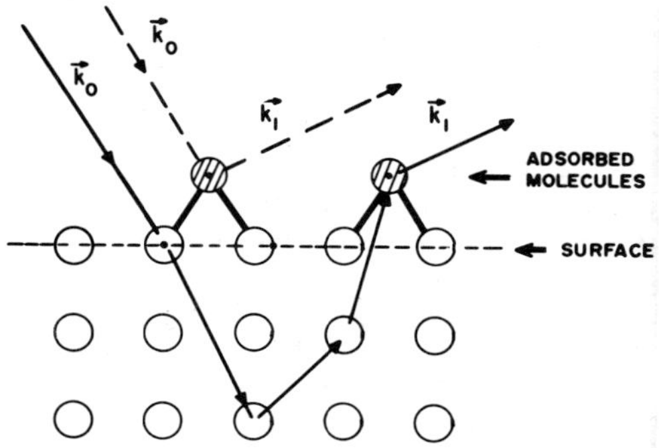

Fig. 6. More detailed picture of the short range scattering depicted in paths #1 and #4 of Fig. 3.

$$\int \frac{d\sigma}{d\Omega}\, d\Omega \Big|_{\substack{\text{short}\\ \text{range}}} \gtrsim \int \frac{d\sigma}{d\Omega}\, d\Omega \Big|_{\text{dipole}}$$

The dipole is peaked in the specular direction so that in general

$$\frac{d\sigma}{d\Omega}\Big|_{\substack{\text{short}\\ \text{range}\\ \text{specular}}} < \frac{d\sigma}{d\Omega}\Big|_{\substack{\text{dipole}\\ \text{specular}}}$$

The exception to the last statement occurs when a negative ion state is formed[15]. In this case, the electron is temporarily trapped in a resonant state of the molecular complex.
(2) The inelastic signal is dominated by processes like those shown in Fig. 6. Therefore

$$\frac{I(\text{loss})}{I(\text{elastic})} \propto \frac{\sigma(E)}{A} \qquad (15)$$

and an energy plot like that shown in Fig. 5 could be very complicated.
(3) There are no strict selection rules but at resonance modes of compatible symmetry will be enhanced. In many gas phase experiments overtones are common so that we would expect to see stronger signals form $0 \rightarrow 2$, $0 \rightarrow 3$, etc. excitations.

III EXPERIMENTAL EXAMPLES

Experimentally, we would expect to see dipole allowed modes in the specular direction, and "all modes" out of specular. If we happened to hit a negative ion resonance, we might see all modes in the specular direction. We would immediately be able to identify this situation by looking at the angle and energy dependence of the intensity of these modes. The first experimental confirmation of these concepts came from a study of a very old and often studied system, β_1, hydrogen on W(100)[19]. The proposed bonding geometry of this system is shown in Fig. 7. There are two hydrogen atoms per W surface atom with the H atoms bridge bonded. Naively, we would expect three vibrational modes: (1) A dipole allowed mode which is the symmetric stretch with the H atom moving perpendicular to the surface; (2) The asymmetric stretch where the hydrogen moves along the axis of the two W atoms to which it is bonded. This mode is parallel to the surface and presumably is "dipole forbidden"; (3) A frustrated rotational mode where the H atom moves perpendicular to the two adjacent W atoms, but parallel to the surface. This mode is "dipole forbidden". Therefore, we would expect to see one mode in the specular direction and three modes away from the specular direction.

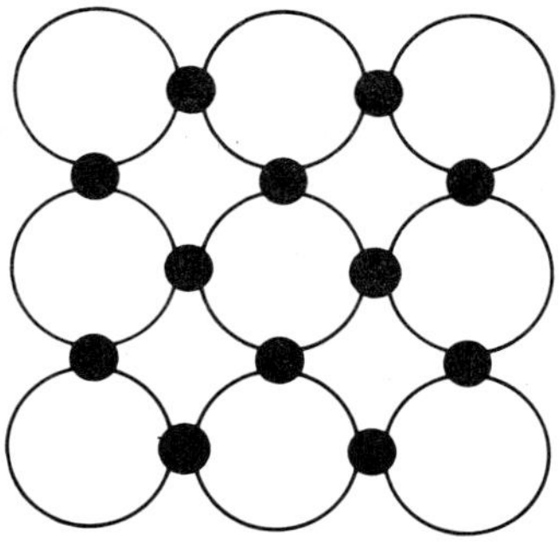

Fig. 7. Model structure for saturation H adsorption of W (100). Open circles are W atoms.

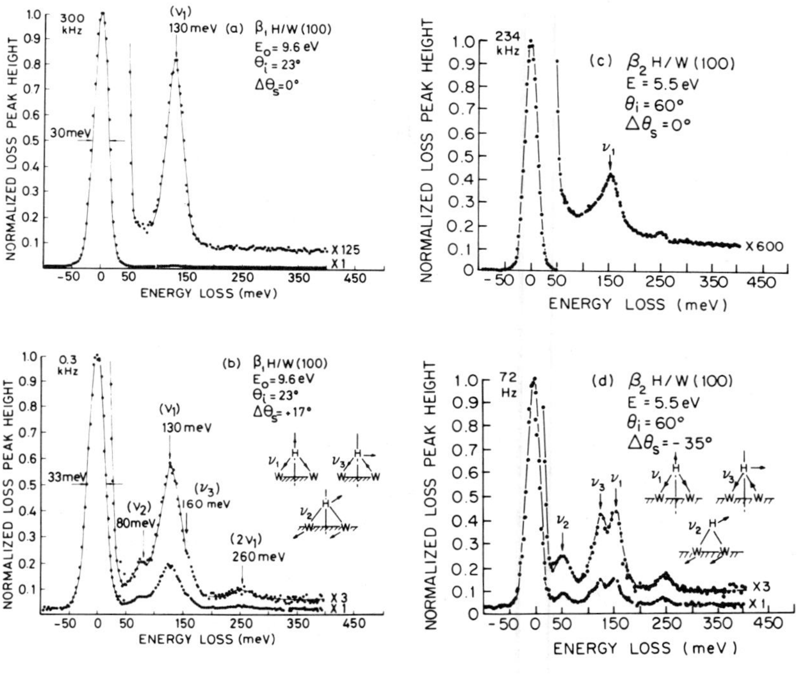

Fig. 8. Electron loss spectra for H adsorbed on W(100). Panels (a) and (c) are collection in the specular direction while (b) and (d) are for collection off of the specular direction. (a) and (b) are for saturated H adsorption, (1x1) structure called β_1, ($\sim 17 \times 10^{14}$ atoms/cm^2)[19], (c) and (d) are for β_2 adsorption ($\sim 4 \times 10$ atoms/cm^2) with a c(2x2) structure[20].

Fig. 8. (a) shows the loss spectrum for β_1 hydrogen in the specular direction. There is only one mode visible in the spectrum at 130 meV energy. According to our previous guidelines, if this modes peaks in the forward direction, then it is a "dipole allowed" mode. Fig. 9 shows that the 130 meV mode peaks in the forward direction[21]. Fig. 9 is for a different incidence energy and angle than Fig. 8 to illustrate that the behavior is general. We now have proven that the 130 meV mode is the symmetric stretch. Fig. 8b shows the spectrum from β_1 H collected 17° away from the specular direction. Four modes are visible (160 meV mode is more pronounced at other angles[21,22]); (1) The 130 meV symmetric stretch; (2) A 160 meV asymmetic stretch; (3) A 80 meV frustrated rotation and; (4) A 260 meV overtone. The asymmetric stretch is at a higher energy than the symmetric stretch because the subtended angle between W-H-W is larger than 90°. This is a clear example of the utility of angle dependent inelastic electron spectroscopy. The "dipole allowed" and "forbidden" modes were identified, and the geometry could have been deduced from their respective energies.

Figures 8(c) and (d) show an example where this procedure was used to determine a structure that was not already known[20]. This is the case of β_2 H adsorption on (100)W. The H density is ~4 x 10^{14} atoms/cm^2 and the structure is c(2 x 2). Previous inelastic electron spectroscopy studies had shown that the single mode seen in the specular direction at 130 meV for the β_1H state shifted to 155 meV for the low coverage β_2 state[22]. The logical interpretation of this shift was that the H atoms in the β_2 state were bonded to a single W atom in the on top position. Fig. 8(c) shows the loss spectrum in the specular direction[20]. The intensity of this single mode at 155 meV peaks in the specular direction so it is the "dipole allowed" mode with the hydrogen moving perpendicular to the surface. Fig. 8(d) shows the spectrum for β_2 hydrogen away from the specular direction. Three modes plus an over tone are seen. The assignment of the modes is shown in the figure. H is still bridge bonded, but the asymmetric mode is at a lower energy than the symmetric mode. This can happen if the angle between W-H-W becomes smaller than ~90° [20]. The angle can decrease if the H-W distance increases or the W-W distance decreases. Barnes and Willis[20] interpreted these measurements as showing that the W substrate was reconstructed in the β_2H system, with the two W atoms ~0.2 Å closer together in the β_2 phase than in the β_1 phase.

Lets return briefly to the angular dependence shown in Fig. 9 for the modes of the β_1 phase of H adsorption on (100)W. The top curve is the quasi-elastic signal strength. The 130 meV mode clearly peaks in the specular direction. The intensity falls off rapidly for the first 5° to 10° in general agreement with dipole theory. For angles larger than 10° the intensity is larger than predicted by dipole theory. This large angle scattering is probably dominated by short range scattering. The 80 meV frustrated rotational mode never peaks in the specular direction, for

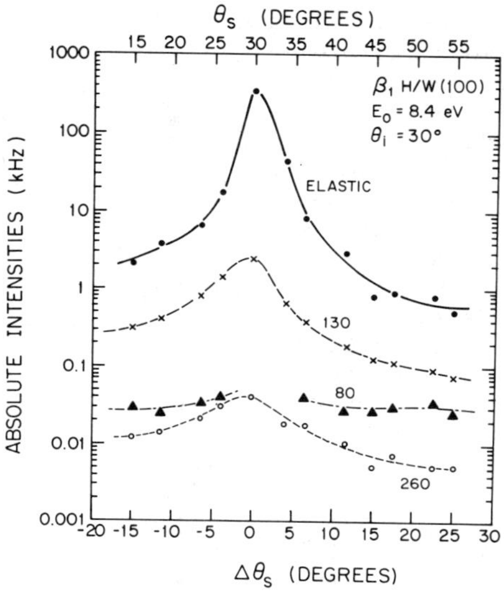

Fig. 9. Angle dependence of the absolute intensities as a function of the angle $\Delta\theta_s$ away from the specular direction. The intensity of the elastic, 130 meV, 80 meV, and 260 meV modes shown in Fig. 8(b) are plotted.

Fig. 10. Normalized loss intensity for 130 meV mode of β_1-H adsorbed on W(100)[21].

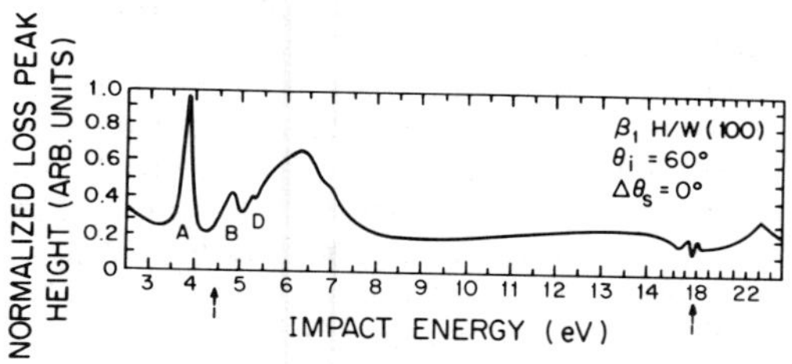

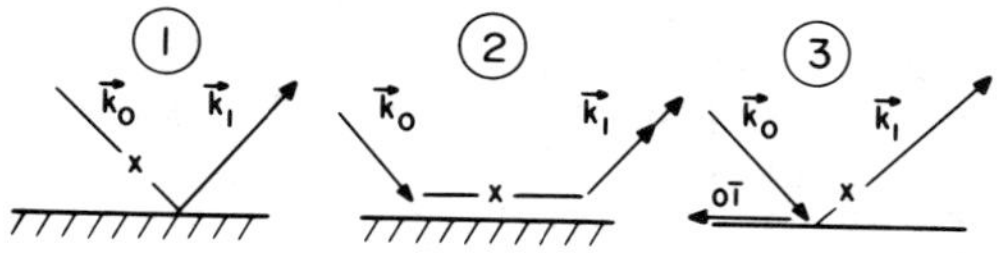

Fig. 11. Schematic drawing of surface resonances and interaction with adsorbate.

any incident angle or impact energy. The 160 meV asymmetric mode is hard to separate from the 130 meV mode with the resolution of this experiment, but it does not appear to peak in the specular direction. These two modes are dipole forbidden. Fig. 9 shows that the overtone at 260 meV has a small maximum in the specular direction at this incident angle and impact energy. In general, it does not peak in the specular direction[21], so it must be excited by the short range scattering. It is this observation that leads to the assignment of this loss to a $0 \rightarrow 2$ excitation instead of two $0 \rightarrow 1$ excitations.

Every experiment with H adsorption that we have discussed, worked exactly as we would have anticipated from the general theoretical discussion outlined in this section. The only remaining test is the energy dependence of the "dipole allowed" mode measured in the specular direction. Since we have proven that the 130 meV loss mode is excited by the dipole field, when the measurement is in the specular direction, we should measure a normalized loss intensity $\frac{I(\text{loss})}{I(\text{elastic})}$ which looks like Fig. 5. The normalized loss signal is very complicated as seen in Fig. 10. The preliminary conclusion is that the excitation mechanism is non-dipole. But in this case, one of our original assumptions in deriving Eqn.14 is incorrect. The reflectivity from W(100) changes rapidly with energy so that the reflectivity for $E_o - \hbar w$ is not the same as that for E_o. It can be shown qualitatively that much of the structure can be explained as changes in the relectivity[21]. The arrows in Fig.10 show another effect. The emergance of new beams, at the energy shown the new beam is just coming out of the crystal, causing rapid changes in the reflectivity. There is yet a third phenomena which makes the normalized loss intensity different than the simple dipole theory would predict; surface resonances[24]. The relevant scattering diagram is shown in Fig.11. Process #1 is the ordinary dipole scattering; in process #2 the incident electron has the correct k to scatter from a reciprical lattice vector exchanging it perpendicular energy for parallel energy. If there is a band gap in the solid at this energy, the electron is trapped in a one dimensional well perpendicular to the surface. These states are quantized [23b] so that E_o and k must be exactly correct. When you tune into one of these states, the incident electron propagates for a long distance down the surface enhancing the probability for inelastic loss. The amplitudes for processes #1 and #2 must be summed and then squared leading to interference terms. The low energy structure in Fig.10 is believed to originate from this mechanism[21]. It should be pointed out that these surface resonances are the two dimensional analoges of the negative ion resonance in the gas phase[21]. The normalized loss intensity as a function of impact energy is not as definitive a test of dipole allowed modes as the angular dependence was. This is especially true when the reflectively is changing rapidly with energy.

There is one major feature in Fig. 10 which can not be explained by any of the qualitative arguments given in the preceeding paragraph. The broad peak near 6.5 eV can not be attributed to rapid changes in reflectivity, surface resonances (of clean W(100)) or to new beam emergence. The energy position of this peak depends upon the incident beam angle so it is not a simple "atomic like" negative ion resonance. It may be associated with two dimensional H-W surface bands.

A quite different situation has been observed for OH groups adsorbed on a thin NiO(111) film, which was grown epitaxially on a Ni(100) substrate[17]. The oxide was about 6 Å thick and was assumed to be half covered with adsorbed OH groups. The energy dependence of the O-H stretch mode intensity is shown in Fig. 5. It is clear that the experimental data does not agree with dipole calculations. This may be the first and only recorded observation of a "negative ion resonance". It is likely that negative ion resonances will be observed for adsorption on semiconductors or insulators. The molecule must be kept far enough from the surface so that the resonance has a reasonably long life time.

We conclude this section with two other examples to prove that the behavior described for CO on Cu or H on W are not special examples. Fig. 12 shows the angular dependence of the C-O stretching mode for CO adsorbed on W(100) at ~150°K. At this temperature, atomic O and C are also present on the surface, but Fig. 12 shows that the 250 meV C-O stretching mode is "dipole allowed". Fig. 13 shows a loss spectrum for N_2 adsorption on W(100) at 125°K[25]. There are four modes observed in this set of data; 265 meV, 180 meV, ~120 meV and at 77 meV[25]. If we had terminally bonded N_2 we would expect only two modes like those shown in Fig. 1 for CO adsorption. The 265 meV mode is at an energy where the N-N stretching mode would be for terminally bonded N_2. The 77 meV mode is approximately correct for the W-N_2 stretch for terminally bonded N_2 and the ~120 meV mode is atomic nitrogen[25]. The unassigned mode is at 180 meV. The first question to be answered is which of these modes are "dipole allowed". Fig. 14 shows via an angular plot that they are all dipole modes. This observation leads to the assignment of the 180 meV mode to the N-N stretching mode for bridge bonded Ni[25]. This is based on the observation that a N-N double bond has a stretching frequency in the range 190-200 meV[26].

IV EXPERIMENTAL EXAMPLES THAT DON'T WORK

All of the previous examples point to a simple conclusion. Any vibrational mode seen in the specular direction is "dipole allowed". If this is true, then there is no need to check the angle or energy dependence to confirm that a given mode seen in the specular direction is in fact "dipole allowed". Fig. 15 shows a loss spectrum for H adsorption at 150°K on Ni(111)[27]. Two modes are observed in the specular direction at ~90 and ~ 140 meV. Given that these modes are seen in the specular direction, it would seem safe to conclude that they are dipole allowed. This would lead to the conclusion that H adsorbs into two different sites on Ni(111). The 90 meV

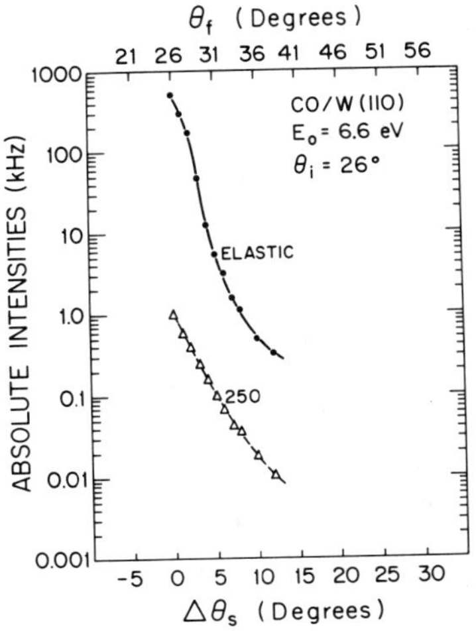

Fig. 12. Angular dependence of CO stretching frequency for CO adsorbed on W(100) at -150 K.[22]

Fig. 13. Electron loss spectrum of N_2 adsorbed on W(100) at 125 K. The collection is in the specular direction.[25]

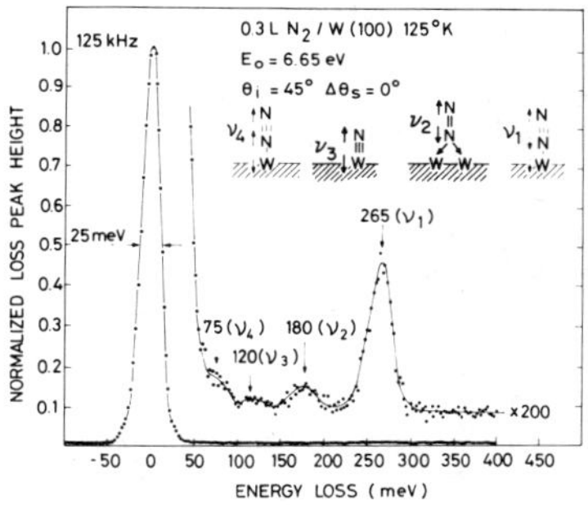

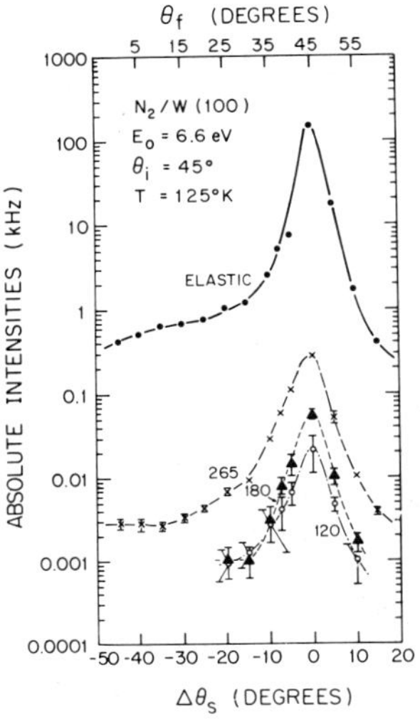

Fig. 14. Angular dependence of the N_2 vibrational modes observed in the spectrum shown in Fig. 13.

mode would probably correspond to the three fold hollow site with a Ni atom below in the next plane and the 140 meV mode is the three fold hollow site without a Ni atom below. The observed frequency of 140 meV corresponds closely to the calculated frequency of 150 meV[28]. The problem with this interpretation is shown in Fig. 16. Neither mode peaks in the specular direction. Do we conclude that neither mode is dipole allowed or that the short range scattering dominates the dipole scattering? One reasonable explanation is that the H id deep in a three fold hollow site so there is no long range dipole field. The 90 meV mode would be the mode with the H vibrating perpendicular to the surface (symmetric stretch) and the 140 meV mode is the asymmetric mode.

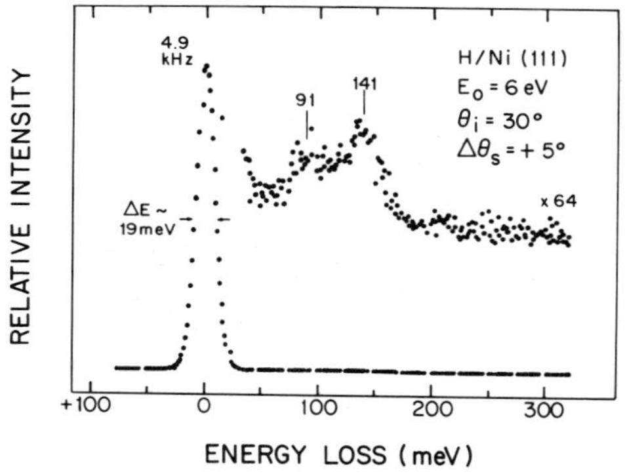

Fig. 15. Electron loss spectrum of H adsorbed on Ni(111) at 150 K[27].

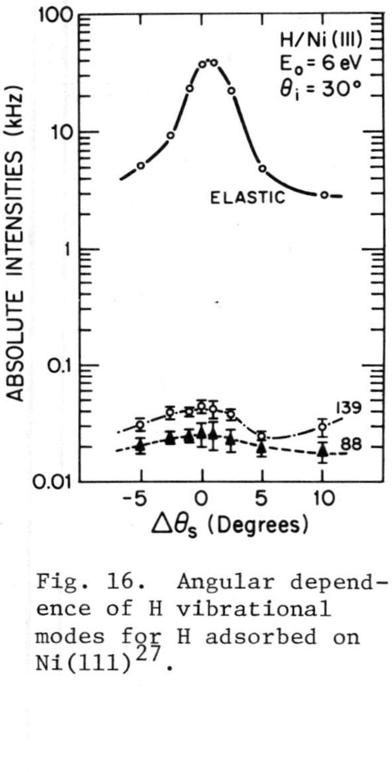

Fig. 16. Angular dependence of H vibrational modes for H adsorbed on Ni(111)[27].

This same behavior has been observed by Baro, Ibach and Bruchmann[29] for hydrogen adsorbed on Pt(111). In this case the two modes have energies of 68 and 153 meV. Neither mode peaks in the specular direction, which leads the authors to conclude that the hydrogen was buried in the surface. The 68 meV mode is perpendicular to the surface and the 153 meV mode is parallel to the surface.

The adsorption of H on either Ni(111) or Pt(111) could be an example of no dipole scattering due to the adsorbed atom being to deep in the surface. One arguenet for this interpretation is that the inelastic signal in the specular direction is very small. An example where the inelastic signal is large in the specular direction but there still is no peaking of the signal in the specular direction is the adsorption of C_2H_2 on Ni(100)[22]. In this case there is no rational explanation for the results.

IV FUTURE

In this final section we will try briefly to outline which problems need to be solved and what new physics may come from inelastic electron scattering. First consider the theoretical questions that have been raised in the previous discussions.

1. What causes the apparent break down of dipole scattering? We intentional use apparent break down, because dipole scattering may not break down. Our understanding of dipole scattering may be imcomplete. There are specific questions that need to be addressed.

(i) What is the effect of order or disorder upon dipole scattering?

(ii) What is the contribution of bulk phonons to the off specular scattering?

(iii) How do you treat an adsorbed atom when it is inside of the surface?

2. What is the nature of the electron-vibrational or electron-phonon coupling? What is needed is a unified theory that can treat short range and long range scattering simultaneously. The substrate electron-phonon interaction must also be considered in this theory, and it must address the excitation of localized molecular vibrations vs. phonon modes. Finally this theoretical description should include overtone or coupled mode excitation.

Experimentally there are several obvious questions that should be answered as well as new experimental developements needed. We need to work on new spectrometers with better resolution and higher signal levels. There is a need for more and better angle and energy dependent spectra on simple systems. For example the following list is almost obvious:

(1) Angular profiles for systems which have both an ordered and disordered phase. This would allows one experimentally evaluate the importance of disorder.

(2) Angular profiles as a function of incident energy, specifically at energies of high and low reflectivities. The normal procedure is to tune the spectrometer in the specular direction to an energy of high reflectivity so

that the signal is large. It would be very useful to tune to a region of low reflectivity to see if the short range scattering can be appreciable in the specular direction.

(3) Incident energy dependent spectra (Fig. 5) of more systems. This will be needed as a data base if the generalized theory discussed in the theoretical discussion is to be checked.

(4) Spectra for ordered systems vs. $q_{||}$. At some stage this spectroscopy will begin measuring phonon modes of ordered overlayers.

We will end this paper with a simple example of what will be possible with phonon modes. We illustrate this measurement by using the simplest overlayer, a (1x1) structure. Fig. 17 shows the optical surface phonon modes for the (1x1) structure calculated by Armand and Theeten[30]. The dispersion curve on the left is for very strong damping where there isn't much interaction with the substrate. The right hand curve is for a weaker damping where the substrate mixes the modes. For example the longintudinal optical phonon (displacement in X direction) mixes with the transverse optical mode whose displacement is in the Z direction (Z is perpendicular to surface). The transverse optical mode whose displacement is in the Y direction doesn't mix with either of the other two modes. Therefore we have one pure transverse optical mode (Y displacement) and two mixed modes (X,Z displacement). At the zone boundaries these modes are pure.

We now want to discuss what would be seen in an inelastic scattering experiment preformed on this hypothetical system. Lets start by considering the form of the matrix element when the incident and final electronic states can be treated as plane waves[8,16].

$$M_f \propto \Gamma_o(E_o, E^1, \theta)(\vec{Q} \cdot \vec{e}_q)^2 \, f(n,q) \, \delta(k_{||}' - k_{||} - q) \, \delta(E - E' - \hbar\omega(q)) \qquad (16)$$

Where Γ_o is a scattering factor and f is an occupancy function[8]. The important term is $\bar{Q} \cdot \bar{e}_q$ where

$$Q = (k_{||}' - k_{||}\,,\; k_z' - k_z) = (q, \Delta k_z)$$

and $\vec{e}_q$ is the displacement direction of a specific mode at q. For the excitation of a three dimensional phonon Eqn. 16 predicts that transverse optical phonon can not be excited because Q is perpendicular to $\vec{e}_q$, unless an Umklapp process is involved[16]. For two dimensional phonon modes this is not the case. $q = k_{||} - k_{||}$ is in the direction of phonon propagation which would be perpendicular to $\vec{e}_q$ for a transverse phonon but Δk_z can always be used to give a finite value to $\vec{Q} \cdot \vec{e}_q$. For dipole allowed modes the coupling is to the z component of e_q so the only term that contributes is $\Delta k_z \times e_{qz}$. The short range

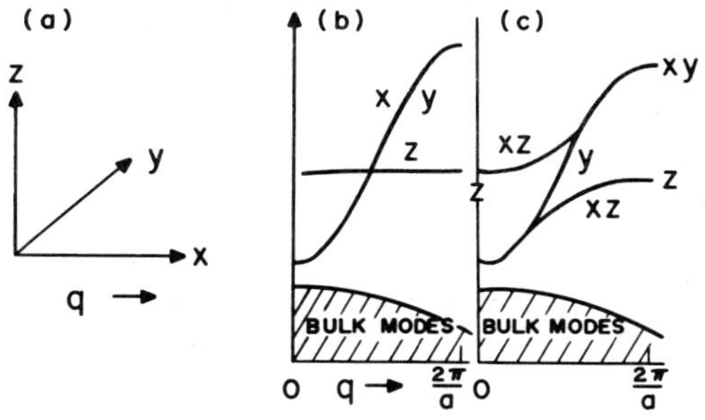

Fig. 17. Optical surface phonon modes for a (1x1) overlayer. (a) Coordinate system, surface is in xy plane with propagation direction in x direction. (b) Modes for very strong damping approximation and (c) weaker damping approximation[24]. X,Y,Z demoter direction of displacement vector.

Fig. 18. Schematic drawing of the intensity vs. momentum transfer for the three phonon modes shown in Fig. 17. The left hand column is for pure modes in the strong damping approximation[24]. The right hand side is for the weaker damping approximation.[30]

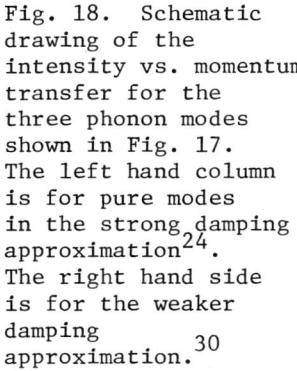

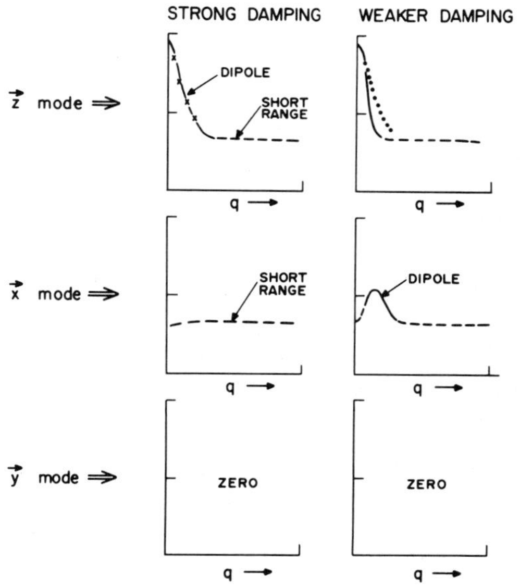

scattering can excite modes with displacement vectors in the surface (X and Y), but Eqn. 16 predicts that the pure mode with Y displacement can never be excited since Q · y = 0. Table 1 lists the different modes in both the strong and weaker damping limits. The excitation mode is checked off for each mode. In this table as well as in Figure 17 the phonon propagates in the X direction, and the incident beam and collection are in the XZ plane.

It is even possible to estimate what the intensity of each mode will be as a function of q. Fig. 18 shows the intensity for the three modes as a function of q for the two phonon dispersion curves shown in Fig. 17. For the strong damping case we can guess the intensity profile from figures like 9, 12, or 14. The z mode (#1 in Table 1) will peak near the specular direction, fall off rapidly until the short range scattering takes over. The X mode (#2) is not dipole allowed so it will have a small nearly isotropic intensity vs. q. The y mode (#3) is forbidden by Eqn. 16 so there is no intensity. When the damping is reduced the X x Z modes mix and the intensity may look like what is shown on the right of Fig. 18 . The Z mode starts off pure at q = 0 but mixes with the x mode as q increases. This means that the

Table 1 PHONON MODE EXCITATION - PROBOGATION IN X DIRECTION

	Mode eq	Type	Strong Damping "Dipole"	Strong Damping "Short" Range	Weaker Damping "Dipole"	Weaker Damping "Short" Range
#1	z	TO	Yes	Yes	Yes	Yes
#2	X	LO	No	Yes	Yes (Mixed)	Yes (Mixed)
#3	Y	TO	No	No	No	No

the component of the dipole perpendicular to the surface is decreased. The figure shows that the intensity will fall off faster than the pure z mode. At the zone boundary this mode is now pure X. In contrast the "X" mode starts off at q = 0 as pure X which is dipole forbidden, but for q ≠ 0 it mixes with the z mode causing it to be dipole allowed. The intensity will increase as the figure shows. The y mode doesn't mix so it will have no intensity.

There are two complications that must be considered. First the general picture presented in Fig. 18 and Table 1 is based on Eqn. 16 which treats the incident and loss electron wave functions as plane waves. It is easy to show using symmetry arguements that if the collection is in a mirror plane of the crystal that the product of the terms in Eqn. 4 is odd for the y mode. This means that the excitation probability is zero. The second complication is not as easy to overcome. The strong signal of the z mode in the specular direction means that there is a large signal forward scattered from the adsorbed layer. If this forward beam is inelastically scattered by the substrate phonons then the signal at large q would contain signal from the specific q plus phonon scattered signal from q ~ 0.

ACKNOWLEDGEMENT

The work of Plummer and Ho was supported by NSF Grant #DMR - 78 - 08520.

REFERENCES

1. F.M. Probst and T.C. Piper, J. Vac. Sci. and Tech. 4, 53 (1967).
2. H. Ibach, Proc. 7th Intern. Vac. Congr. and 3rd Intern. Conf. Solid Surface, Vienna p. 743 (1977).
3. S. Andersson, Topics in Surface Chemistry, edited by E. Kay and P. Bagus (Plenum Pub. Co. 1978)p. 291
4. S. Andersson, ECOSS 2 (Surf. Sci)
5. A. Crossley and D. King, Surf. Sci. 68, 528 (1977).
6. P. Hollins and J. Pritchard, ECOSS 2 (Surf. Sci.)
7. D.L. Mills, Progress in Surface Science 8, 143 (1977).
8. E. Evans and D.L. Mills, Phys. Rev. B5, 4126 (1971).
9. D.M. Newns, Phys. Letters 60A, 461 (1977).
10. (a) B.N.J. Persson, Solid State Comm. 24, 573 (1977).
    (b) B.N.J. Persson, private communication.
11. D. Sokevic, A. Lenac, R. Brako, and M. Sunjic, Zeit. fur Physik B28, 273 (1977).
12. F. Delanaye, A. Lucas, and G.D. Mahan, Surf. Sci. 70, 629 (1978).
13. Z. Lenac, M. Sunjic, D. Sokcevic, and R. Brako, Surf. Sci. 80, 602 (1979).
14. V. Roundy and D.L. Mills, Phys. Rev. B5, 1347 (1972).
15. J. W. Davenport, W. Ho, and J.R. Schrieffer, Phys. Rev. B17, 3115 (1978).
16. Ziman Electrons and Phonons, Oxford Press (1962).
17. S. Andersson and J.W. Davenport, Solid State Commun. 28, 677 (1978).
18. C. Allyn, T. Gustafsson and E.W. Plummer, Solid State Commun. 24, 531 (1977).
19. W. Ho, R.F. Willis, and E.W. Plummer, Phys. Rev. Letter 40, 1463 (1978).
20. M.R. Barnes and R.F. Willis, Phys. Rev. Letter 41, 1729
21. W. Ho, R.F. Willis, and E.W. Plummer, Phys. Rev. (to be published).
22. W. Ho, Ph.D. Thesis, Univ. of Pennsylvania.
23. (a) H. Froitzhelm, H. Ibach, and S. Lehwald, Phys. Rev. Letters 36, 1549 (1976).
    (b) A. Adnot and J.D. Carette, Phys. Rev. Letter 39, 209 (1977).
24. E.G. McRae and G.H. Wheatley, Surf. Sci. 29, 342
25. W. Ho, R.F. Willis and E.W. Plummer, to be published.
26. L.J. Bellamy, The Infra-red Spectra of Complex Molecules 3rd ed., Chapman and Hall, London 1975
27. W. Ho, J. DiNardo and E.W. Plummer, to be published in J. Vac. Sci. and Tech.
28. T.H. Upton and W.A. Goddard, III, Phys. Rev. Letter 42, 472 (1979).
29. A.M. Baro, H. Ibach and H.P. Bruchmann, to be published.
30. G. Armand and J.B. Theaten, Phys. Rev. B9, 3969 (1974).

TWO-DIMENSIONAL PHASE SEPARATION: CO-ADSORPTION OF HYDROGEN AND CARBON MONOXIDE ON THE (111) SURFACE OF RHODIUM*

Ellen D. Williams**, Patricia A. Thiel**,
W. Henry Weinberg† and John T. Yates, Jr.‡
Division of Chemistry and Chemical Engineering
California Institute of Technology
Pasadena, CA 91125

ABSTRACT

The co-adsorption of CO and H_2 on Rh(111) at low temperature ($\sim$ 100 K) has been studied using thermal desorption mass spectrometry (TDS) and Low-Energy Electron Diffraction (LEED). The probability of adsorption of CO on rhodium pretreated with hydrogen has been found to decrease slowly with increasing amounts of hydrogen on the surface. In addition, the effect of surface hydrogen on the CO LEED patterns indicates segregation of hydrogen and CO. These results can be explained in terms of a strong repulsive CO-H interaction and a mobile precursor model of CO adsorption.

INTRODUCTION

Studies of the co-adsorption of hydrogen and CO on transition metals are of practical importance due to their relevance to the formation of hydrocarbons from the products of coal gasification. The two commonly posited intermediates in hydrocarbon formation are the following: (1) An active surface carbon species formed by dissociation of CO, and (2) A hydrogen-CO complex formed by interaction of chemisorbed hydrogen and CO.[1,2] Recent experimental work on both pure and supported transition metals has shown that CO dissociation is the more important step of the two in methane formation.[3-9] At the same time, there is increasing evidence to suggest that surface complexes of CO and hydrogen do exist[10-12] and that direct hydrogenation of molecular CO is an alternative route to hydrocarbon formation.[3]

In this study, the co-adsorption of hydrogen and CO on the atomically smooth (111) surface of rhodium has been investigated. The adsorption experiments were performed under ultra-high vacuum and at a surface temperature of approximately 100 K. No evidence for reaction of hydrogen and CO under these conditions was observed.

The individual chemisorption of hydrogen and CO has been studied previously and is well characterized.[13-15]

---

*Research supported by the National Science Foundation under Grant No. CHE77-16314.

**National Science Foundation Predoctoral Fellow.

†Camille and Henry Dreyfus Foundation Teacher-Scholar.

‡Sherman Fairchild Distinguished Scholar, 1977-78. Permanent Address: National Bureau of Standards, Washington, D.C. 20234.

ISSN:0094-243X/80/610275-09$1.50

RESULTS

Thermal desorption mass spectrometry (TDS) was used to determine activation energies for desorption, coverage-exposure relationships and adsorption kinetics for both hydrogen and CO in the co-adsorbed system. For brevity, these results are only summarized here.

The activation energy for desorption of hydrogen decreases markedly upon co-adsorption with CO. This decrease can be from 1.5 to 3 kcal/mole depending upon the coverage. The desorption energy for CO does not change when it is co-adsorbed with hydrogen. This is not surprising since all the hydrogen has been desorbed from the surface by the temperature at which desorption of CO begins.

At low coverages of CO, hydrogen adsorption occurs readily, reaching a lower saturation value than on the clean surface. At mid- to high-coverages of CO, hydrogen adsorption is efficiently blocked. The adsorption of CO onto a hydrogen covered surface demonstrates some interesting behavior as shown in Fig. 1. The open and filled circles represent CO adsorption onto two different clean surfaces, and the stars represent CO adsorption onto a surface precovered with 0.58 monolayer of hydrogen. The experimental data have been compared with coverage-exposure behavior predicted using first-order Langmuir adsorption kinetics and adsorption via a mobile precursor.[16,17] The parameters used in the calculations were chosen to give good agreement with the clean surface data. The values used were, for the

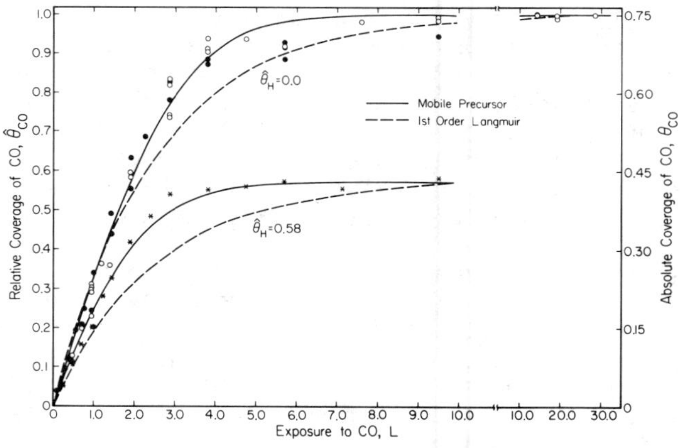

Figure 1. Comparison of coverage-exposure data with curves predicted by first-order Langmuir and mobile precursor adsorption kinetics.

Langmuir model,

$$\frac{d\hat{\theta}}{d\varepsilon} = 0.4(\hat{\theta}_s - \hat{\theta}), \qquad (1)$$

and for the mobile precursor

$$\frac{d\hat{\theta}}{d\varepsilon} = 0.525 \frac{\hat{\theta}_s - \hat{\theta}}{0.5 + \hat{\theta}_s - \hat{\theta}} \qquad (2)$$

where $\hat{\theta}$ is the coverage of CO normalized to unity at saturation on the clean surface, $\hat{\theta}_s$ is the measured relative saturation coverage of CO for the given initial coverage of hydrogen, and ε is the exposure in Langmuirs.

It is clear from Fig. 1 that the mobile precursor model fits the experimental data quite well, whereas the Langmuir model fails badly. The same level of agreement was found for other initial coverages of hydrogen. A major factor in the poor agreement of the Langmuir model is that it predicts too rapid a decrease in the initial slope of the coverage-exposure curve with increasing hydrogen coverage. The reason for this is that sites occupied by hydrogen atoms are considered to be blocked completely for CO adsorption. The mobile precursor model, on the other hand, allows physical adsorption of a CO molecule above a hydrogen atom, as well as above another CO molecule or an empty site prior to chemisorption. For this reason the mobile precursor model predicts a less rapid decrease in initial slope with increasing hydrogen coverage.

Low-Energy Electron Diffraction (LEED) was used to study the geometrical structure of the co-adsorbed system. The LEED patterns due to individual adsorption of hydrogen and CO have been investigated previously.[13-15] No ordered structures observable by LEED form during adsorption of hydrogen. Upon adsorption at 100 K, CO forms three ordered structures, at different coverages. At low coverage, CO forms a weak p(2x2) structure which increases in intensity steadily from zero coverage, suggesting island formation. At a coverage slightly less than one-quarter monolayer, local formation of a $(\sqrt{3} \times \sqrt{3})R30°$ structure ($\sqrt{3}$ structure) begins. This structure increases rapidly in intensity, reaching a maximum at the optimum coverage of one-third monolayer. Further exposure to CO causes compression of the overlayer until a (2x2) structure with three molecules per unit cell is reached at saturation.

No new LEED patterns were observed for co-adsorbed hydrogen and CO. However, the clean surface CO structures do form, under some circumstances, with modified intensity during co-adsorption. The effect of exposing the different CO superstructures to hydrogen and the formation of the CO superstructures on hydrogen covered surfaces have been investigated by monitoring the intensity of the LEED beams as a function of exposure to either hydrogen or CO.

Addition of hydrogen to the fully ordered ($\theta_{CO} = 0.23$) p(2x2) structure causes a rapid decrease in intensity of the p(2x2) pattern.

Similarly, addition of hydrogen to the fully ordered (θ_{CO} = 0.35) $\sqrt{3}$ structure causes a loss in intensity of the $\sqrt{3}$ pattern. Clearly, hydrogen atoms cannot occupy a position within either the p(2x2) or the $\sqrt{3}$ unit cell without perturbing the neighboring CO molecules.

The effect of hydrogen on the $\sqrt{3}$ structure for coverages of CO less than one-third is illustrated in Fig. 2. The dashed line shows

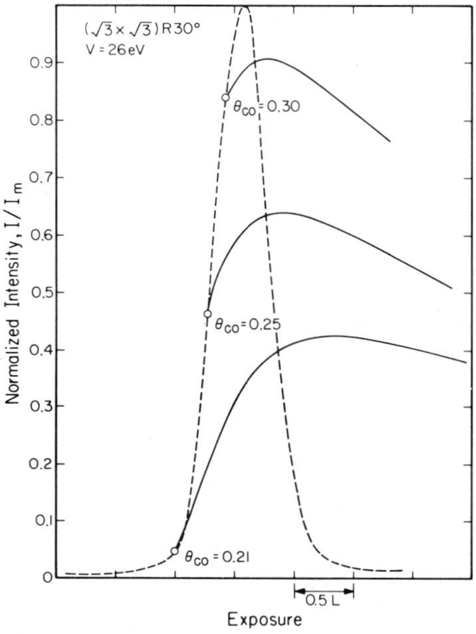

Fig. 2: LEED intensity as a function of exposure for the $\sqrt{3}$ structure. Dashed line shows the result of exposing the clean surface to CO. Solid lines show the result of terminating CO exposure at various coverages and initiating exposure to hydrogen. The surface temperature is approximately 100 K. Intensities are normalized to unity at the maximum intensity of the $\sqrt{3}$ structure on the clean surface.

the intensity-exposure behavior for CO adsorption on the clean surface. If exposure to CO is terminated before maximum ordering occurs, and followed by exposure to hydrogen, the solid curves result. At θ_{CO} = 0.25 and 0.30, hydrogen increases the amount of the ordered $\sqrt{3}$ structure on the surface. At θ_{CO} = 0.21, where only the p(2x2) CO structure is present, addition of hydrogen causes a transformation from the p(2x2) to the $\sqrt{3}$ structure. The maximum intensities obtained by addition of hydrogen are less than the maximum reached for CO adsorption alone. The asymmetric decrease in intensity with hydrogen exposure beyond the maxima is due to the decreasing probability of adsorption of hydrogen (18).

Addition of hydrogen to the partially ordered (θ_{CO} < 0.23) p(2x2) structure does not increase the order as for the $\sqrt{3}$ structure. Rather, there is a steady loss of intensity which is slower for lower coverages of CO.

Exposure to hydrogen caused no change in intensity of the partially ordered (θ_{CO} = 0.5 and 0.7) (2x2) structure. This may be due to the small probability of adsorption of hydrogen when a high coverage of CO is present.

The adsorption of CO onto a hydrogen precovered surface mimics adsorption onto the clean surface, as shown in Figs. 3 and 4. In both figures, curve (a) represents the intensity-exposure behavior for adsorption onto the clean surface. Curves (b) - (e) are for CO

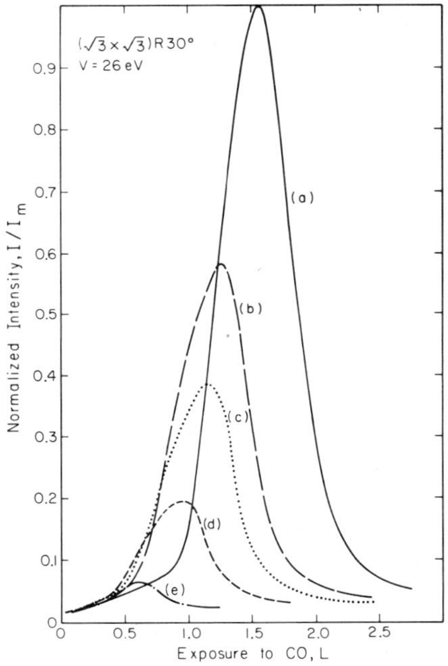

Fig. 3: LEED intensity as a function of exposure to CO for the $\sqrt{3}$ structure on a surface covered with varying amounts of hydrogen.

(a) $\hat{\theta}_H = 0$ (d) $\hat{\theta}_H = 0.75$

(b) $\hat{\theta}_H = 0.42$ (e) $\theta_H = 0.92$

(c) $\hat{\theta}_H = 0.58$

Intensities are normalized to unity at the maximum intensity of the $\sqrt{3}$ structure on the clean surface.

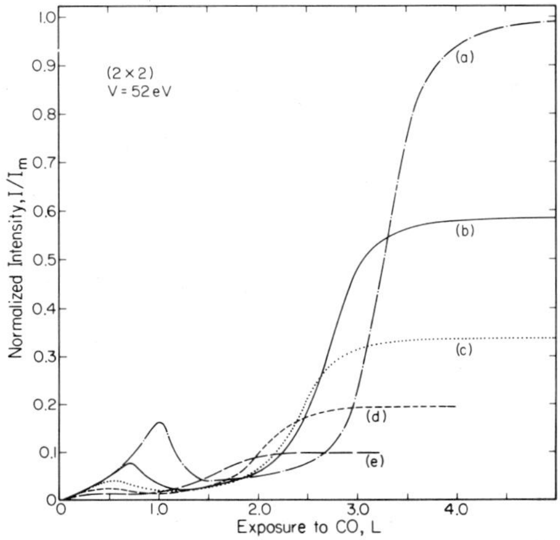

Fig. 4. LEED intensity as a function of exposure to CO for the two (2x2) structures on a surface covered with varying amounts of hydrogen.

(a) $\hat{\theta}_H = 0$ (d) $\hat{\theta}_H = 0.58$

(b) $\hat{\theta}_H = 0.26$ (e) $\hat{\theta}_H = 0.75$

(c) $\hat{\theta}_H = 0.42$

Intensities are normalized to unity at maximum intensity of the high coverage (2x2) structure.

adsorption onto a surface covered with increasing amounts of hydrogen. As on the clean surface, on a hydrogen-covered surface, CO first forms a p(2x2) structure which is transformed quickly to a $\sqrt{3}$ structure. Then the final (2x2) structure is formed by continuous compression of the $\sqrt{3}$ structure. However, on the hydrogen covered surface, the CO structures form at lower coverage and with lower intensity than on the clean surface.

DISCUSSION

The LEED results indicate that the structure of the mixed overlayer is different depending on the order of adsorption of hydrogen and CO. The results of the thermal desorption experiments, together with the intensity-exposure behavior give a consistent picture of the co-adsorbed system. As discussed in more detail elsewhere (18), the effect of adding hydrogen to the fully or partially ordered p(2x2) and $\sqrt{3}$ structures indicates clearly that there is a long range repulsive interaction between hydrogen atoms and CO molecules.

The nature of the CO-hydrogen repulsive interaction is of considerable interest. The disruption of the p(2x2) CO structure by addition of hydrogen shows that this interaction occurs over a hydrogen-CO distance as large as 3.1 Å. The hard core radius of CO, determined from the distance of closest approach in the saturation structure is 1.55 Å. The radius of a chemisorbed hydrogen atom is probably less than 0.75 Å.[19] Thus, the repulsive interaction between hydrogen and CO must be a through-metal effect rather than a result of orbital overlap.[20,21]

A comparison of LEED intensities for the formation of the p(2x2) structure, in the presence and in the absence of hydrogen, shows that little more CO is disordered in the presence of hydrogen than on the clean surface.[18] Since the p(2x2) structure cannot coexist with intermingled hydrogen atoms, the hydrogen and CO must be segregated on the surface. The $\sqrt{3}$ structure and the (2x2) structure follow the same intensity coverage behavior, indicating that the same degree of segregation of hydrogen and CO persists over the entire accessible coverage range.

The segregation of the co-adsorbed species requires some mechanism by which hydrogen and CO "find" others of their own kind. Chemisorbed hydrogen is thought to be relatively mobile even at low temperature. Hydrogen atoms could thus simply migrate away from the region of influence of a CO molecule. On the other hand, it is difficult to imagine migration of CO molecules on the scale necessary to form relatively large islands. Fortunately, the possibility of a physically adsorbed mobile precursor to adsorption for CO provides an elegant alternative to large scale migration. The probability for the precursor to chemisorb near a hydrogen atom would be decreased by the repulsive CO-hydrogen interaction. The precursor would then be likely to move away from sites near hydrogen atoms, and to chemisorb in sites near other CO molecules where the interactions are more favorable. The growth of islands of CO would be a straightforward result of this mechanism.

The mechanism for formation and growth of CO islands on a hydrogen covered surface deduced from these measurements is illustrated schematically in Fig. 5. The nucleation of an island is shown in Fig. 5(a). There, the initial chemisorption of a CO molecule causes

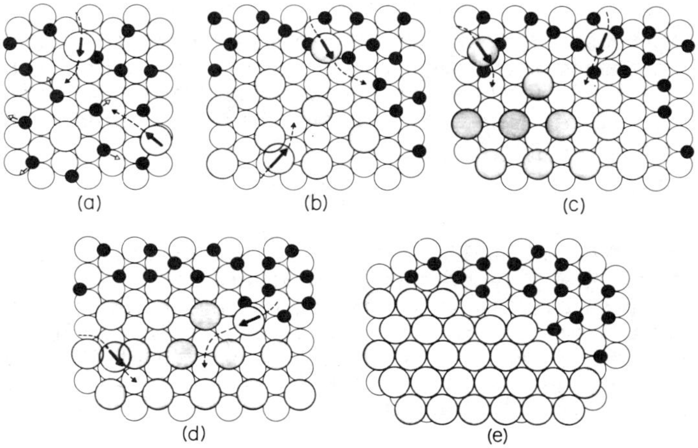

Fig. 5: Schematic illustration of adsorption of CO onto a surface covered with hydrogen. Open circles are Rh atoms with d = 2.7 Å. Shaded circles are CO molecules with d = 3.1 Å. Shaded circles with arrow are physically adsorbed, mobile precursor CO molecules. Solid circles are H atoms with d = 1.5 Å (estimated). Adsorption sites shown have been chosen arbitrarily. Hydrogen and CO trapped in the "wrong" regions are not shown for clarity. See text for discussion.

migration of hydrogen atoms away from its vicinity due to the repulsive hydrogen-CO interaction. Physically adsorbed CO molecules adsorb preferentially near the original CO molecule, minimizing the CO-hydrogen interaction and initiating island formation. As more CO adsorbs, statistically some nucleation sites will fail to develop. Pockets of CO, either lone molecules or small clusters, will then be trapped in regions of high hydrogen density. Similarly, some hydrogen atoms may be trapped in the growing CO islands. These trapped species are not shown in Fig. 5. In Fig. 5(b), the p(2x2) island has reached its maximum size. At this point, addition of a CO molecule at the edge of the island, which would cause either a decreased CO-H distance or a decreased H-H distance becomes energetically less favorable than addition of a CO to the interior of the island causing a decreased CO-CO distance. Transformation of the island to the $\sqrt{3}$ structure begins, and this is shown after completion in Fig. 5(c). A simple calculation using the experimental results indicates that the islands of $\sqrt{3}$ structure occupy three-halves as much surface area at their maximum intensity as the islands of p(2x2) structure at their maximum intensity. Therefore, once transformation of the p(2x2) islands to a $\sqrt{3}$ structure is complete, further growth by addition of CO molecules to the edges of the islands must occur. This process is shown beginning in Fig. 5(c) and at completion in Fig. 5(d). When the islands occupy so much area that addition of CO at the edges causes too great an increase in CO-H and H-H repulsive energy, island growth stops; and further addition of CO occurs within the island, causing compression of the CO overlayer to the high coverage (2x2) structure shown in Fig. 5(e). At this point, no further adsorption of CO occurs and saturation is reached.

CONCLUSIONS

(1) Adsorption of CO on Rh(111) proceeds via a physically adsorbed intermediate, a mobile precursor to adsorption.

(2) There is a repulsive interaction between co-adsorbed CO molecules and hydrogen atoms on rhodium. This results in complete segregation of hydrogen and CO following adsorption of CO onto an adlayer of hydrogen at 100 K. The repulsive interaction is a through-metal effect rather than a result of orbital overlap.

REFERENCES

1. M. A. Vannice, Catal. Rev. 14, 153 (1976).
2. G. C. Bond, Catalysis by Metals, Academic Press, New York, 1962.
3. J. A. Rabo, A. P. Risch and M. L. Pontsma, J. Catal. 53, 295 (1978).
4. M. Araki and V. Ponec, J. Catal. 44, 439 (1976).
5. R. D. Kelley, T. E. Madey and J. T. Yates, Jr., J. Catal. 50, 301 (1977).
6. T. E. Madey, D. W. Goodman and R. D. Kelley, J. Vacuum Sci. Technol. 16, 433 (1978).
7. R. D. Kelley, T. E. Madey, K. Revesz and J. T. Yates, Jr., Appl. Surface Sci. 1, 266 (1978).
8. B. A. Sexton and G. A. Somorjai, J. Catal. 46, 167 (1977).
9. D. J. Dwyer and G. A. Somorjai, J. Catal. 52, 291 (1978).
10. K. Kraemer and D. Menzel, Ber. Buns. Ges. 79, 649 (1975).
11. V. H. Baldwin and J. B. Hudson, J. Vacuum Sci. Technol. 8, 49 (1971).
12. J. C. Bertolini and B. Imelik, Surface Sci. 80, 586 (1979).
13. J. T. Yates, Jr., P. A. Thiel and W. H. Weinberg, Surface Sci. 84, 427 (1979).
14. P. A. Thiel, E. D. Williams, J. T. Yates, Jr. and W. H. Weinberg, Surface Sci. 84, 54 (1979).
15. D. G. Gastner, B. A. Sexton and G. A. Somorjai, Surface Sci. 71, 519 (1978).
16. P. Kisliuk, J. Phys. Chem. Solids 3, 95 (1957).
17. R. Gorte and L. D. Schmidt, Surface Sci. 76, 559 (1978).
18. E. D. Williams, P. A. Thiel, W. H. Weinberg and J. T. Yates, Jr., J. Chem. Phys. 72, 0000 (1980).

19. K. Christmann, R. J. Behm, G. Ertl. M. A. Van Hove and W. H. Weinberg, J. Chem. Phys. 70, 4168 (1979).
20. T. B. Grimley and M. Torrini, J. Phys. C 6, 868 (1973).
21. T. Einstein and J. R. Schrieffer, Phys. Rev. B 7, 3629 (1973).

MAGNETIC PHASE DEPENDENCE OF THE NICKEL CARBONYL REACTION RATE[&]

R.S. Mehta[†#], M.S. Dresselhaus[†],
G. Dresselhaus[‡], and H.J. Zeiger[*]
Massachusetts Institute of Technology, Cambridge, MA 02139, USA

ABSTRACT

The activation energy for $Ni(CO)_4$ formation on surfaces of $Ni_{1-x}Cu_x$ alloys is found to be greater when the metallic substrate is ferromagnetic than when it is paramagnetic. This difference in activation energy is identified with the bulk exchange energy of the substrate and depends on the bulk Ni concentration. On the basis of the experimental results, a model for the step associated with rupture of the metallic bond is presented.

INTRODUCTION

The influence of the bulk ferro- to paramagnetic phase transition on the rate of a surface chemical reaction is reported. This effect, known as the Hedvall effect, was discovered in 1934 in the context of catalytic chemical reactions.[1,2] Two variations of this effect have been noted, one corresponding to an anomaly in the reaction rate at the Curie temperature T_c (Hedvall Effect I) and the other to a change in activation energy at T_c (Hedvall Effect II). We consider here the influence of the magnetic state of Ni (whether ferro- or paramagnetic) on its reaction with CO gas at atmospheric pressure to form diamagnetic $Ni(CO)_4$ in the reaction: $Ni + 4CO \rightarrow Ni(CO)_4$. We show here that the nickel carbonylation reaction provides an excellent system for the study of the Hedvall Effect.

The nickel carbonylation reaction is conveniently carried out in the temperature range $25 < T < 180C$, the upper limit imposed by the rapid spontaneous decomposition of $Ni(CO)_4$ for $T > 180C$,[3] and the lower limit imposed by the slowness of the reaction rate at low T. An understanding of the correlation between bulk magnetization and chemical reactivity could be important for modelling electronic interactions in catalytic chemical reactions which frequently employ catalysts exhibiting magnetically ordered phases. Nickel is an example of an extensively used catalyst which is normally ferromagnetic.

†Department of Electrical Engineering and Computer Science and Center for Materials Science and Engineering.
‡Francis Bitter National Magnet Laboratory, supported by NSF.
*MIT Lincoln Laboratory. The Lincoln Lab portion of this work was sponsored by the Department of the Air Force.
#IBM predoctoral fellow during 1977-78.
&Paper read by G. Dresselhaus.

ISSN:0094-243X/80/610285-11$1.50

The etching reaction on the magnetic metal caused by CO gas is ideally suited for the study of the correlation between the magnetic state of the reacting material and its chemical reactivity for several reasons: (i) The final reaction product $Ni(CO)_4$ is in the gas phase and is carried off the Ni surface leaving behind the unreacted Ni atoms. Hence the incoming CO gas always sees a magnetic metal surface with exposed Ni atoms. (ii) Nickel and copper form substitutional alloys $Ni_{1-x}Cu_x$ over the entire composition range x, and the Curie temperature, T_c, for this alloy system is a linear function of x. Thus it is possible to obtain $Ni_{1-x}Cu_x$ alloys with T_c in the range $0 < T_c < 631K$; this range of Curie temperatures is achieved with Cu concentrations in the range $0.6 \geqslant x \geqslant 0$.[4] Hence the nickel carbonylation reaction can be carried out using $Ni_{1-x}Cu_x$ alloys with T_c within the temperature range where the reaction is conveniently carried out. (iii) The interaction of CO with $Ni_{1-x}Cu_x$ alloys is limited to reaction of CO with Ni to form $Ni(CO)_4$ since Cu does not react with CO under the conditions in this experiment.

We report here a study of $Ni(CO)_4$ formation using polycrystalline, single crystal and powdered Ni samples and polycrystalline Ni-Cu alloy samples. For each Ni and $Ni_{1-x}Cu_x$ sample, an Arrhenius plot of ℓnR vs $1/T$ shows a linear dependence in the range $300 < T < 400K$ where R is the $Ni(CO)_4$ formation rate. From the slopes of these plots we obtain the activation energy, ΔE, for this reaction. We find that ΔE changes from ΔE_{ferro} to ΔE_{para} when the samples are cycled through the ferro- to paramagnetic transition, and that these changes in ΔE at T_c correlate with the bulk exchange interaction of the substrate.

EXPERIMENTAL DETAILS

A schematic diagram of the system[5] used to study the magnetic phase dependence of the nickel carbonylation reaction is shown in Fig. 1. This reaction is carried out in a Vycor glass flow reactor, which is temperature controlled by an oil bath operating in the range $25 < T < 180C$. The trace amounts of $Ni(CO)_4$ in the CO gas emerging from the reactor are continuously detected by a photoionization detector (PID) before being absorbed in an alcohol/iodine solution for disposal or calibration tests.[6] Gas stream impurities such as trace oxygen, water and hydro-carbons are carefully controlled and monitored with appropriate instrumentation.[5] The most important experimental considerations include the preparation and characterization of the Ni and $Ni_{1-x}Cu_x$ samples, the purity of the gas streams,[7,8] the control of the reaction conditions such as temperature and flow rate[7,9] and the detection system for $Ni(CO)_4$ [10,11]. The experimental details are described in more detail elsewhere.[5,11]

The aspect of the system that was critical to the success of the experiment was the PID on-line detection system for quantitative determination of trace amounts of $Ni(CO)_4$ in a large CO background (relative $Ni(CO)_4$ concentration $\simeq 10^{-6}$).[5,11] The detection problem was complicated by the extreme toxicity of $Ni(CO)_4$ which constrained us to work in the low reaction rate regime ($<4\mu g/min$).[12] The PID (shown in Fig. 2) contains a UV light source of photons ($h\nu \simeq 10.2eV$)

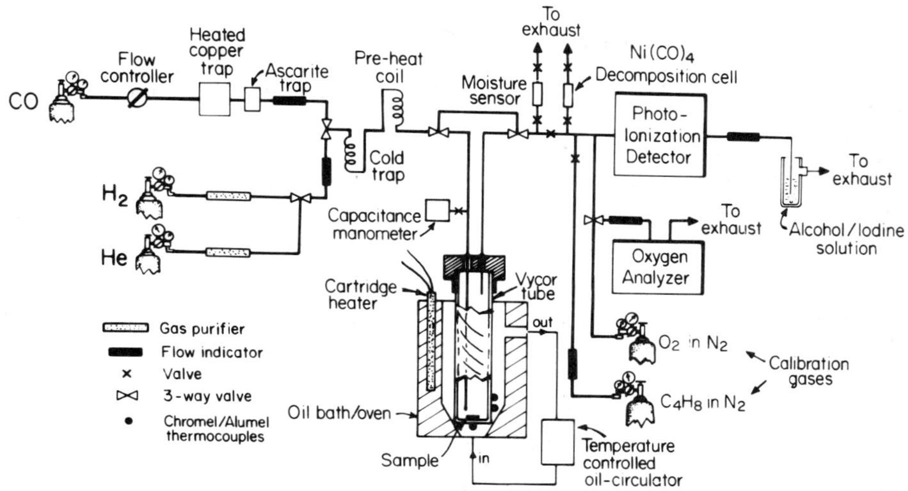

Fig. 1 Schematic representation of the experimental system used to study the $Ni(CO)_4$ reaction.

incident into a photoionization chamber. Since the photoionization threshold for $Ni(CO)_4$ is 8.28eV[13], its presence in the chamber gives rise to a collection current, whereas other gases which are present in the gas stream at various times (CO, H_2, He, CH_4, C_2H_6 and other hydrocarbons) have photoionization thresholds in excess of 10 eV,[13] thereby resulting in negligible photoionization current.

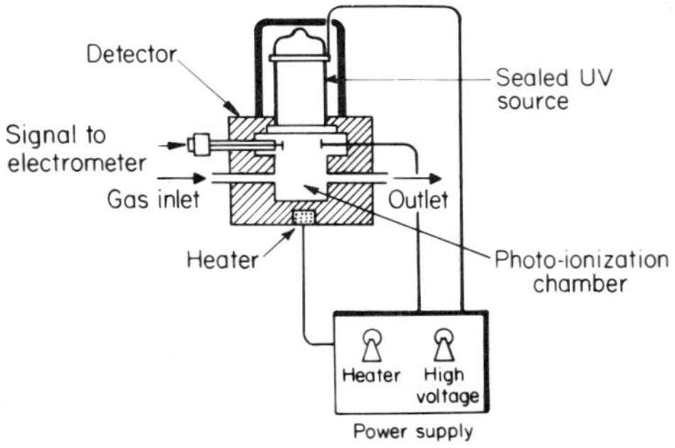

Fig. 2 Schematic diagram of the photoionization detector (PID). The UV source ionizes all molecules with an ionization potential less than 10.2 eV, which enables the detection of $Ni(CO)_4$ with an ionization potential of 8.28 eV in a large background of CO (ionization potential of 14 eV).

Prior to initiation of the carbonylation reaction, the samples underwent several procedures which were somewhat different for the Ni and $Ni_{1-x}Cu_x$ samples The Ni samples were mechanically and chemically polished as described elsewhere [5,11] before insertion into the reaction chamber. Each samples was then heated for 4 hours in a stream of H_2 gas at atmospheric pressure and at a temperature of 400C and subsequently cooled to room temprature in the H_2 gas stream. The same heat treatment was used for the $Ni_{1-x}Cu_x$ alloy samples prior to initiation of the carbonylation reaction.

For the $Ni_{1-x}Cu_x$ samples several additional steps were employed between the initial chemical polishing and the heat treatment process Each alloy was homogenized at 1050C for about 100 hours in an argon filled quartz ampoule. The ampoule was then removed from the high temperature furnace, rapidly quenched in cold water, and the sample was then characterized for the bulk and surface concentration of Ni and Cu. The bulk concentration was found at several points on the sample using an electron microprobe operating at 15 keV for which the electron penetration depth is estimated[14] to be 1μm, and the data analysis was corrected for matrix effects.[14] The surface composition was determined using Auger spectroscopy. Because of the different erosion rates for Ni and Cu from the alloy surface[15] when subjected to argon-ion bombardment, the alloy surfaces were not sputtered prior to measurement. The Auger spectrum for $Ni_{.6}Cu_{.4}$ is shown in Fig. 3 for the newly prepared sample after the heat treatment process (top) and for the sample after reacting with CO in the reaction chamber for about 4 hours.

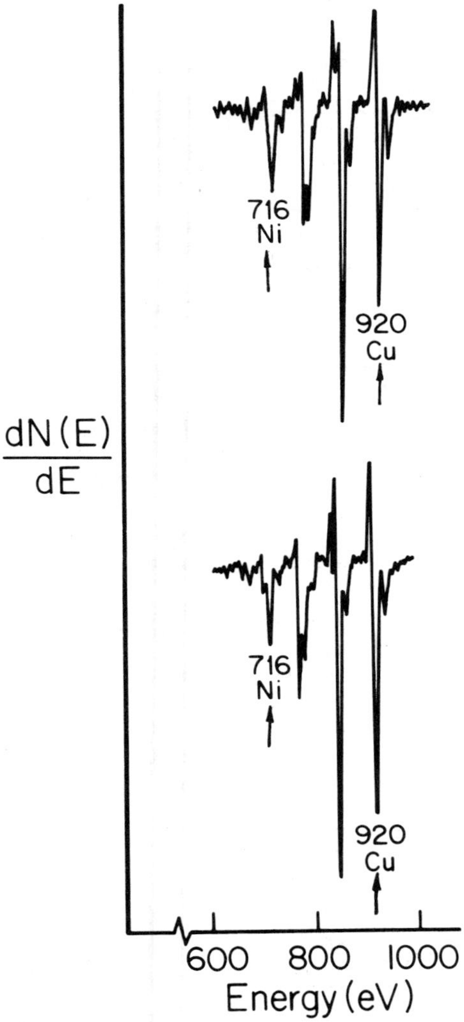

Fig. 3 Auger spectrum of the $Ni_{.6}Cu_{.4}$ alloy taken before the reaction (top) and after reaction with CO for about 4 hours in the reaction chamber. The ratio of the Ni "peak" at 716 eV to the Cu "peak" at 920 eV exhibits approximately the same time dependence as the $Ni(CO)_4$ reaction rate. The build up of Cu on the surface decreases the reaction rate, but does not alter the activation energy.

The ratios of the Ni to Cu Auger peaks at 716 eV and 920 eV

respectively was used to infer the change of surface conditions during the $Ni(CO)_4$ production. The ratio of the Auger peaks decayed exponentially, $[Ni/Cu]_{Auger} \sim e^{-t/\tau}$, with approximately the same decay time τ as in the rate of $Ni(CO)_4$ formation, indicating that the presence of Cu on the surface blocks the formation of $Ni(CO)_4$. In relating the Auger peaks to the surface properties, we note that the Auger electron escape depth is estimated to be 5-8 atomic layers[15] for these Auger electron energies.

Because of the high sensitivity of the nickel carbonylation reaction to trace impurities, high purity CO (99.99%) and H_2 (99.9999%) gases and special gas purification systems were used (see Fig. 1). Since the carbonylation reaction is especially sensitive to Ni oxidation,[7] the oxygen level was specially monitored by a suitable gas analyzer.[5] Clean helium gas was used for flushing out the system, and high purity C_4H_8 gas diluted in N_2 was used to periodically calibrate the PID. The use of stainless steel in the system was minimized, especially in critical areas, to avoid the possibility of iron carbonyl formation which would also be detected by the PID. The temperature was measured by chromel-alumel thermocouples attached externally to the reaction chamber.

On first reacting a freshly hydrogen-treated Ni sample with CO, a large transient reaction rate is observed, followed by a rapid decrease to a reproducible steady-state value (steady for tens of minutes) as shown in Fig. 2 of Ref. 11. For the $Ni_{1-x}Cu_x$ samples, the initial transient is followed by a slow decay characterized by the decay time τ. This time dependence (t) for the reaction rate R is conveniently displayed by plotting $\log_{10}R$ vs t, as shown in Fig. 4

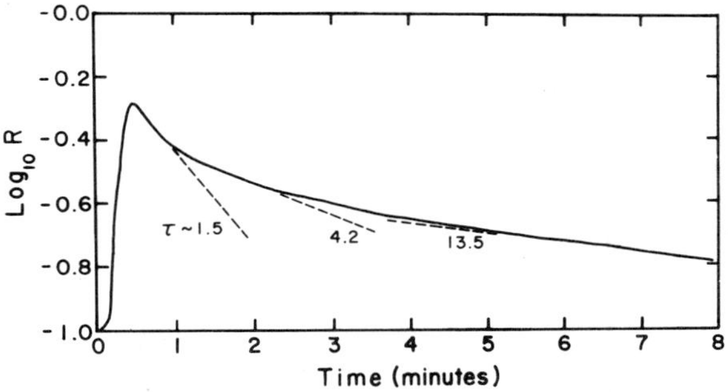

Fig. 4 Log (base 10) plot of $Ni(CO)_4$ production rate vs time ($t \leq 8$ min) for a $Ni_{.72}Cu_{.28}$ alloy sample. The initial build up occurs in less than 1 minute which is the time to sweep the H_2 out of the reaction chamber. The intial production has two exponential dependences with decay times of 1.5 min and 4.2 min respectively followed by the "steady state" region with a 13.5 min. decay time. Within 10 minutes all measured samples were found to settle down to a steady state behavior, which was characterized by single decay time for a time period of several hours. The activation energies reported in this article were measured under steady state conditions.

for a $Ni_{.72}Cu_{.28}$ alloy sample. The time dependence of Fig. 4 suggests that just after the heat treatment procedure, there are a certain number of highly reactive surface sites on the magnetic substrate (presumably with a smaller activation energy). After these sites have been depleted by the reaction, surface etching, characterized by a long decay time τ, sets in. This slow decrease in reaction rate (referred to as the "steady state" condition) is associated with a change in the exposed nickel surface area resulting from the presence of Cu atoms that are not removed by the reaction. Surface contamination caused by reaction with gas impurities (e.g. NiO formation) is responsible for a very slow decrease in reaction rate (time constant of several hours) which occurs for both Ni and $Ni_{1-x}Cu_x$ alloy substrates. The steady-state reaction rates R are temperature dependent and the temperature dependence is reproducible for each sample during a given run. The dependence of R on time and 1/T for Ni and the $Ni_{1-x}Cu_x$ alloys is found to be exponential. Thus, the rate data at various values of T can be analyzed to provide an accurate fit to the relation $R=R_0 \exp(-\Delta E/kT) \exp(-t/\tau)$, thereby yielding ΔE. In analyzing the experimental results, ΔE is most conveniently found from the slope of an Arrhenius plot of ℓnR vs 1/T. The points on this plot are obtained making use of the exponential time dependence of the steady state reaction rates to compare R at constant time when the temperature is changed from T_1 to T_2. Another possible approach to the determination of ΔE could be through a plot of $\ell n\ 1/\tau$ vs 1/kT.

Provided that the temperature range of the measurement does not include the Curie temperature, T_c, the carbonylation rate data follow a simple Arrhenius plot yielding a well-defined activation energy. An example of such a simple Arrhenius plot is shown in Fig. 5 for a

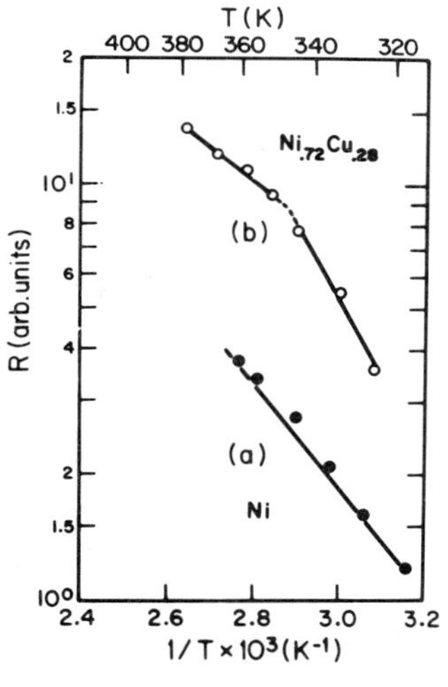

Fig. 5 Arrhenius plots for the temperature dependent reaction rate R (of the $Ni(CO)_4$ reaction) for (a) a pure nickel sample and (b) a $Ni_{1-x}Cu_x$ sample with x=.28. From the slope of ℓnR vs (1/T) the activation energies ΔE are determined. We note that for $Ni_{1-x}Cu_x$ alloy sample, ΔE changes within a temperature range of 10K about T_c from $\Delta E_{para} = 0.15 \pm .05$ eV in the paramagnetic phase to $\Delta E_{ferro} = 0.35 \pm 0.05$ eV in the ferromagnetic phase. For pure Ni, $\Delta E_{ferro} = 0.23 \pm .03$ eV. In the figure R is plotted in arbitrary units on a $\log_{10}$ scale.

pure Ni sample, yielding $\Delta E_{ferro} = 0.23 \pm 0.03$ eV.[16] However, for Cu concentrations such that T_c is contained within the temperature range of measurement, a change in the slope of the Arrhenius plot is found at T_c, as shown in Fig. 5 for a $Ni_{.72}Cu_{.28}$ sample, where ΔE changes from $\Delta E_{ferro} = 0.35 \pm 0.05$ eV to $\Delta E_{para} = 0.15 \pm 0.05$ eV in a temperature interval of <10K about T_c. It is significant that despite the large decrease in R during a typical run on a $Ni_{1-x}Cu_x$ sample (by as much as one order of magnitude), indicating a major change in the number of exposed Ni atoms, ΔE associated with that sample remained invariant during the course of the run and from one run to the next. It is also significant that by treatment for 4 hours at 400C in an H_2 atmosphere, the reaction rate of a $Ni_{1-x}Cu_x$ sample could be increased without change of ΔE. From this analysis we conclude that ΔE for a given sample is determined by the bulk composition and is relatively insensitive to the number of exposed surface Ni atoms. From Arrhenius plots such as in Fig. 5, values for ΔE_{para} and ΔE_{ferro} are obtained for the various Ni and $Ni_{1-x}Cu_x$ samples and the results are presented in Fig. 6 as a function of Cu

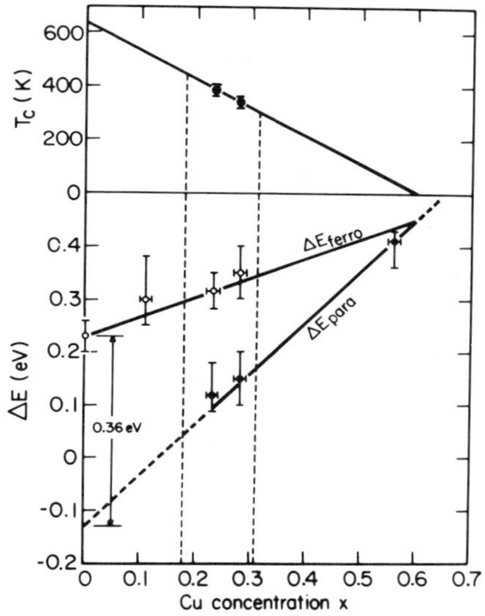

Fig. 6 Plot of ΔE for the nickel carbonylation reaction on ferromagnetic and paramagnetic substrates vs Cu concentration x for the $Ni_{1-x}Cu_x$ alloy system. The experimental points are consistent with the indicated lines. The result for ΔE_{ferro} for pure Ni is given at x=0. Also plotted are T_c vs x for the $Ni_{1-x}Cu_x$ system showing that $T_c \to 0$ for x=0.6. The experimental points for T_c vs x are obtained from the break in the slopes of the Arrhenius plots (see Fig. 5). Indicated by dashed vertical lines is the range of concentrations for which T_c falls within the temperature range 300<T<440K over which the $Ni(CO)_4$ reaction is conveniently studied. The difference ($\Delta E_{ferro} - \Delta E_{para}$) indicated on the diagram for pure Ni is identified with the exchange energy for Ni.

concentration x. Also shown in this figure is the dependence of T_c on x for the $Ni_{1-x}Cu_x$ system, and the range of x over which T_c falls within the temperature range of the nickel carbonylation reaction.

A plausible outline for the steps in the $Ni(CO)_4$ formation reaction is sketched within the framework of the schematic reaction coordinate diagram of Fig. 7. The maximum energy of the Ni-CO system is represented by point A corresponding to a Ni atom on the surface and 4 CO gas molecules: $[Ni]_s + 4[CO]_g$. Point F represents the gaseous $[Ni(CO)_4]_g$ molecule which is lower in energy than $[Ni]_s + 4[CO]_g$ by $\sim$1.7 eV representing the heat of formation of the

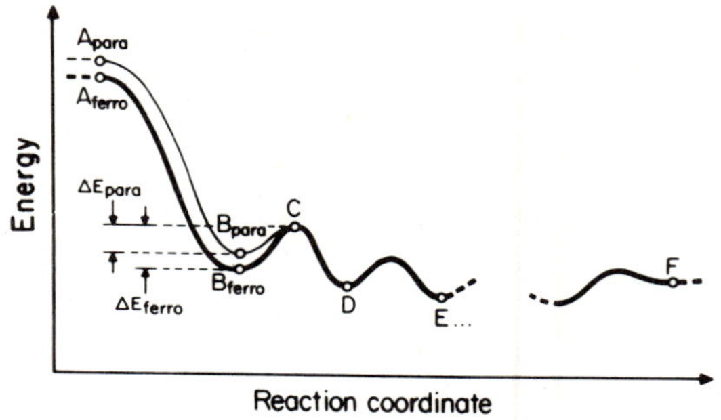

Fig. 7 Schematic reaction coordinate diagram for $Ni(CO)_4$ formation. The energy separation between points A and B represents the energy of chemisorption of CO on a clean nickel metal surface while the energy difference between points A and F refers to the formation energy of $Ni(CO)_4$ from Ni metal and CO gas. The activation energies ΔE_{para} and ΔE_{ferro} for the step at which the Ni bond to the surface is ruptured are indicated by the potential barriers between points B and C. According to our model, when the substrate is ferromagnetic (heavy line), points A_{ferro} and B_{ferro} are both lowered relative to A_{para} and B_{para} (light line) by the exchange energy, thereby causing ΔE_{ferro} to be greater than ΔE_{para}. Our model also assumes that the addition of Cu lowers points A and B by an equal amount relative to point C leading to an increase in ΔE with addition of Cu. Once $[Ni(CO)_x]_s'$ is formed on the surface at point D, the energy becomes independent of the amount of Ni exposed on the surface and of its magnetic state. Points E... denote the addition of successive CO molecules to $[Ni(CO)_x]_s'$.

free $Ni(CO)_4$ molecule.[3] Since the energy to remove a nickel atom from the surface to the free atom state is about ∿4.4 eV,[13] the total bond energy in the free $Ni(CO)_4$ molecule is about 6 eV. Studies of the temperature dependence of desorption of CO from Ni yields an adsorption energy[17] of ∿1.2 eV at coverage $\theta \sim 0.5$ which is indicated by point B on the figure and represents $[Ni(CO)]_s + 3[CO]_g$. In order to form a molecule of gaseous $Ni(CO)_4$, a Ni atom must be dislodged from the surface. We argue that this step gives rise to a potential barrier of height ΔE between points B and C of the figure. Point D at lower energy represents $[Ni(CO)_x]_s' + (4-x)[CO]_g$, where $[Ni(CO)_x]_s'$ denotes a molecular unit bound weakly to the surface, and x is between 1 and 3. The binding of the $Ni(CO)_x$ unit in the state $[Ni(CO)_x]_s'$ contains a large fraction of the bond energy of the $Ni(CO)_4$ molecule. Points E...in Fig. 7 are identified with the addition of successive CO molecules to the $[Ni(CO)_x]_s'$ unit with small gains in energy, and correspondingly small activation barriers and rapid reaction rates. The final step in the chain is reached at point F, representing $[Ni(CO)_4]_g$.

The identification of point B with the process of rupturing the Ni atom bond to the surface is consistent with the observation that the activation energy for $Ni(CO)_4$ formation is larger in the ferro- than in the paramagnetic state by approximately the bulk exchange energy (∿.4 eV)[18] (see heavy and light curves in Fig. 7). This interpretation that ΔE is associated with the rupture of the Ni atom bond to the surface further suggests that $[\Delta E_{ferro} - \Delta E_{para}]$ for the $Ni_{1-x}Cu_x$ alloys is also identified with the exchange energy of these alloys. This interpretation is also consistent with a decrease in the exchange energy and in $[\Delta E_{ferro} - \Delta E_{para}]$ for these alloys with increasing Cu concentration, and a vanishing of these quantities as $T_c \to 0$.

The bonding energy of a Ni atom in a $Ni_{1-x}Cu_x$ alloys can be estimated from the heats of sublimation of Ni and Cu,[19] and is found to decrease with the addition of Cu, thereby increasing the energy difference between points A and F in Fig. 7. There is evidence from thermal desorption measurements[20] that the addition of Cu atoms does not affect the energy of adsorption of CO on Ni. Thus, the addition of Cu does not affect the energy difference between points A and B. The present experiments however show that the addition of Cu results in an increase in ΔE. This implies that the activated complex at point C must be raised in energy more than at point B in order to produce a net increase in ΔE with Cu alloying. Once the surface bond has been ruptured, the energy becomes independent of the amount of exposed Ni on the surface or of its magnetic state (see Fig. 7).

Sales and Maple have previously reported observation of a Hedvall Effect in the oxidation of Ni.[21] For the NiO reaction the magnitude of the activation energy is much larger than for $Ni(CO)_4$ formation, with $\Delta E_{ferro} \sim 2.7$ eV and $\Delta E_{para} \sim 1.7$ eV. Although the Hedvall Effect for the $Ni(CO)_4$ and NiO reactions has the same sign, a direct comparison of the magnitudes of $\Delta E_{ferro} - \Delta E_{para}$ is not possible because in one case the reaction product $Ni(CO)_4$ comes off as a gas while in the other case NiO remains on the surface after the reaction is completed. Thermally-activated diffusion of Ni ions through NiO appears to be important in the oxidation of Ni and may be important in other surface reactions. Because of the large activation energy (∿1.5 eV) for bulk diffusion of nickel through metallic Ni and $Ni_{1-x}Cu_x$[22] relative to the values of ΔE in Fig. 6, we further conclude that nickel diffusion is not the principal mechanism controlling the $Ni(CO)_4$ reaction rate.

Suhl[23] has discussed theoretically the role of long range magnetic fluctuations near a second order phase transition in chemical reaction rates and has predicted anomalies in the reaction rate near T_c. D'Agliano and Huberman[24] have developed a theoretical model for the sublimation of Co atoms from the surface of Co metal which predicts an anomaly at T_c in addition to a change in activation energy. In contrast, the change in activation energy at T_c for the $Ni(CO)_4$ reaction (see Fig. 5) does not, within experimental error, give evidence for an additional anomaly. Further careful experimental measurements in the region of T_c for $Ni(CO)_4$ reaction rates are necessary to detect possible deviations from Arrhenius behavior which is assumed for $T < T_c$ and $T > T_c$ in Fig. 5.

We suggest that a diamagnetic complex is formed when the Ni atom breaks away from the surface and becomes part of an adsorbed molecule (at point D in Fig. 7). The $Ni(CO)_4$ molecule in the gas phase is known to be diamagnetic.[3] The observation that the activation energy for $Ni(CO)_4$ formation is dependent on the magnetic state of the Ni argues strongly that the initial adsorption complex $[Ni(CO)]_s$ at point B is still magnetic, and that the metallic bonding of a Ni atom to its neighbors inhibits the formation of the diamagnetic $[Ni(CO)_x]_s$' complex. It appears highly probable that after the rupturing of the Ni bonds to the metallic surface, the Ni-CO complexes are diamagnetic as described by our model. This interpretation indicates a need for calculations of the magnetic properties of the intermediate species in the nickel carbonylation reaction.

ACKNOWLEDGMENTS

We are grateful to Profs. E.I. Solomon and H. Suhl, and Drs. T. Collins and G. Wolken, Jr. for valuable and illuminating discussions. We also wish to express our thanks to Prof. J.S. Kouvel and Dr. J.W. Cable for generous gifts of some of the $Ni_{1-x}Cu_x$ samples used in this study. We gratefully acknowledge support for this work from the Air Force Office of Scientific Research, Grant #77-3130.

REFERENCES

1. J.A. Hedvall, R. Hedin and O. Persson, Z. Physik. Chem. B27, 196 (1934).
2. G. Cohn and J.A. Hedvall, J. Phys. Chem. 46, 841 (1942).
3. P.W. Jolly and G. Wilke, The Organic Chemistry of Nickel, Vol. 1, Academic Press, NY (1974), p. 2.
4. R.M. Bozorth, Ferromagnetism, D. Van Nostrand Co., Inc., Princeton, NJ, 1961.
5. R.S. Mehta, Ph.D. Thesis, M.I.T., 1979 (unpublished).
6. J.G. Kincaid, E.L. Stanley, C.H. Beckworth and F.W. Sunderman, Am. J. Clin. Pathol. 26, 107 (1956).
7. G.S. Krinchik and R.A. Shvartsman, Sov. Phys. JETP 40, 1153 (1975).
8. W. Goldberger, D. Ch. E. Thesis, Polytechnic Inst. Brooklyn, 1961 (unpublished).
9. W.M. Goldberger and D.F. Othmer, Ind. Eng. Chem. Process Des. Develop. 2, 202 (1963).
10. The PID instrument model no. PI 201 is manufactured by HNU Systems, Inc., Newton, MA.
11. R.S. Mehta, M.S. Dresselhaus, G. Dresselhaus and H.J. Zeiger, Surface Science L681, 78 (1978).
12. Hygienic Guide Series, Nickel Carbonyl, Am. Ind. Hyg. Assoc. J. 29, 304 (1968).
13. R.C. Weast, ed., CRC Handbook of Physics and Chemistry, 55th edition, CRC Press, Cleveland, Ohio, 1974.

14. S.J.B. Reed, Electron Microprobe Analysis, Cambridge University Press, Cambridge, 1975.
15. C.R. Helms and K.Y. Yu, Vac. Sci. Technol. 12, 276 (1975).
16. The present determination of ΔE is based on more data points and more samples than were used in the previous determination ($\Delta E = 0.21 \pm 0.05$ eV), Ref. 11, so that the present result has a smaller experimental error.
17. G. Wedler, H. Papp and G. Schroll, Surface Science 44, 463 (1974). Although the bonding energy of CO on Ni depends on CO coverage, the value pertinent to the $Ni(CO)_4$ formation is identified with the broad plateau ($\theta \approx 0.5$).
18. C. Herring, Magnetism Vol. IV, edited by G.T. Rado and H. Suhl, Academic Press, NY, 1966.
19. J.J. Burton, E. Hyman and D.G. Fedak, J. Catal. 37, 106 (1975).
20. Y. Soma and W.M.H. Sachtler, Japan J. Appl. Phys. Suppl. 2 Pt. 2, 241 (1974).
21. B.C. Sales and M.B. Maple, Phys. Rev. Lett. 39, 1636 (1977).
22. A. Ya. Kipnis and N.F. Mikhailova, Kinetics and Catalysis 12, 1240 (1971).
23. H. Suhl, The Physical Basis for Heterogeneous Catalysis, eds. E. Drauglis and R.I. Jaffee (Plenum Press, New York, 1975) p. 427; and Phys. Rev. B11, 2077 (1975).
24. E.G. d'Agliano and B.A. Huberman (to be published in Solid State Commun.).

LASER DIAGNOSTICS OF HO RADICAL FORMATION IN THE $H_2 + N_2O$ REACTION ON PLATINUM

L. D. Talley* and M. C. Lin
Chemistry Division
Naval Research Laboratory
Washington, DC 20375

ABSTRACT

Catalytic oxidation of H_2 by N_2O on Pt has been investigated in a fast flow system. HO radicals were detected by means of the laser-induced fluorescence technique using a tunable dye laser. The kinetics of the HO radical production have been measured at different reactant pressures and catalyst temperatures. These results indicate that the desorption process

$$HO* \longrightarrow HO + *$$

takes place with an activation energy of 28 ± 2 kcal/mole.

INTRODUCTION

Due to its intensity, monochromaticity and its ever expanding tunability, the laser has been most successfully applied to excite and detect atomic and molecular species at concentration levels of $\geq 10^8$ particles/cm^3. For some metal atoms with large optical absorption cross-sections, single-atom detection[1,2] with the laser was achieved several years ago.

We have recently employed a flash-lamp-pumped tunable dye laser, frequency-doubled to about 300 nm to detect the production of the HO radical from several reactions involving ground state[3] and electronically excited state[4] oxygen atom reactions using the laser-induced fluorescence (LIF) technique. The principle and application of this technique have been reviewed by Kinsey.[5] Very recently, we applied this very sensitive probing technique to monitor the formation of HO radicals produced in the catalytic oxidation of H_2 by O_2.[6,7]

In this work, we investigate the kinetics and mechanism of the $H_2 + N_2O$ reaction on Pt surfaces. We employed the LIF

*NRC/NRL Postdoctoral Research Associate

ISSN:0094-243X/80/610297-09$1.50

technique to probe the production of HO radicals as a function of both reactant pressure and catalyst temperature. The results of this study are reported herein.

EXPERIMENTAL

A. Samples

One percent mixtures of H_2 (Matheson, Ultrahigh purity) and N_2O (Matheson, CP) in argon (Matheson, UHP) were prepared separately for use as samples by mixing at the time of the experiment. O_2 (Matheson, extra dry) was used to prepare the catalyst. The H_2-Ar sample mixture was also used for the catalyst preparation.

B. Catalyst Preparation

The production of HO was found to be sensitive to the method of preparing the Pt catalyst. The pretreatment of the catalyst in these experiments is described as follows:

1. burning of the Pt wire at 900°C for about 30 minutes in the presence of 5 torr O_2.

2. vacuum degassing of the catalyst at 950°C for 30-45 minutes.

3. reduction by H_2 with 50 mTorr of the slowly flowing 1% H_2-Ar mixture at around 900°C.

C. LIF Measurements of HO Production

The experimental apparatus used to detect the production of HO from the Pt catalyst in the gas phase is shown schematically in Fig. 1. The reaction cell consisted of a black-anodized cylinder equipped with the following: a stainless steel reactant-gas mixing nozzle in the top and a vacuum port in the bottom for continuous pumping; sidearms with light baffles and Brewster-angle windows for passage of the excitation laser pulses; a large-aperture window for collection of fluorescence at 90° to the laser propagation axis and to the gas flow direction.

A Pt (99.999% pure, 0.50 mm diameter) wire was coiled to form a grid which was suspended in the horizontal plane just below the gas-mixing nozzle. The catalyst wire was heated resistively and the wire temperature was determined by means of the measured resistance. Gas pressures were measured *in situ* with an MKS Model 145 Baratron capacitance manometer.

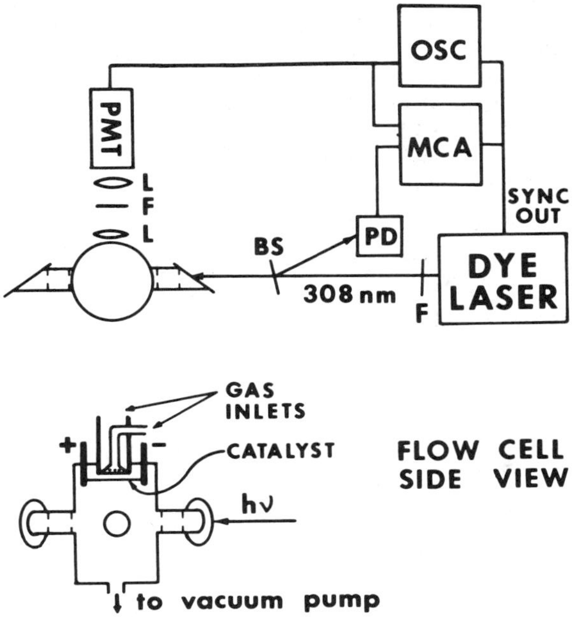

Fig. 1. Diagram of the apparatus used for HO LIF measurements in the gas phase. The second harmonic (307.8 nm) of the dye laser frequency is filtered (F) and sampled by a beam-splitter (BS) to a photodiode (PD). The HO fluorescence is collected by lenses (L) and imaged onto a photomultiplier tube (PMT). The PMT and PD signals are processed by an oscilloscope (OSC) and a multichannel analyzer (MCA).

The Chromatix CMX-4 dye laser was used to excite the $A^2\Sigma \leftarrow X^2\Pi$ transition of HO in the gas phase. The wavelength of the second harmonic laser output was tuned to 307.8 nm corresponding to the Q_1 rotational bandhead frequency of the (0-0) vibronic transition of HO. The HO resonance fluorescence was detected with an EMI G13D411 photomultiplier tube. A dielectric interference filter with transmission maximum at 309 nm was used in the collection optics to discriminate against stray light at other wavelengths. The signal from the photomultiplier tube was sent to an oscilloscope for real-time display and to a signal averager (Nicolet model 1072) for intensity accumulation.

The laser intensity at 307.8 nm was monitored with a diode whose output was also accumulated on the signal averager. The accumulated photomultiplier signal was normalized to the accumulated laser intensity signal for typically 256 shots.

It was observed that changing experimental variables such as catalyst temperature and reactant gas pressures resulted in a change in HO fluorescence intensity which required only a few seconds for the signal equilibration to be effected. Data obtained at different Pt catalyst temperatures were collected randomly to assure that stability of the apparatus was being maintained. The signal at each temperature was corrected for background noise as determined in separate runs.

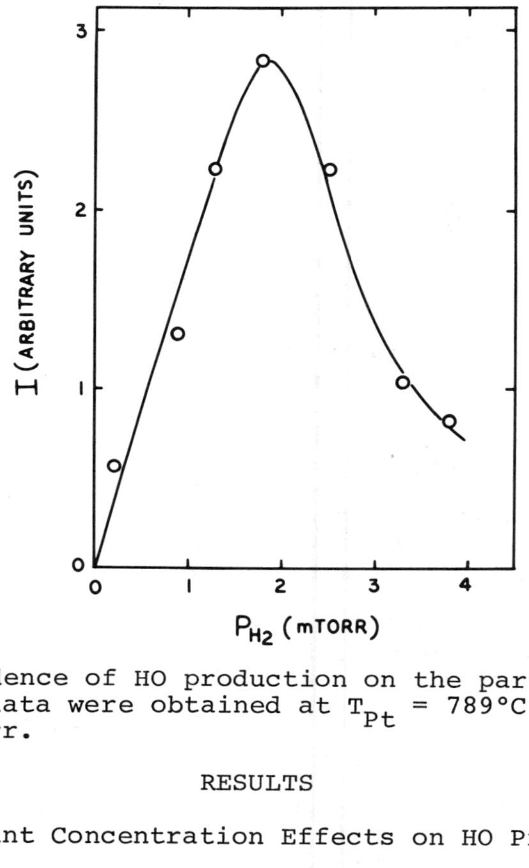

Fig. 2. Dependence of HO production on the partial pressure of H_2. These data were obtained at T_{Pt} = 789°C and P_{N_2O} = 9.6 mTorr.

RESULTS

A. Reactant Concentration Effects on HO Production

The dependence of HO production on the partial pressure of H_2 and N_2O is shown in Figs. 2 and 3, respectively. These data were obtained by varying the pressure of one of the reactants while keeping the partial pressure of the other constant. The result shown in Fig. 2 indicates that the concentration of the HO radical increases rapidly and almost linearly with the partial pressure of H_2 below 2 mTorr, beyond that pressure it drops precipitously. This result, as will be discussed later, suggests the presence of two competing channels for the consumption of H atoms which affect HO formation. This is in contrast to the result shown in Fig. 3, which demonstrates a monotonic increase in HO concentration with the partial pressure of N_2O. These findings are similar to those observed in the oxidation of H_2 by O_2 on the same Pt catalyst.[8]

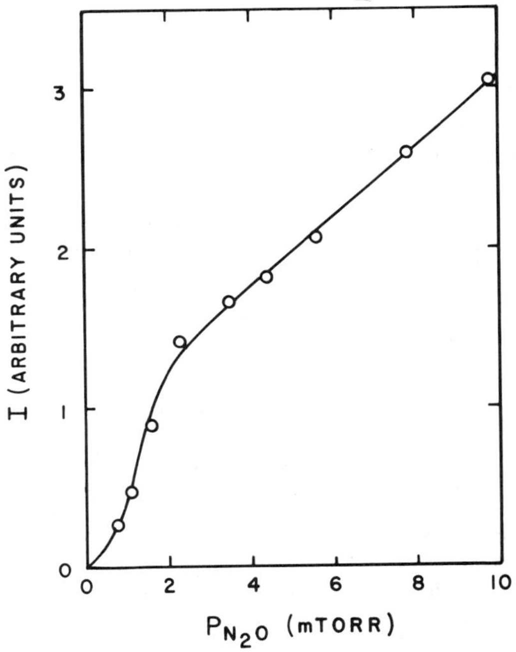

Fig. 3. Dependence of HO production on the partial pressure of N_2O. These results were taken from experiments at T_{Pt} = 797°C and P_{H_2} = 1.2 mTorr.

B. Catalyst Temperature Effect

We have also examined the effect of catalyst temperature on the production of HO in this system. The result of this experiment is plotted in the Arrhenius form as shown in Fig. 4. From the slope of this plot, we have evaluated an activation energy of 28±2 kcal/mole. This value for HO production compares closely with those observed in the decomposition of H_2O[9] and in the oxidation of H_2 by O_2[7,8] as summarized in Table I. These findings will be discussed in the following section in terms of the proposed reaction mechanism.

DISCUSSION

The results of previous studies on the decomposition of N_2O on noble metal surfaces indicate that the decomposition occurs initially via the production of adsorbed O atoms;[10]

$$N_2O \quad + \quad * \quad \rightarrow \quad N_2 \quad + \quad O*$$

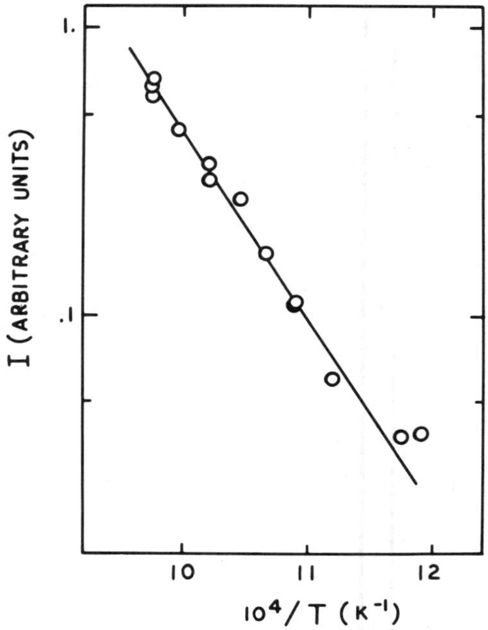

Fig. 4. Arrhenius plot of the observed HO LIF intensity vs. the reciprocal Pt catalyst temperature for the $H_2 + N_2O$ reaction. The reactant pressures are $P_{H_2} = 1.0$ mTorr and $P_{N_2O} = 9.4$ mTorr, diluted in 1030 mTorr of Ar.

where the asterisk represents an active site. The reaction is strongly inhibited by O_2 but not by N_2. The results of isotope-labeling experiments[11] show conclusively that N_2 derives entirely from a single N_2O molecule. This rules out the importance of the path involving the breaking of the N=N bond.

The production of HO radicals in the present system can be accounted for by the following reaction mechanism involving dissociatively adsorbed H and O atoms, similar to that proposed for the $H_2 + O_2$ reaction occurring on the same Pt catalyst.[8]

$$N_2O + * \rightarrow N_2 + O* \quad (1)$$

$$H_2 + 2* \rightarrow 2H* \quad (2)$$

$$H* + O* \rightarrow HO* + * \quad (3)$$

$$H* + HO* \rightarrow H_2O + 2* \quad (4)$$

$$HO* \rightarrow HO + * \quad (5)$$

H_2 is known to be readily adsorbed dissociatively on Pt surfaces[12] as represented by reaction (2). Although we can not directly detect adsorbed HO species (HO*), there is strong evidence that it is stable at room temperature or

Table I. Observed activation energies for HO radical production in different reactions

Reaction	Catalyst	Temperature(°C)	E_a(Kcal/mole)	Reference
$H_2 + N_2O$	Pt	566 - 750	28±2	This work
$H_2 + O_2$	Pt	609 - 776	32±3	8
$H_2 + O_2$	Pt-Rh(10%)	660 - 1100	29±1	7
H_2O decomp.	Pt	596 - 811	30±1	9

colder.[13-15] The observed large activation energy in the present and other experiments as listed in Table I suggests that the rapid disappearance of HO* on Pt surfaces observed in EELS[15] and XPS[16‡] experiments at low temperatures results perhaps mainly from the occurrence of reaction (4) rather than (5). These two reactions, however, become competitive at high temperatures (i.e., ≥600°C, as observed in this and several other experiments summarized in Table I using the HO radical LIF method). Due to the very high sensitivity of this method, we are able to monitor its presence at relatively low concentration levels ($\geq 10^9$ particles/cm^3).

On the basis of the above mechanism, the addition of H_2, or equivalently, the increase in H*, is expected to enhance the steady-state concentration of HO* and that of desorbed HO radicals via reactions (3) and (5), respectively, when N_2O pressure is kept constant. Since reaction (4) occurs concurrently, too much H_2 or H* tends to remove HO* proportionately and thus results in the observed rapid decrease in HO* and HO-radical LIF signals as demonstrated by the result shown in Fig. 2. Contrary to this behavior, the addition of an excess amount of N_2O leads, instead, to a monotonic increase in HO* and HO radical concentrations. This may be attributed either to a small affinity between HO* and O* or, perhaps more plausibly, to the instability of the nascent HO_2* species which immediately decomposes to H* + O_2* (or O_2 + *); viz.,

$$HO* + O* \rightarrow HO_2* + *$$
$$\rightarrow H* + O_2* \quad (\text{or } O_2 + *)$$

‡Norton[16] observed a shift in the characteristic O 1s line from 530.2 ±0.2 ev to 533 ±0.5 ev when H_2 was added to an O_2-covered Pt foil.

Very similar behaviors were observed in the oxidation of H_2 by O_2 and the decomposition of H_2O[9] on Pt when excess amounts of H_2 and O_2 were introduced.

The activation energy for HO production in the present system, 28 ±2 kcal/mole, is in close agreement with the values measured in other systems, centering around 30 kcal/mole (see Table I). This value is believed to be associated with the amount of average energy required to desorb the HO radical from different active Pt sites. Since the reactivity of a catalyst (particularly the one which has not been physically characterized before each run such as the present case) depends on the procedure of its pretreatment, the observed desorption energy is expected to vary somewhat when different procedures are used. The results obtained from this series of studies, nevertheless, agree closely and lie in the vicinity of 30 kcal/mole.

CONCLUDING REMARKS

In this work, we have investigated the reaction of H_2 with N_2O on Pt surfaces using the laser-induced fluorescence technique to monitor the formation of HO radicals. The production of the HO radical was found to have an activation energy of 28 ±2 kcal/mole which agrees closely with the values observed for HO production from $H_2 + O_2$ and the decomposition of H_2O on the same catalyst. A mechanism, which involves dissociatively adsorbed H and O atoms, has been proposed to account for the production of HO in the present system.

The present work demonstrates again the usefulness of LIF technique for detecting low concentrations of free radical species escaping from catalyst surfaces which are otherwise not noticed. The present work will soon be extended to the study of the reduction of metal oxides (such as CuO, NiO, PtO_2 etc.) by H_2 or H atoms, and to the determination of the internal energy content (i.e., rotational and vibrational energies) of the HO radical formed in various catalytic oxidation reactions.

1. W. M. Fairbank, Jr., T. W. Hänsch and A. L. Schawlow, J. Opt. Soc. Am. 65, 199 (1975); for review see, for example, ref. 2.
2. W. M. Fairbank, Jr. and C. Y. She, Opt. News 5, 4 (1979)
3. J. E. Butler, J. W. Hudgens, M. C. Lin and G. K. Smith, Chem. Phys. Lett. 58, 216 (1978)
4. G. K. Smith, J. E. Butler and M. C. Lin, Chem. Phys. Lett. 65, 115 (1979)
5. J. L. Kinsey, Ann. Rev. Phys. Chem. 28, 349 (1977)
6. L. D. Talley, D. E. Tevault and M. C. Lin, Chem. Phys. Lett., in press
7. M. E. Umstead, L. D. Talley, D. E. Tevault and M. C. Lin, Laser Applications to Heterogeneous Catalysis:

Reactant Excitation and Product Diagnostics, in a special (Jan./Feb., 1979) issue of Optical Engineering on Laser Applications to Chemistry, in press.

8. D. E. Tevault, L. D. Talley and M. C. Lin, to be published.
9. L. D. Talley and M. C. Lin, to be published.
10. E. K. Rideal, Concepts in Catalysis (Academic Press, London and New York, 1968), p. 31.
11. Y. H. Hu and J. W. Hightower, Kinet. Katal. 18, 545 (1977).
12. For review, see G. A. Somorjai, Acc. Chem. Res. 9, 248 (1976) and Adv. Catal. 26, 1 (1977).
13. B. A. Morrow and P. Ramamurthy, Can. J. Chem. 49, 3409 (1971).
14. B. A. Morrow and P. Ramamurthy, J. Phys. Chem. 77, 3052 (1973).
15. B. A. Sexton, results reported in this workshop.
16. P. R. Norton, J. Catal. 36, 211 (1975).

ATOM EJECTION STUDIES BY CLASSICAL TRAJECTORY SIMULATION

D.E. Harrison, Jr.
Naval Postgraduate School, Monterey, CA 93940

INTRODUCTION

This paper discusses the use of classical trajectory simulations in the study of low energy ion bombardment of clean and chemically reacted single crystal surfaces. After a brief discussion of the relationship between this work and other theoretical efforts in this area the method is described in some detail. Then a specific example is presented to illustrate the type of information obtainable from this method of investigation.

SIMULATION AND THEORY

Computer simulation is a form of theoretical investigation which is psychologically closely akin to experimentation. It is best viewed as experimentation in which the research is performed in a mathematical apparatus. The investigator works in a universe which is totally under his control; so the thrust of his effort must be to obtain insight and physically useful results from the simplest possible model. Numerical agreement with any particular set of experimental data can always be forced; so the investigations must be interpretive and predictive.

BACKGROUND

Full lattice (FL) classical trajectory simulations of atom ejection from the surface of single crystals by ion bombardment were first undertaken to test the assumptions of analytic sputtering theory. Recently their use in a synergistic relationship with Secondary Ion Mass Spectrometry (SIMS) experiments has led to plausible interpretations of previously inexplicable results.

The analytic theories of sputtering owe much of their formalism to neutron transport theory[1,2]. These early attempts have been greatly extended by Brandt and Laubert[3], Thompson[4] and most notably by Sigmund[5]. Recently Winters[6] has published an excellent and impartial review of analytic physical sputtering theory in which he outlines the assumptions and approximations which must be made to develop an analytic theory.

The analytic theories suffer from three major limitations: 1) they are inherently statistical; so the calculated results must pertain to averages over many collision events; 2) they cannot include a description of the ordered nature of a crystalline target; so their results have validity only in amorphous materials; and 3) to be analytically tractible they must deal with binary collisions in which one of the partners is initially at rest.

Mention must also be made of the binary collision (BC) classical trajectory simulations. The best known of these codes is MARLOWE[7],

ISSN:0094-243X/80/610 307-11$1.50

which artificially introduces the lattice. Other amorphous target codes have also been written[8]. The advantage of the BC simulations is that they can complete the calculation of a single trajectory much more rapidly than can the FL simulation described below. Where thousands of trajectories are required the savings in computer costs is very significant. However, in most cases judicious use of sampling negates the necessity for such large samples[9] and comparable information can be obtained with a limited number of trajectories.

The full lattice calculations are slower because they compute with timesteps, while the BC codes are "event stored" models in which collisions are the events. The timestep computations are slower, because they compute the interaction of many atoms at each step, thus they allow simultaneous events.

This is not the place to give a detailed discussion of the relative merits and limitations of the analytic theories, or the BC and FL simulations. Clearly the analytic theories provide formulas for average quantities such as the total yield of atoms per ion. The BC codes can deal with potential functions which are beyond the capability of closed form analysis, and can in some cases, introduce some lattice effects. The FL model retains all of the geometrical characteristics of the target and the full power of classical dynamics without resorting to approximations or averaging. It is apparent that the other methods are at best approximations to the FL model, but that there may be situations in which such approximations could be desirable. Present indications are that the FL approach is best suited to provide the sort of detailed information required for direct comparison with modern surface experiments.

The major strength of the FL simulations lies in their ability to elucidate the details of the trajectories of individual atoms. It is possible to study the formation of multimers in detail, and to interpret the anisotropies of atom ejection from single crystals into information about the original location and lattice environment of the ejected atom. The computations have suggested experiments[10], and the predictions have been confirmed[11].

THE SIMULATION

A FL simulation is a very complex classical dynamics problem evaluated numerically. The dissipation of momentum when an energetic ion strikes a solid is modeled using classical mechanics. The positions and velocities of the primary ion and lattice atoms develop in time during the trajectory or collision cascade. The trajectory advances by timesteps. In each timestep the total force on every particle is determined and new positions and velocities are calculated using Hamilton's equations of motion. The cascade is terminated when the momentum has dissipated through the microcrystallite and no more atoms can be ejected from the surface. The final positions and velocities are used to determine yields, ejected atom energy distributions and the angular distribution of ejected atoms. Further analysis with suitably chosen diatomic potential functions provides information about dimer formation.

All interactions are described by pair-wise interatomic potential functions; so the total force on any particle is the sum of many such interactions. Complete potential functions, in the sense that they describe both the attractive and repulsive portion of the interaction, are used. The details have been published previously[12]. The potential functions used to detect dimer formation are discussed in the same reference.

The calculations are relatively insensitive to the mathematical form of the potential functions. The published results use a Born-Mayer "wall" and Morse "well", joined with a cubic spline. At various times Coulomb, Thomas-Fermi and Moliere potential functions have also been tested. Because the Born-Mayer function is not suitable for high energy ions, the calculations have been limited to ion energies less than 20.0 KeV. Higher energy calculations will use the Moliere potential.

The size of the target microcrystallite has grown steadily as program and computer speeds have increased. At the moment most of the calculations are done with ~300 target atoms. This size microcrystallite is adequate for the calculations now being undertaken, but it does not completely contain the cascade from a single ion in the 3.0 to 20.0 KeV energy range for Cu/Ar^+. The error in total yield is not more than a few percent, and the other observables are unaffected, because they are ratios.

Almost all calculations have been done with a perfect static lattice. "Warming" the lattice to finite temperature has been explored, but the cost, in terms of badly deteriorated data which require many more trajectories for statistical confidence, has not justified the return in physical understanding.

Except for the fcc (110) orientation, the target microcrystallite is only four atom layers thick. This number was chosen after extensive testing, and is under continuous reexamination, because it is very much smaller than the depths required in the analytic theories and the BC simulations. The standard test is to deepen the microcrystallite for a few selected trajectories, broaden it for the same trajectories, and then repeat the runs in a target crystal which has been both broadened and deepened. Particularly in the 3.0 to 8.0 KeV energy range, each broadening of the area of the target increases the yield by a few percent in some cases, although sometimes it decreases it. The decrease occurs because atoms ejected from the edges of the smaller crystal are retained when they are moved back from the edge by the addition of more atoms. Deepening the crystal sometimes increase the yield by about the same amount, but often leaves it unchanged. This increase usually comes from the center of the surface area, not from the edges. There is no increase of yield from the second and third layers of the target.

The yield from the second and third layers of the target provide an interesting example of the differences between the various models. The analytic theories and BC computations predict significant yield from atoms whose original sites lie more than five layers below the surface, and that the yield from the second layer may approach that from the first. The FL computations show that the probability that a second or third layer atom will be ejected is quite small, see Figure 1. The second layer contribution peaks around 10 percent for the fcc (111) surface, and the third layer contribution remains even smaller for ion energies <20.0 KeV. On the (001) surface the second layer contribution peaks around 15 percent (remember that the first layer is more open) at 8.0 KeV, and the third layer also peaks at the same ion energy.

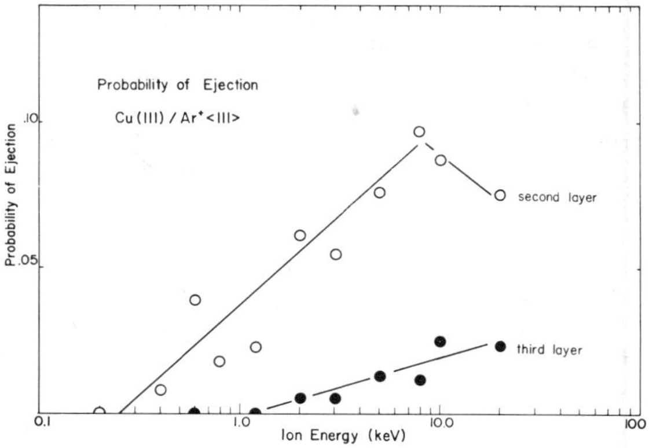

Figure 1. The probability of ejection is the yield from that layer divided by the total yield at that energy. The third layer has peaked at 10.0 KeV on the (001) surface; so it probably has already begun to decline here also.

Careful analysis of individual trajectories indicates the source of the difference: multiply scattered atoms in the FL trajectories tend to be moving approximately in channeling directions. After several collisions the probability of a hard collision is significantly reduced by the correlated collisions produced by the full lattice interactions. Thus the FL calculations predict that ion bombardment will be a more precise surface analysis tool than one might expect from other models.

A typical analysis of a single surface orientation with a normally incident beam at one ion energy requires 80 to 100 trajectories aimed into a symmetry zone on the target surface. All impact points could be mapped into this zone. Oblique ion incidence may double the number of trajectories required because the zone size is increased. Rather good agreement with the total yield can often be obtained with samples as small as 40 trajectories if the surface does not contain channels, but larger samples are required for multimer studies and angle-of-ejection analysis. In either case the impact points for the various trajectories on the target surface are chosen on a grid which samples that surface, rather than randomly, as is done in the BC simulations. Surfaces which contain channels are more difficult to sample because the yield from trajectories which enter the channel is effectively zero.

Most of the investigations completed to date have used the Cu/Ar^+ system. Some studies have been done with Ni/Ar^+, and on copper and nickel surface which have been reacted with oxygen and carbon monoxide. Studies are now underway to determine procedures for systematic changes of potential functions for different bombarding ions. With this information in hand the more difficult problems introduced when the target species is changed can be addressed. Some preliminary work has also been done on copper-nickel alloys.

ATOM EJECTION FROM OXYGEN REACTED COPPER

As an example of the sort of information obtainable from these computations, consider a simulation study of the atoms ejected from oxygen reacted copper. Three recent publications from our group bear on this problem. The first was a prediction that angle resolved SIMS could be used to determine the registry of an ordered oxygen overlayer on the (100) surface of copper[10]. The second is the experimental confirmation of this prediction[11]. Much of the background for these two papers is contained in a more general study of Cu(001)+O/Ar^+<001> published somewhat earlier[12]. The computations for refs. 10 and 12 were done at 600 eV ion energy, while the experiments were performed at 1500 eV. Now we have performed additional calculations at 3.0 KeV to determine whether more information could be gleamed from experiments at higher energies.

The target microcrystallite for this investigation consisted of 338 Cu atoms arranged in a (001) oriented four layer structure and 36 O atoms in the four-fold symmetry positions placed to form an ordered c(2x2) structure on the surface. The calculations were also performed with 13 O atoms, in the four-fold symmetry positions, placed to form a p(2x2) structure. The identical clean copper microcrystallite had previously been run at 3.0 KeV as part of another investigation.

The O atoms were placed 1.74 Å above the surface layer, which coincides with the value used in the earlier calculations. The experiments suggest a somewhat smaller value[11]. With this configuration the static surface binding energy between an O atom and the surface is 0.76 eV. The effects introduced by changing the elevation and the well depth of the attractive potential functions are discussed in ref. 10, which should be consulted for more information on the computations and potential functions. Note that the crystal size has been enlarged because of the higher ion energy.

The new results are compared to the 600 eV data in Table I. As previously reported, the clean target Cu yields are too large, which indicates that our Cu/Ar^+ potential function must be modified. Fortunately the error is in the direction of too great a yield, because large yields encourage multimer production. The impact point sets for the clean and reacted surfaces are not identical, but the differences should not influence the results reported here.

At 600 eV the Cu yield is little affected by the presence of a $\theta = 0.25$ coverage of oxygen but the $\theta = 0.50$ coverage is beginning to cause an effect. At 3.0 KeV the reduction is apparent at both coverages. Although the oxygen yield from c(2x2) is very large the O/Cu ratio is larger at 600 eV. Apparently the lower energy encourages a relative increase in the fraction of momentum transferred to the oxygen layer because the Ar^+ scattering angle is larger.

The presence of oxygen atoms enhances the formation of dimers at both energies because the CuO bond is stronger than the Cu-Cu bond. Note that the CuO dimers form at the expense of the Cu_2 dimers, and that CuO is strongly preferred when sufficient oxygen is present.

TABLE I

$Cu(001) + O$ {4-Fold}/Ar^+<001>

	600 eV[12]			3.0 KeV		
	Clean	p(2x2)	c(2x2)	Clean	p(2x2)	c(2x2)
All Atoms						
Trajectories	111	110	121	89	95	95
Number Cu Ejected	436	435	416	1102	1028	962
Yield Cu (Atoms/Ion)	3.93	3.96	3.44	12.38	10.82	10.13
Number O Ejected	---	267	607	---	341	1158
Yield O (Atoms/Ion)	---	2.43	5.02	---	3.59	12.19
From Second Layer	2	Not Available		95	116	120
From Third Layer	0	Not Available		11	9	6
Multimers						
Cu_2	17	20	13	57	56	38
O_2	---	2	19	---	0	3
CuO	---	23	34	---	34	81
Cu_3	3	2	2	16	13	3
Cu_2O	---	0	5	---	10	14
CuO_2	---	0	4	---	3	13
O_3	---	0	1	---	0	1
Cu_2O_2	---	0	0	---	1	9
Cu_2O_3	---	0	0	---	0	1
Cu_3O_3	---	0	0	---	0	3
All Tetramers	0	0	1	7	3	18
All Pentamers	0	0	1	3	0	4
All Hexamers	0	0	1	2	0	3
All Septamers	0	0	0	0	1	1

The O_2 dimer rarely forms at 3.0 KeV, even though it contains the strongest bond of the three. This is explained by the large yield of large multimers which involve more than one oxygen atom. These multimers cannot form when the system is oxygen deficient. The presence of oxygen strongly suppresses the formation of the large pure copper multimers. Cu_4 is the largest observed with the c(2x2) coverage at 3.0 KeV.

The oxygen layer increases the yield of Cu atoms from the second layer, but reduces it from the third. The recoiling oxygen atoms behave like additional light projectiles in the second copper layer, but they break up the subtle mechanisms required to eject atoms from the third layer, depressing its yield.

One oxygen atom, which is in a very favorable momentum transfer position, is ejected 88 percent of the time. The ejection probability of the others is startling uniform. The actual numbers vary from approximately 20 percent to less than 50 percent. The oxygen atom most likely to link is on the left edge of the target, approximately in the <100> direction from the impact zone, in a position where ejected Cu atoms can carry it along as they leave. The next highest is also on the edge, just below it. The next highest is near the center of the surface in a good position to benefit from the operation of the up-down ejection mechanism[13].

As reported in the 600 eV study[12], there is no evidence that the ion lifts large patches of material from the surface. Next-nearest neighbor Cu-Cu links are more common that nearest neighbor links, and even nearest neighbor O-O links do not seem to be preferred over larger separations. Large multimers tend to contain one nearest neighbor or one next-nearest neighbor link. Neither appears twice in the same multimer, and only one multimer contains one of each. Thus the higher energy studies confirm the picture of molecular recombination above the surface after the individual atoms have been ejected.

A recent investigation[14] suggests that it should be possible to obtain surface structural information from angle resolved SIMS by judicious tuning of the ion's angle of incidence and the ejected atom energy. The present investigation confirms the earlier impression that surface analysis by angle resolved SIMS should be performed at the lowest feasible ion energy. The large number of heavy clusters obtained in this study indicate that more energetic ions will confuse the picture by encouraging the formation of many multimers species.

A schematic drawing of an angle resolved SIMS apparatus[11] is shown in Figure 2. The azimuthal angle, ϕ , is measured from a <100> direction around the Ar^+ "axis". The experimental signals are conveniently analyzed by measuring the peak to valley ratio of the anisotropy, or by subtracting out the minimum intensity and averaging over the four peaks[11,15,16].

The results of a similar computer simulation are shown in Figure 3. These results differ from the experiments because the "collector area" varies for different values of the polar angle, θ. For the purposes of this preliminary study there is no need to convert to a form equivalent to a constant experimental aperture. In all of the

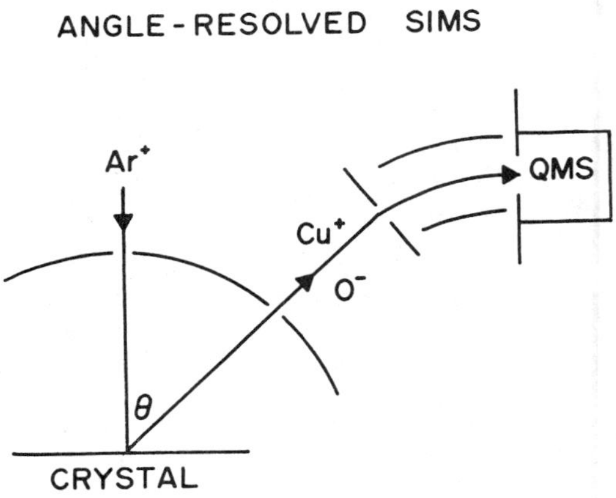

Figure 2. Schematic drawing of the angle resolved SIMS apparatus[11], with the permission of the authors. Consult the reference for further information about the experiments.

remaining figures, the topmost curve, labeled a in Figure 3, is from all ejected atoms, the next, b, is from atoms with energy greater than 5.0 eV, the next, c, greater than 10.0 eV, and the bottom, d, greater than 20.0 eV. At time the labels have been omitted, but the convention applies. The a and b curves contain information which is not available in the current angle resolved SIMS experiments; the c and d curves can be directly compared with experiments taken at a constant polar angle. The ordinates are actual counts after the symmetry of the impact zone has been used to develop the full four-fold symmetry. All of the available symmetry transformations have not been included; so the variations in the tops of the peaks give a measure of the uncertainty of the numbers.

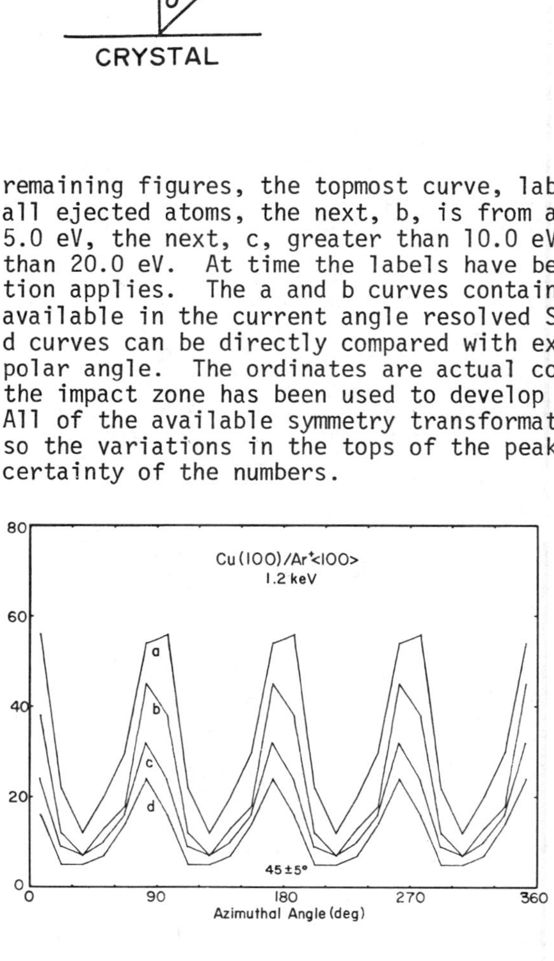

Figure 3. Full azimuthal angle scan at 45° polar angle as computed for four ejected atom energies. See the text for definitions of curves a, b, c, and d. The ordering convention is maintained in those figures which are not labeled. The ordinate is actual counts obtained after the symmetry transformations for the impact zone have been applied.

Figure 3 contains excess redundancy; so it is convenient to use half azimuthal scans, as in Figure 4. Figure 4 shows the results for $\Delta\phi = 15°$ and $\Delta\theta = 10°$ at two different polar angles. Note that the oxygen has not suppressed the anisotropy of ejection completely, but it has been reduced. With the oxygen present the higher energy plots at 45° suggest that recognizable subsidiary peaks at 45° and 135° may be detectable from the reacted surface in addition to the main peak at 90° reported previously.

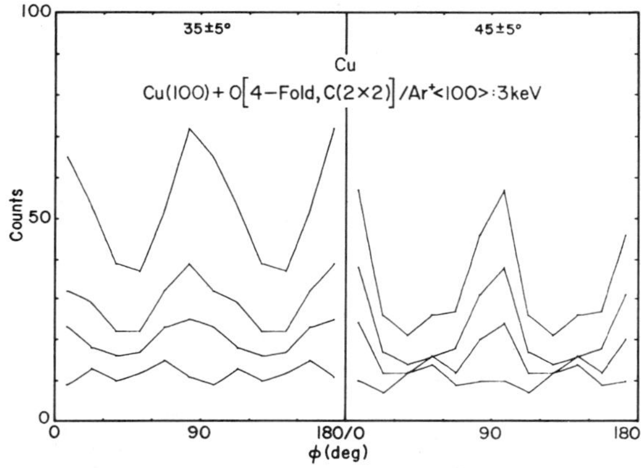

Figure 4. This, and succeeding figures, presents half azimuthal scans for two polar angles. Magnitudes cannot be compared between different polar angles because the "detector apertures" are different. These are the computed Cu monomer data.

Figure 5 shows that the anisotropy for the O atoms is a strong function of the ejected atom energy. Apparently there is a strong 90° peak of low energy atoms which is almost completely gone for atoms above 10.0 eV. The 45° and 135° peaks are present, and a third peak at 90° may be detectable. Contrast the structure at polar angles of 35° and 45°.

Figure 6 shows the Cu_2 dimer data. At $\theta = 25°$ there is 45° and 135° structure, while at 35° there is a "90° peak", but shifted to

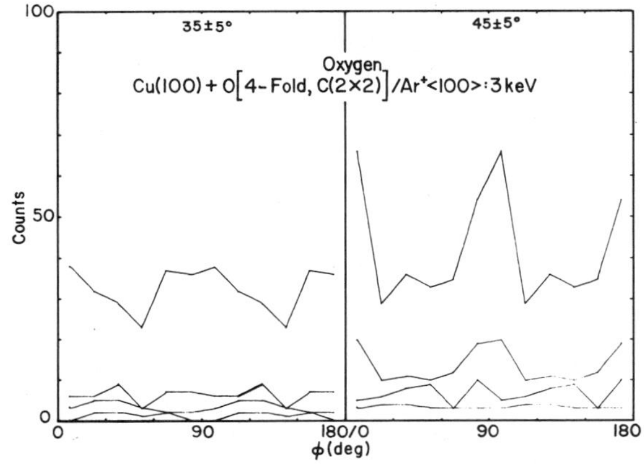

Figure 5. These are the oxygen monomer data from the same computations.

lower azimuthal angles. In general dimer anisotropy tends to appear at smaller polar angles than monomer anisotropy. Unfortunately, there were insufficient O_2 dimers to make an angular plot.

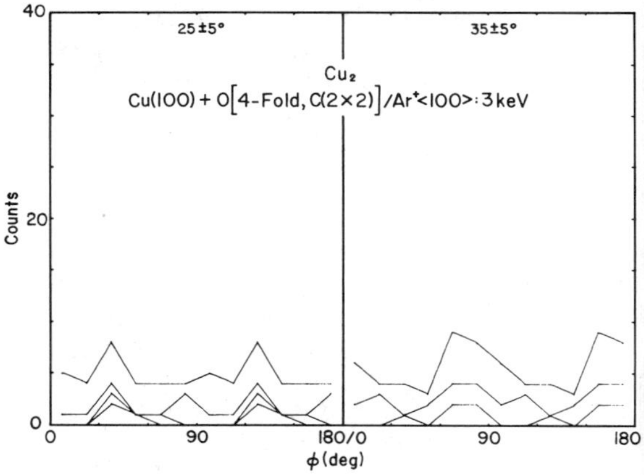

Figure 6. These are the linked Cu-Cu data from the same computations. Some of these dimers would be combined into higher multimers in the laboratory experiment.

Figure 7 predicts structure in the CuO emission. This effect has not been reported previously. Comparison with the monomer curves indicates that the structure is determined by the Cu atom in the partnership. The appearance of this type of ejection, in which an outgoing Cu atom picks up a neighboring O atom as it is ejected, has already been mentioned above in connection with the formation of dimers Unfortunately, because of the smaller polar angle, this prediction cannot be tested experimentally until a more flexible angle resolved SIMS apparatus is available[11].

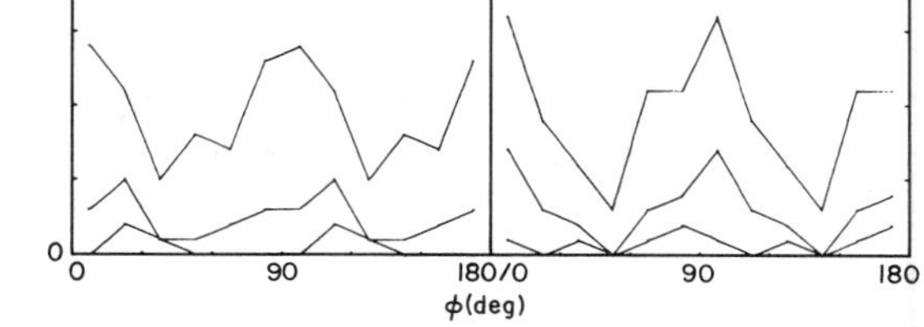

Figure 7. These are the linked Cu-O data from the same computations. Some of these dimers would be combined into higher multimers in the laboratory experiment.

SUMMARY

This paper has been written to give a general introduction to classical trajectory simulation methods, to give some indications of the considerations which influence these calculations and to give a limited example of the type of detailed information produced. The close synergism with angle resolved SIMS is apparent. There is every reason to believe that this partnership has become a major new experimental tool which gives promise of the ability to answer many major questions in surface physics.

ACKNOWLEDGEMENTS

These investigations were supported by the National Science Foundation and by the Foundation research program of the Naval Postgraduate School.

REFERENCES

1. F. Keywell, Phys. Rev. 87, 160 (1952); 97, 1611 (1955).
2. D.E. Harrison, Jr., Phys. Rev. 102, 1473 (1956); 105, 1202 (1957).
3. W. Brandt, R. Laubert, Nucl. Instrum. Methods 47, 201 (1967).
4. M.W. Thompson, Phil. Mag. 18, 377 (1968).
5. P. Sigmund, Phys. Rev. 184, 383 (1969).
6. H.F. Winters, Advances in Chemistry Series, No. 158, Radiation Effects in Solids, M. Kaminsky, Ed., American Chem. Soc., 1976.
7. M.T. Robinson and I.M. Torrens, Phys. Rev. B9, 5008 (1974).
   M.T. Robinson, Sputtering by Ion Bombardment: Physics and Applications (Ed. R. Behrish; Springer-Verlag, New York)(to be published).
   M. Hou and M.T. Robinson, Appl. Phys. 18, 381 (1979).
8. R. Shimizu, Proc. Int. Vac. Congr., 7th, 2, 1417 (1977).
9. B.J. Garrison, N. Winograd and D.E. Harrison, Jr., J. Chem. Phys. 69(4), 1440 (1978).
10. N. Winograd, B.J. Garrison, and D.E. Harrison, Jr., Phys. Rev. Lett. 41(16), 1120 (1978).
11. S.P. Holland, B.J. Garrison, and N. Winograd, Phys. Rev. Lett. 43(3), 220 (1979).
12. B.J. Garrison, N. Winograd, and D.E. Harrison, Jr., Phys. Rev. B18(11), 6000 (1978).
13. D.E. Harrison, Jr., and C.B. Delaplain, J. Appl. Phys. 47(6), 2252 (1976).
14. K.E. Foley and B.J. Garrison, J. Chem. Phys., in press.
15. S. Kono, S.M. Goldberg, N.F.T. Hall, and C.S. Fadley, Phys. Rev. Lett. 41, 1831 (1978).
16. D.P. Woodruff, D. Norman, B.W. Holland, N.V. Smith, H.H. Farrell, and M.M. Traum, Phys. Rev. Lett. 41, 1130 (1978).

THE RELATIONSHIP OF THE GEOMETRY OF THE OBSERVED STEADY STATE CHEMICAL CONVERSION RATE TO THE BASIC SURFACE REACTION PROCESS IN ELECTROCHEMISTRY†

D. G. Retzloff, B. DeFacio, J. E. Bauman and P. H. Ragatz
University of Missouri-Columbia, Columbia, Mo. 65211

ABSTRACT

The steady state surface reaction rate for electrochemical reactions is frequently measured by the polarization curve. In the anodic region of this curve the dynamics of the reaction process are generally described by a semiflow map. The fixed points of this map describe the steady state dynamics of the surface reactions. These generic singularities are classified according to their bifurcation structure. For the corrosion of aluminum this steady state bifurcation structure is found experimentally to be given by a butterfly catastrophy surface. Each structure has a characteristic signature which can be experimentally observed. This is most readily seen in the polarizations curves obtained in the study of surface reactions in electrochemistry and corrosion. Each characteristic signature is described by two variables: the codimension (cod) and k-determinacy (σ). These variables are obtained directly from the polarization curve and are determined by its geometric structure. The codimension and k-determinacy can be calculated from any detailed physical model of the surface reactions that occur. Thus the experimental results constrain the possible physical models that can be considered to describe the reaction process because the codimension and k-determinacy values must agree for both the experiment results and the physical model. Because cod and σ are measures of the geometric shape of polarization curve what is being required is that any physical model be capable of representing the geometry of the experimental results. In the study of electrochemical corrosion reactions this analysis completely explains why the usual Tafel Law and its associated model of the surface reactions involved is entirely inadequate to describe the physics of the anodic region of the polarization curve. Results of this analysis as applied to electrochemical reaction studies of Aluminum will be described.

†Work supported in part by the Army Research Office, Grant Number DAAG29-77-G-0216 , a grant from the Research Council of the Graduate School, University of Missouri-Columbia and the Applied Mathematics Group, Ames Laboratory USDOE, Ames, Iowa 50011.

ISSN:0094-243X/80/610319-14$1.50

INTRODUCTION

The nonlinear features inherent in chemical reactions occuring at a metal interface during anodic corrosion provide a key to understanding the underlying chemical processes. The importance of these nonlinearities in anodic corrosion is well known from the work of Stern[1] and Oldham and Mansfield[2] but their fundamental significance was not recognized. The work of Turner[3] established that chemical reactions controlled by nonlinear effects exhibit a wide range of behavior which include limit cycle instabilities, hysteresis, multiple homogeneous steady states and bifurcation points displaying a variety of dissipative structures. The type of behavior observed is related to the chemical processes that occur. However the work of Thom[4], Poston and Stewart[5] and Retzloff[6] shows that this relationship is not one to one because of the local nature of the observed phenomena. Retzloff further reported the classification and identification of the characteristic structures associated with certain nonlinear effects in terms of catastrophy theory. It is these characteristic signatures that are observed in anodic polarization curves.

The corrosion process we consider here is restricted to aluminum corroding in contact with a 0.1 M solution of $LiClO_4$ in H_2O with the pH adjusted to the appropriate value by the addition of suitable amounts of either $HClO_4$ or NaOH. It is well known for this process that pH, overpotential E, and current i constitute the most important variables. By examining the generic singularity structure of the anodic polarization region we will demonstrate that there exists two additional variables that must be added to the list. Furthermore, these generic singularities are classified by their steady state bifurcation (catastrophy) structures. Each structure has a unique signature determined by the two variables: k-determinacy (σ) and codimension (cod). Once these variables are determined from the experimental polarization curve, they provide a constraint for any detailed physical model of the surface reactions that occur. Although it has been proved that σ and cod do not uniquely determine the kinetic processes that occur, we will show that the constraints are sufficiently restrictive to eliminate a large number of potentially possible reaction mechanisms.

Finally for aluminum corroded by water at a fixed temperature of 25°C the potential E versus pH has an interesting passive region which is shown in Fig. 1. We will give a model which explains the "phase-like diagram" depicted in Fig. 1. The model employed here is nonlinear and utilizes Thom's catastrophy theory to analyze the corrosion behavior. The butterfly catastrophy will be shown to be entirely adequate to provide the first successful model of the anodic region.

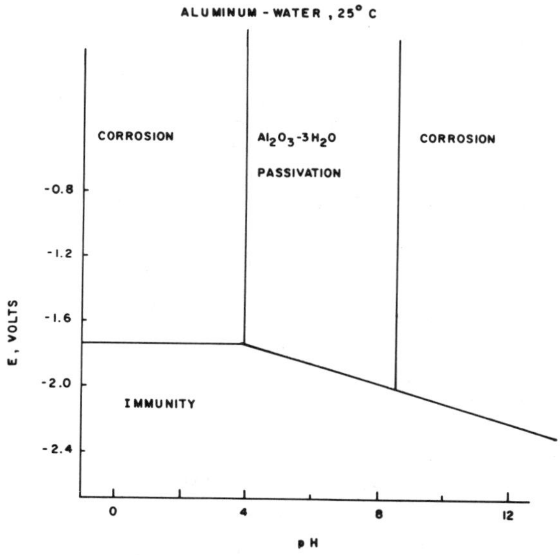

Figure 1. A "phase-diagram" for the corrosion of Aluminum by water at T = 25°C.

THE ORIGIN AND CLASSIFICATION OF BIFURCATION STRUCTURES IN CORROSION PROCESSES

To understand the origin of the dynamical bifurcation structures in corrosion kinetics and its relationship to catastrophy theory, we observe that the dynamics of the corrosion process are generically described by an equation of the form:

$$\frac{dx}{dt} = g(x,\omega) \ . \tag{1}$$

Here g is a C^k mapping with

$$g: X \times \Omega \to Y \tag{2}$$

where $X \subset Y$ are manifolds and Ω is the parameter space. Marsden[7] has shown that (1) defines a semiflow

$$G_t^{\omega}: X \to X \tag{3}$$

such that $G_t^{\omega}(x_0)$ is the solution of

$$\dot{x} = g(x,\omega);\ x(0) = x_0\ . \tag{4}$$

The fixed point (x_0,ω) determined by $g(x_0,\omega) = 0$ defines x_0 as a steady state point of the dynamics, i.e.

$$G_t^{\omega}(x_0) = x_0\ . \tag{5}$$

Kiehn and Retzloff[8] have shown that

$$(g(x_0,\omega) = 0) \cap (H = 0) = \text{steady state bifurcation set}, \tag{6}$$

where H is the mean curvature of the solution manifold, defines the points of marginal stability separating the regions of stability and instability in the steady state case. The classification of the generic singularities of (5) contained in (6) in terms of σ and cod determines the characteristic signature of each type of steady state bifurcation. It is this signature which is experimentally observed in polarization curves. Because σ and cod are measures of the geometric shape of the polarization curve we are characterizing the geometry of the steady state corrosion process. By requiring that the experimentally determined values of σ and cod agree with those calculated for any physical model we are in fact demanding that the physical model be capable of representing the geometry of the experimental results. The generic classification of the steady state bifurcation structures relevant to the anodic polarization curves via σ and cod have been carried through by Thom[4] in the subject of catastrophy theory. We will discuss this classification as it relates to the anodic polarization curve when we consider the experimental results. Before doing this we need to identify the physical source of the nonlinearities that give rise to the bifurcation structures at steady state.

The polarization current in electrochemical corrosion experiments is given by

$$i = \overleftarrow{i} - \overrightarrow{i} = F\,\overleftarrow{r} - F\,\overrightarrow{r};\ \overrightarrow{i} = F\,\overrightarrow{r},\ \overleftarrow{i} = F\,\overrightarrow{r} \tag{7}$$

where $\overleftarrow{i}$ is the de-electronation current, $\overrightarrow{i}$ is the electronation current, F is Faraday's constant, $\overleftarrow{r}$ is the overall de-electronation reaction rate and $\overrightarrow{r}$ is the overall electronation reaction rate. At equilibrium $\overrightarrow{i} = \overleftarrow{i} = i_0$ and the polarization current can be written for a sequence of elementary reactions with a rate controlling step as[9]

$$i = i_0\,[e^{\overleftarrow{\alpha}FE/RT} - e^{-\overrightarrow{\alpha}FE/RT}]\ , \tag{8}$$

in terms of the transfer coefficients $\overrightarrow{\alpha}$, $\overleftarrow{\alpha}$ and the overpotential E. At sufficiently high applied overpotential (8) reduces to

$$E = \frac{RT}{\overleftarrow{\alpha}FE}\ln i - \frac{RT}{\overleftarrow{\alpha}FE}\ln i_0\ , \tag{9}$$

the usual Tafel law. However (7), (8) and (9) mask the more interesting structure that arises from the nonlinear electrochemical reaction regime. A further examination of (7) reveals that $\overrightarrow{r}$ and $\overleftarrow{r}$ are the sources of the nonlinear structure for they represent the effect of five major processes. These processes are: 1) diffusion from the bulk solution to the corroding electrode, 2) absorption, 3) "surface" reaction, 4) desorption and 5) diffusion from the electrode to the bulk solution. In addition diffusion from the interior of the electrode to its surface can be important and the "surface" reaction can be and usually is a complicated set of series, parallel or series-parallel electron transfer reactions with resulting non-linear kinetics. Hence the mathematical representation of the anodic polarization curve via $\vec{r}$ and $\overleftarrow{r}$ is inherently nonlinear. We now consider the experimental realization of these ideas.

EXPERIMENTAL ANODIC POLARIZATION CURVES FOR ALUMINUM

An experimental polarization curve for Al-1100 at a pH of 4.025 is given by the solid line of Fig. 2. The geometric shape is suggestive of a cross-section of the butterfly catastrophy surface

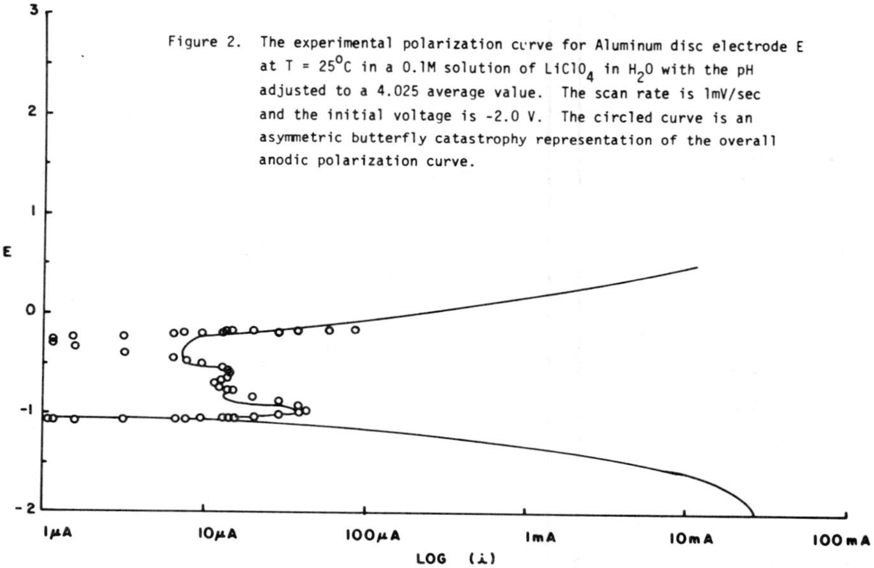

Figure 2. The experimental polarization curve for Aluminum disc electrode E at T = 25°C in a 0.1M solution of $LiClO_4$ in H_2O with the pH adjusted to a 4.025 average value. The scan rate is 1mV/sec and the initial voltage is -2.0 V. The circled curve is an asymmetric butterfly catastrophy representation of the overall anodic polarization curve.

shown in Fig. 3. The mathematical expression for this surface in terms of the relevant corrosion variables is

$$\eta(E-E_b)^5-\beta(\log i-\log i_b)-\xi(pH-pH_b)(E-E_b)-\gamma(E-E_b)^2-\delta(E-E_b)^3=0 \qquad (10)$$

where η, β and ξ are scale factors, γ and δ are products of scale factors and as yet unidentified control variables, and i_b, E_b

and pH_b locate the origin of the catastrophy surface. This arises from a dissipative Hamiltonian[10] of the form

$$H = \frac{\eta}{6}(E-E_b)^6-\beta(\log i-\log i_b)(E-E_b)-\frac{\xi}{2}(pH-pH_b)(E-E_b)^2 -\frac{\gamma}{3}(E-E_b)^3-\frac{\delta}{4}(E-E_b)^4 \tag{11}$$

with the anodic polarization curve being given by

$$\frac{\partial H}{\partial E} = 0 \ . \tag{12}$$

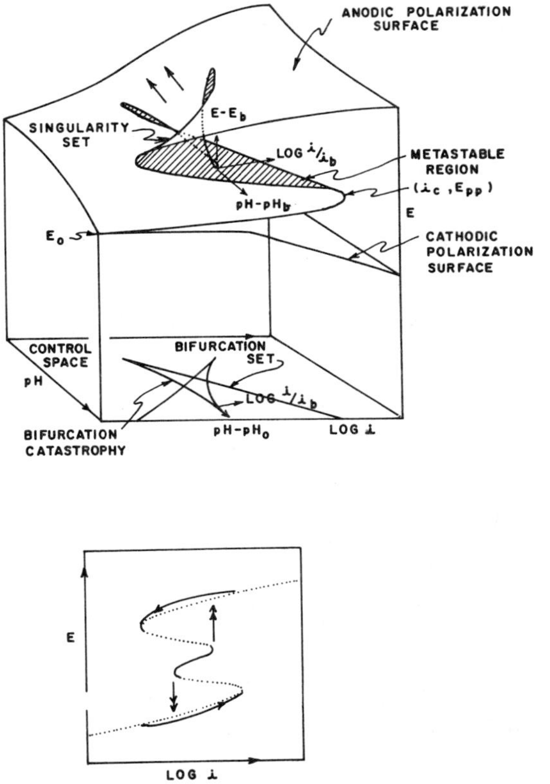

Figure 3. A butterfly catastrophy surface calculated from (10) of the text with parameters chosen to fit the experimental data of Fig. 2.

By taking $\eta = 1$ and using <u>only</u> the four singular points of the experimental polarization curve to determine β, ξ, γ and δ the "best" butterfly representation of the anodic polarization region is asymmetric and given by the circled curve of Fig. 2. Using this representation the butterfly catastrophy surface of Fig. 3 predicts that the passive region (fold over surface region) will disappear under acidic condition for a $pH < 3$. The experimental results are shown as the solid line in Fig. 4 and the calculated values (with no adjustable parameters) are represented by the circled

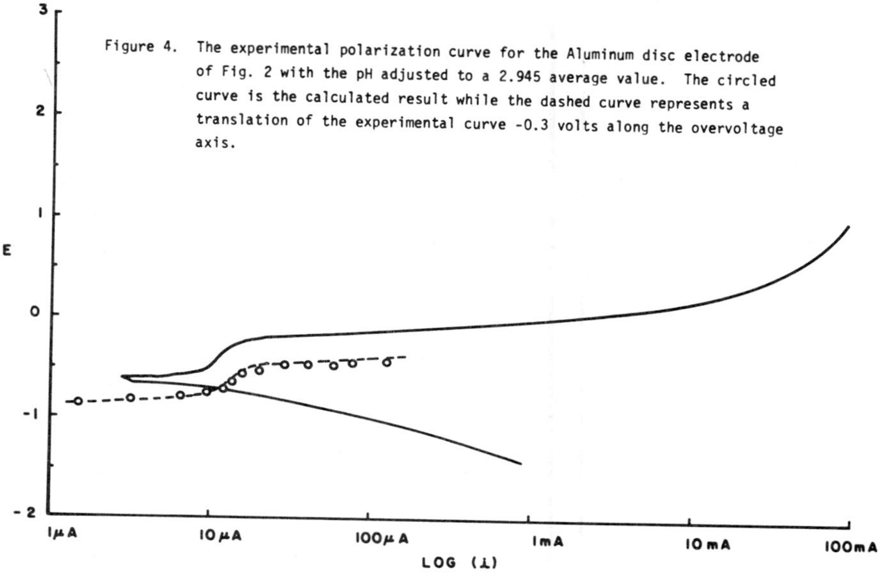

Figure 4. The experimental polarization curve for the Aluminum disc electrode of Fig. 2 with the pH adjusted to a 2.945 average value. The circled curve is the calculated result while the dashed curve represents a translation of the experimental curve -0.3 volts along the overvoltage axis.

curve. The dashed line represents a translation of the experimental curve -0.3 volts along the overvoltage axis. The agreement between calculation and the translated experimental curve is excellent. The origin of this translation of the experimental results is believed to be due to a change in the character of the surface of the aluminum electrode due to the polishing that is required to renew the surface between experimental runs. This idea is currently being investigated. A second aluminum sample was studied at a pH of 11.75. The experimental results are shown as the solid curve of Fig. 5. The circled curve is the calculated result of the butterfly catastrophy model using only the critical points of the experimental curve to determine β, ξ, δ and γ. To further test the model for this sample the polarization curve for a pH of 12.31 was calculated and is shown as the circled curve in Fig. 6. The experimental result is given by the solid curve. Once again the agreement is reasonably good. Finally the disappearance of the

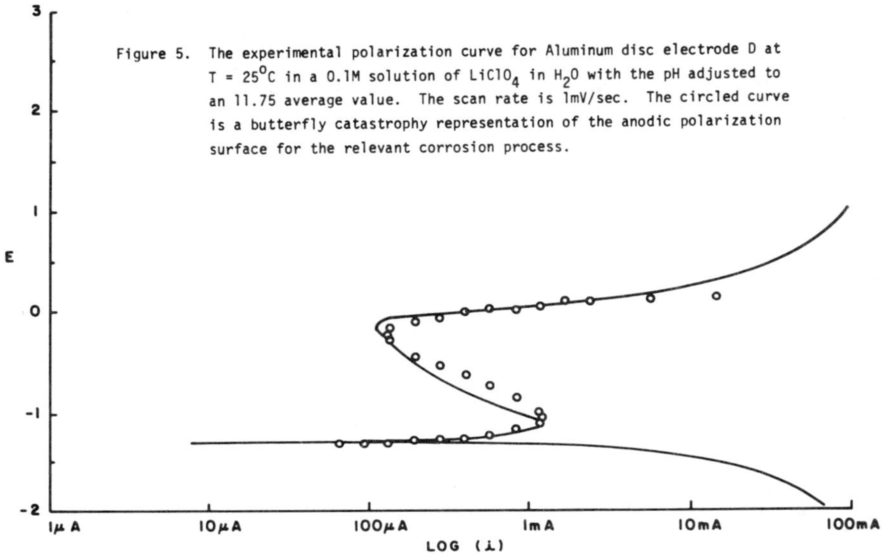

Figure 5. The experimental polarization curve for Aluminum disc electrode D at $T = 25^{o}C$ in a 0.1M solution of $LiClO_4$ in H_2O with the pH adjusted to an 11.75 average value. The scan rate is lmV/sec. The circled curve is a butterfly catastrophy representation of the anodic polarization surface for the relevant corrosion process.

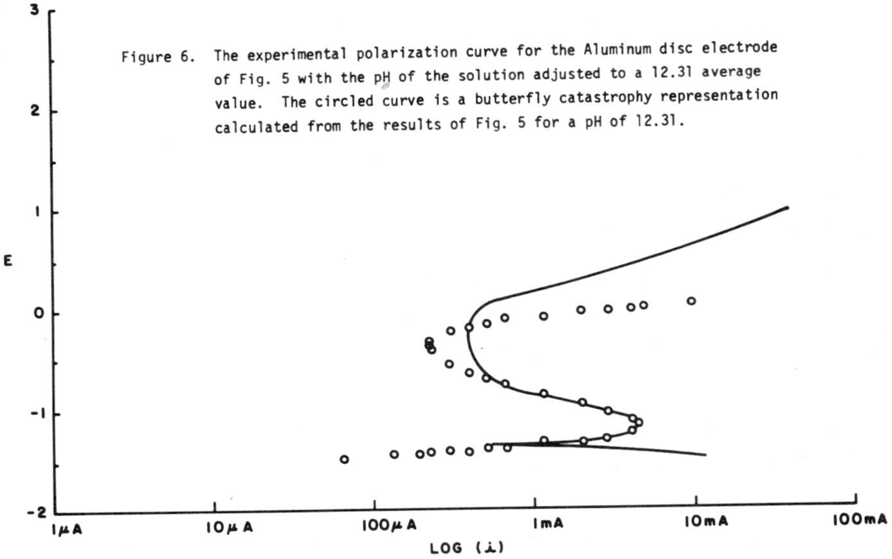

Figure 6. The experimental polarization curve for the Aluminum disc electrode of Fig. 5 with the pH of the solution adjusted to a 12.31 average value. The circled curve is a butterfly catastrophy representation calculated from the results of Fig. 5 for a pH of 12.31.

passive region for this sample was calculated to occur for pH < 2. The experimental result is shown as the solid curve in Fig. 7 and the calculated curve is given by the circles. The dashed curve represents a translation of the experimental curve -0.28 volts along the overvoltage axis and +0.38 mA along the current axis. The origin

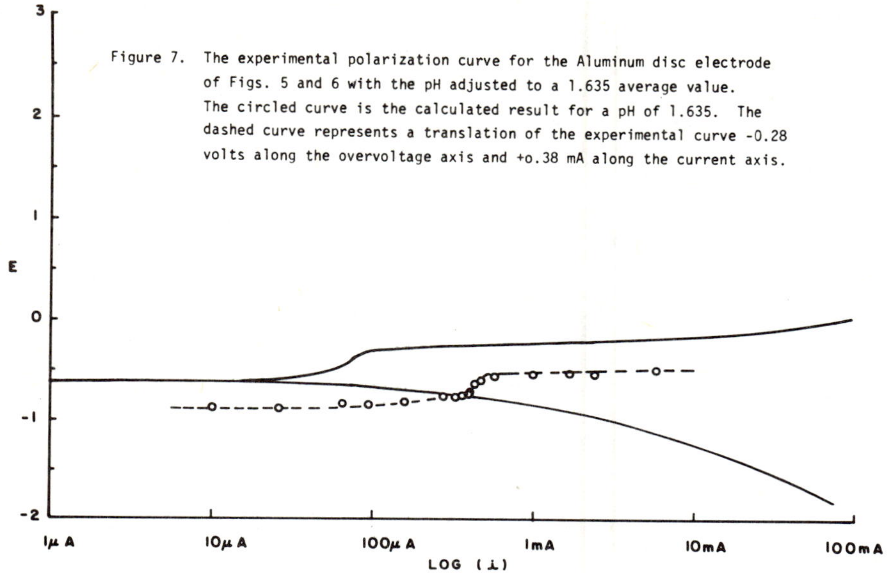

Figure 7. The experimental polarization curve for the Aluminum disc electrode of Figs. 5 and 6 with the pH adjusted to a 1.635 average value. The circled curve is the calculated result for a pH of 1.635. The dashed curve represents a translation of the experimental curve -0.28 volts along the overvoltage axis and +o.38 mA along the current axis.

of this translation is also thought to be associated with polishing and surface renewal effects.

The results given in Figs. 2, 4, 5, 6 and 7 are adequately represented by an asymmetric butterfly catastrophy surface and the disappearance of the passive region for a sufficiently acidic environment is correctly predicted. The geometry of this bifurcation structure is completely determined by σ and cod. Thus these parameters provide constraints on the possible physical descriptions of the ocrrosion process which we will discuss next.

CONSTRAINTS IMPOSED BY THE BIFURCATION STRUCTURE ON MODELS OF THE ANODIC CORROSION CHEMISTRY

The transversal bifurcation structure at the fixed point of the semiflow map describing the dynamics of the corrosion process has now been experimentally identified as a butterfly catastrophy. The work of Thom[4] has established for this bifurcation structure that $\sigma = 6$ and cod = 4. The determination of $\sigma(f)$ and cod(f) at steady state for a given function f which arises from a mathematical model of the corrosion process is a straight forward albeit tedious algebraic calculation which is lucidly explained in the work of Poston and Stewart[5]. Because the integer values of σ and cod identify a specific bifurcation structure, any proposed model for the corrosion process must have values of σ and cod identical to those of the experimentally observed steady state bifurcations in the anodic polarization curve. This requirement provides a fairly restrictive constraint on the possible descriptions of the corrosion process. This is illustrated in Table 1 where several possible descriptions of corrosion kinetics are listed. When each of these models is expressed in terms of overpotential E, pH and current i, the resulting expressions are all found to have $\sigma = 2$ and cod = 1. Thus none of these models are adequate descriptions of the anodic corrosion process. The same is true for the Tafel law given in (9) which also has $\sigma = 2$ and cod = 1. This explains why the Tafel law has been found to be totally inadequate for describing the detailed features of the anodic region. Although the requirements imposed by σ and cod have been useful in eliminating models not consistent with the experimental results, it should be emphasized that they do not <u>uniquely</u> determine the correct description of anodic corrosion. This lack of uniqueness arises because σ and cod determine an equivalence class under nonsingular coordinate transformations. Thus several distinct descriptions of the corrosion kinetics can possess the same <u>local</u> geometric catastrophy structure at their respective fixed points[5,6].

Finally, because cod is numerically equivalent to the number of control parameters required for a transversal description of the corrosion process, the experimental results indicate that four variables are needed for an adequate description of the anodic polarization surface. Two of these have already been identified as pH and i. Two still remain to be identified. They are believed to be related to a characterization of the corroding surface.

Type of Process	Overall Electrochemical Reaction	$\overleftarrow{r}$	$\overrightarrow{r}$
Series of elementary one electron transfer reactions with a rate determining step	$A^{n+} + n\,e = D$	$\overleftarrow{k}_c \prod_{i=n-\gamma-1}^{n} K_i \; C_D e^{(\overleftarrow{\gamma} + 1 - \beta)F\Delta\phi/RT}$	$\overrightarrow{k}_c \prod_{i=1}^{\overrightarrow{\gamma}} K_i \; C_{A^{n+}} e^{-(\overrightarrow{\gamma} + \beta)F\Delta\phi/RT}$
Single elementary one electron transfer reaction	$A^+ + e = D$	$\overleftarrow{k}_c C_D \, e^{(1-\beta)F\Delta\phi/RT}$	$\overrightarrow{k}_c C_{A^+} e^{-\beta F\Delta\phi/RT}$
Blockage by surface absorption	$A^+ + e = D$	$\overleftarrow{k}_c C_D \, \theta \, e^{(1-\beta)F\Delta\phi/RT}$	$\overrightarrow{k}_c C_{A^+}(1-\theta)e^{-\beta F\Delta\phi/RT}$
Surface reaction rate controlling and absorption equilibrium	$A^+ + e = D$	$\dfrac{\overleftarrow{k}_c K_D C_D \, e^{(1-\beta)F\Delta\phi/RT}}{1 + K_{A^+}C_{A^+} + K_D C_D}$	$\dfrac{\overrightarrow{k}_c K_{A^+} C_{A^+} e^{-\beta F\Delta\phi/RT}}{1 + K_{A^+}C_{A^+} + K_D C_D}$
Absorption of acceptor A^+ controlling and surface reaction equilibrium	$A^+ + e = D$	$\dfrac{k_c K_c K_D C_D e^{(1-\beta)F\Delta\phi/RT}}{1 + (K_c K_D + K_{A^+})C_D}$	$\dfrac{k_c C_{A^+} e^{-\beta F\Delta\phi/RT}}{1 + (K_c K_D + K_{A^+})C_D}$
Diffusion of acceptor A^+ controlling			$-D\left(\dfrac{\partial C_{A^+}}{\partial X}\right)_i$

Table 1: Expressions for the electronation ($\overrightarrow{r}$) and de-electronation ($\overleftarrow{r}$) rates for selected rate controlling processes. Here $\overrightarrow{k}_c$ is the chemical rate constant for the electronation reaction and $\overleftarrow{k}_c$ the chemical rate constant for the de-electronation reaction, K_i is the equilibrium constant for reaction i, C_i is the concentration of species i, $\overrightarrow{\gamma}$ is the number of reaction steps preceeding the rate determining reaction step, $\overleftarrow{\gamma}$ is the number of reaction steps following the rate determining reaction step, β is the symmetry factor, $\Delta\phi$ is the potential across the interface, subscript C refers to the chemical step and θ is the fraction of the surface blocked by absorbed species.

SUMMARY

This analysis has developed several concepts that are repeated here for emphasis. We emphasize that it is the nonlinear kinetics which is responsible for the observed polarization curves.

1. The anodic polarization curves are 2-dimensional slices through a surface of higher dimensions.
2. The main features of the bifurcation structure (passive region) present in the anodic polarization curves for aluminum (Al-1100) are adequately represented by a butterfly catastrophy surface.
3. The butterfly catastrophy surface representation of the anodic region correctly predicts the disappearance of the passive region in a sufficiently acidic environment.
4. The butterfly catastrophy representation demonstrates that two (currently unidentified) parameters in addition to pH and current, i, are required to adequately describe the anodic region of the polarization surface.
5. The codimension and k-determancy completely identify a catastrophy structure. These quantities can be calculated for any proposed model of the corrosion process. Potential models whose codimension and k-determancy have values that are not comensurate with the observed anodic polarization behavior can thus be eliminated. This explains, for instance, why the Tafel Law with cod = 1 and $\sigma = 2$ cannot represent the observed anodic behavior for aluminum which has cod = 4 and $\sigma = 6$.

REFERENCES

1. M. Stern, J. Electrochem. Soc. 102, 609 (1957).
2. K. B. Oldham and F. Mansfeld, Corrosion 27, 434 (1971).
3. J. S. Turner, Discrete Simulation Methods for Chemical Kinetics, Preprint of the Proceedings of the Symposium on Reaction Mechanisms, Models, and Computers, 173rd National Meeting, A.C.S., New Orleans, March 20-25, 1977.

4. R. Thom, Structural Stability and Morphogenesis, (W. A. Benjamin, New York, 1975).
5. T. Poston and I. N. Stewart, Catastrophy Theory and Its Applications, (Pitman, San Francisco, 1978).
6. D. G. Retzloff, Stability Analysis and Multiple Solutions in the Design of Chemical Reactors, A.I.Ch.E. Preprint, St Louis Symposium 1978, Ibid, A.I.Ch.E. Free Forum Session 71st National A.I.Ch.E. Meeting, Miami Beach, Florida (1978).
7. J. E. Marsden, Bulletin A.M.S. 84, 1125 (1978).
8. R. M. Kiehn and D. G. Retzloff (to be published).
9. J. O'M Bockris and A. K. N. Reddy, Modern Electrochemistry Vol. 1 and 2 (Plenum, New York, 1973).
10. R. M. Kiehn, J.M.P. 15, 9 (1974).

AN EXPERIMENTAL STUDY OF ADSORPTION INTERFERENCE IN BINARY MIXTURES FLOWING THROUGH ACTIVATED CARBON

R. Madey and P.J. Photinos
Kent State University*
Kent, Ohio 44242

ABSTRACT

We measured the isothermal transmission through activated carbon adsorber beds at 25°C of two binary mixtures (viz., acetaldehyde-propane and acetylene-ethane) in a helium carrier gas. The transmission of each component in the mixture is the ratio of the outlet concentration to the inlet concentration. The inlet concentration of each component was in the range between 10 and 500 ppm. The constant inlet volumetric flow rate was controlled at 200 cc (STP)/min in the acetaldehyde-propane experiments and at 50 cc(STP)/min in the acetaldehyde-ethane experiments. Comparison of the experimental results with the corresponding single-component experiments under similar conditions revealed interference phenomena between the components of the mixtures as evidenced by changes in both the adsorption capacity and the dispersion number; more specifically, throughout the range of concentrations studied here, we observed that propane displaces acetaldehyde from the adsorbed state. This displacement is manifested by the fact that the acetaldehyde transmission exceeds unity at some stage. We observed also that the outlet concentration profiles of propane in the binary mixtures tend to become more diffuse than the corresponding concentration profiles of the one-component experiments. We observed similar features with the binary mixtures of acetylene and ethane; in this case, the displacement of acetylene by ethane is less pronounced because of the low adsorptivity of both compounds.

THEORETICAL BACKGROUND

The adsorption of gases and vapors on porous solids is a process with numerous applications (e.g., in separation techniques). Separations are achieved by allowing a gas stream to flow through an adsorber bed. The performance study of an adsorber bed usually involves the so-called "breakthrough" or "transmission" curve, i.e., the normalized outlet concentration of a particular species versus the time (or the eluted volume). The shape of the transmission curve is strongly determined by (a) the equilibrium relation between the (stationary) solid-phase concentration and the (moving) fluid-phase concentration, and by (b) the overall rate of adsorption.[1,2,3] In this paper, we confine our discussion to the

*Supported in part by the National Aeronautics and Space Administration under NSG-2013.

ISSN:0094-243X/80/610333-11$1.50

isothermal transmission of a step-input concentration flowing steadily through an initially clean adsorber bed. The problem then consists of integrating the mass-balance equation along with the rate equations. Since mathematical difficulties posed by the rate equations preclude a general solution, simplifying assumptions become necessary. The analysis is simplified by assuming a linear equilibrium relation between the solid-phase and liquid-phase concentrations (i.e., a linear isotherm). Several authors have presented analytical solutions based on different diffusion mechanisms for the transmission of a single species; for example, Thomas[4] assumed that the adsorption rate is controlled by the diffusion from the bulk of the fluid to the external surface of the solid (external mass transfer). Lapidus and Amundson[5] obtained solutions for external mass transfer and axial diffusion. Rosen[6] considered both external mass transfer and diffusion in the solid phase, while Masamune and Smith[7] combined external, surface, and intraparticle diffusion.

For concave isotherms, the transmission curves become broader with increasing bed length; while for convex isotherms, the transmission curves tend to attain a constant pattern as the bed length increases. This constant pattern behavior was utilized by several authors[8-12] to simplify the mathematical problem while others utilized powerful numerical techniques.[13-15]

The situation is more complex for multicomponent adsorption because it is necessary to deal with a system of coupled equations describing each component of the mixture. Assuming non-dispersive adsorption, Glueckauf[16] presented analytical results for two-component mixtures. Cooney and Lightfoot[17] and Cooney and Strussi[18] extended Glueckauf's work by including a mass-transfer term together with the constant-pattern approximation. Weber and Crittenden[19] and Hsieh et al[20] developed numerical methods for more general conditions.

EXPERIMENTAL TECHNIQUE

We measured the isothermal transmission through activated carbon adsorber beds of two binary gas mixtures, namely acetaldehyde-propane and acetylene-ethane in a helium carrier gas. The transmission of each component of the mixture is the ratio of the outlet concentration to the inlet concentration. The cylindrical adsorber beds were packed with "Columbia" type 4LXC 12/28 activated carbon. The adsorber bed temperature was 25°C and was controlled to ± 20 millidegrees by means of a dual-bath constant-temperature system. The experimental apparatus is shown in Fig. 1. The flow system is constructed of stainless-steel tubing and valves with teflon seats and gaskets. Flow controllers FCA, FCB and FCC enable setting of the gas flow rate from cylinders A, B and He (usually Helium), while rotameters RA, RB, RC and RD give an approximate value of the flow rate. Three-way ball valves B1, B2, B3, and B4 can divert the flow to the electronic dual-

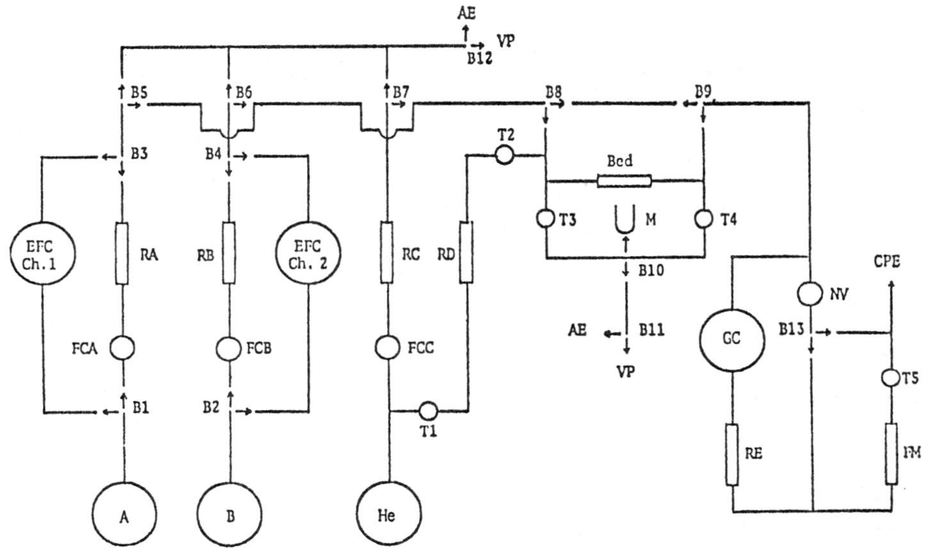

Fig. 1. Schematic diagram of the flow system

channel mass-flow controller, which controls the flow rate to ± 1 percent. The flow meter FM measures the flow rate with an accuracy of ± 0.5 percent. Three-way ball valves B5, B6 and B7 enable mixing of the three gas streams; while similar valves B8 and B9 permit two modes, namely flow through the adsorber bed, and flow bypassing the adsorber bed. The manometer M along with the toggle valves T3 and T4 are used to measure the pressure drop across the bed. Typical values for the pressure drop ranged between 10 and 20 mm Hg, while the total pressure in the adsorber bed ranged between 740 and 780 mm Hg. The automatic data-acquisition system, which is located downstream from the adsorber bed, consists of an automated sampling valve, a gas chromatograph with a flame ionization detector, and a digital integrator. With this apparatus we measured the time-dependent concentration C(t) at the outlet of the adsorber bed. The concentration of each gas was known within ± 1 percent. The concentration of each gas was checked against a primary standard mixture before and after each series of experiments; no appreciable drift was observed. In the acetaldehyde-propane experiments, we used an adsorber bed (10 cm long and 0.454 cm i.d.) packed with 0.569 g of carbon; whereas in the acetylene-ethane experiments, we used an adsorber bed (40 cm long and 0.486 cm i.d.) packed with 2.638 g of carbon. At the end of each experiment, the saturated adsorber bed was desorbed by allowing high-purity helium to flow through the bed at a rate of

100 to 200 cc/min for 14 hours, and simultaneously heating the adsorber to 150°C. Sufficient time was allowed before the beginning of each experiment to ensure thermal equilibrium. In these experiments, we used low-concentration mixtures to minimize non-equilibrium effects caused by the heat of adsorption.

The transmission curves for three binary gas mixtures of acetaldehyde and propane in a helium carrier gas are shown in Fig. 2. The concentrations of acetaldehyde and propane in parts per million (ppm) were (a) 90 and 12, (b) 66 and 34, and (c) 33 and 67, respectively. The volumetric flow rate was 200 cc (STP)/min for all three experiments. We note that the transmission of acetaldehyde exceeds unity over a certain range of eluted volume; or equivalently, the outlet concentration of acetaldehyde exceeds the inlet concentration. This phenomenon (sometimes called "rollover") occurs because propane displaces acetaldehyde from the adsorbed phase to the gas phase. At a sufficiently large eluted volume, the adsorber becomes saturated and the transmission of both components reaches unity. It should be noted also that the overall slope of the propane transmission curves decreases with decreasing propane concentration; or equivalently, the overall dispersion for propane increases with increasing acetaldehyde concentration. This phenomenon should be attributed to the displacement of acetaldehyde by propane; however, the acetaldehyde transmission curve does not show any substantial change in slope.

The adsorption capacity K in cc(STP)/g for each component can be evaluated from the transmission curve by means of the general formula

$$K = \frac{C_o}{m} \int_o^{\infty} [1 - T(V)]dV \tag{1}$$

where C_o is the inlet concentration, m the mass of the adsorber in grams, T(V) is the transmission and V is the eluted volume in cc(STP). This integration was carried out numerically, and the results are summarized in Table I. Also listed in Table I are

Table I. Adsorption capacities in cc(STP)/g of activated carbon at 25°C for single-component gases (K_1) and binary mixtures (K_2).

Binary Mixture	C (ppm)	K_1	K_2
Acetaldehyde	90	3.13	2.87
Propane	12	0.65	0.54
Acetaldehyde	66	2.50	2.30
Propane	34	1.60	1.43
Acetaldehyde	33	1.43	1.21
Propane	67	2.64	2.61

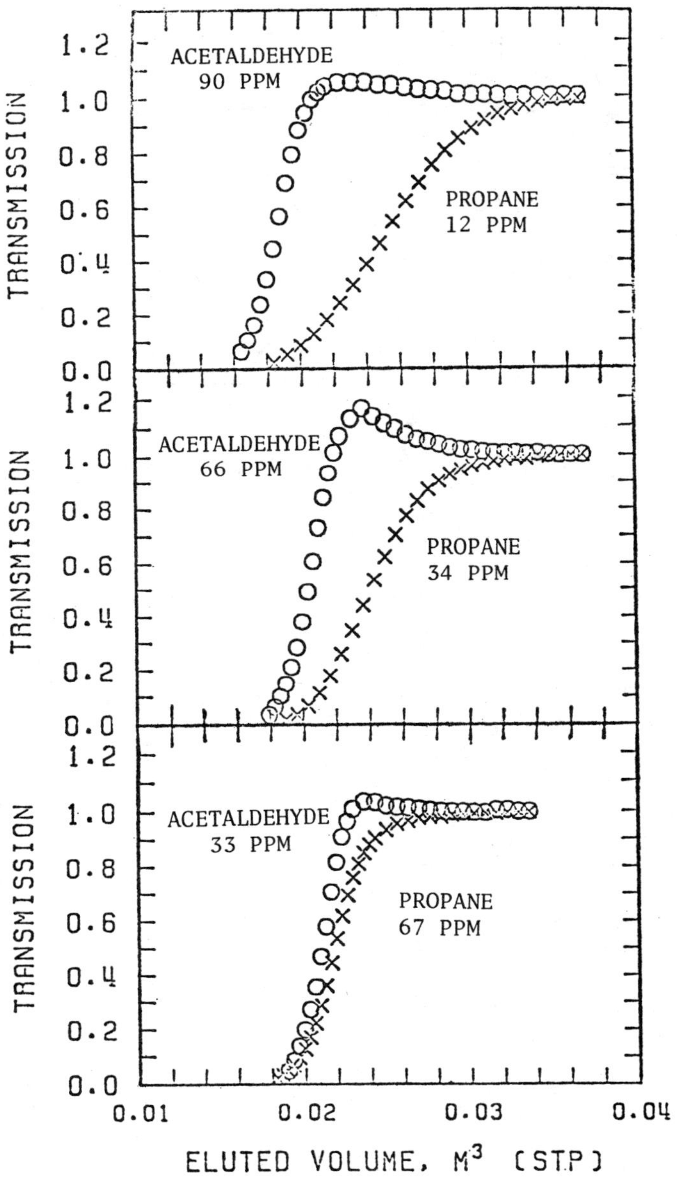

Fig. 2. Transmission of binary mixtures of acetaldehyde and propane in helium through an adsorber bed at 25°C packed with activated carbon.

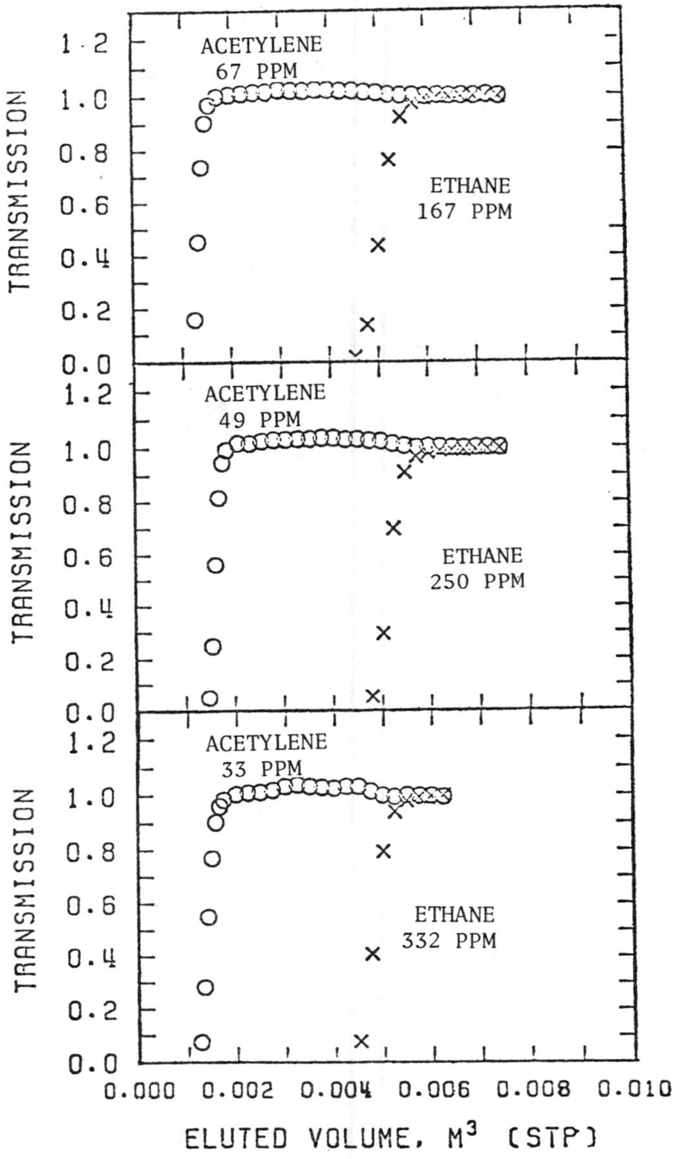

Fig. 3. Transmission of binary mixtures of acetylene and ethane in helium through an adsorber bed at 25°C packed with activated carbon.

the adsorption capacity values obtained from single-component adsorption experiments. Comparison of the corresponding values in the single-component experiments and binary-mixture experiments reveals a substantial reduction of the adsorption capacity for both components in the binary mixture. This reduction is a result of competitive adsorption. The situation is different in the acetylene-ethane binary mixtures. The transmission curves for three mixtures are shown in Fig. 3. The concentrations of acetylene and ethane in ppm are (a) 67 and 167, (b) 49 and 250, and (c) 33 and 332, respectively. The volumetric flow rate was 50 cc(STP)/min for all three experiments. Here acetylene breaks through first, and is displaced by ethane, as indicated by the "rollover"; however, the locations and slopes of the transmission curves do not vary appreciably for the three mixtures. This result should be attributed to the low adsorptivities of both components of the mixture. It should be mentioned also that the overall slope of the ethane transmission curve is equal to that of the single-component transmission curve under the same experimental conditions. This result is compatible with the findings of Shen and Smith,[21] who examined the rates of adsorption in a benzene-hexane binary mixture. They found that in the linear isotherm range the rate of adsorption of benzene from mixtures was the same as in single-component adsorption. It should be pointed out that in the range of concentrations examined here, acetylene obeys a linear isotherm,[22] while the adsorption of ethane does not deviate appreciably from the linear shape.[23]

REFERENCES

1. T. Vermeulen, "Advances in Chemical Engineering," T.B. Drew and J.W. Hoopes, ed., Vol. II, Academic Press (1958).
2. E.N. Lightfoot, R.J. Sanchez-Palma, and D.O. Edwards, "New Chemical Engineering Separation Techniques," H.M. Schoen, ed., Interscience Publishers (1962).
3. T. Vermeulen, G. Klein, and N.K. Hiester, Chemical Engineers' Handbook, 5th ed., R.H. Perry and C.H. Chilton, ed., McGraw-Hill (1973).
4. H.C. Thomas, J. Chem. Phys. 19, 1213 (1951).
5. L. Lapidus and N.R. Amundson, J. Phys. Chem. 56, 984 (1952).
6. J.B. Rosen, J. Chem. Phys. 20, 387 (1952); Ind. Eng. Chem. 46, 1590 (1954).
7. S. Masamune and J.M. Smith, A.I.Ch.E. Journal 11, 34 (1965); A.I.Ch.E. Journal 11, 41 (1965).
8. J.B. Rosen, Ph.D. thesis, Columbia University (1952).
9. D.O. Cooney and E.N. Lightfoot, Ind. Eng. Chem. Fundam. 4, 233 (1965).
10. K.R. Hall, L.C. Eagleton, A. Acrivos, and T. Vermeulen, Ind. Eng. Chem. Fundam. 5, 212 (1966).
11. R.D. Fleck, Jr., D.J. Kirwan and K.R. Hall, Ind. Eng. Chem. Fundam. 12, 95 (1973).

12. D.R. Garg and D.M. Ruthven, Chem. Eng. Sci. 28, 791 (1973); A.I.Ch.E. Journal 21, 200 (1975); Chem. Eng. Sci. 30, 1192 (1975); Chem. Eng. Sci. 28, 799 (1973).
13. C. Tien and G. Thodos, A.I.Ch.E. Journal 5, 373 (1959); A.I.Ch.E. Journal 11, 845 (1965).
14. W.S. Kyte, Chem. Eng. Sci. 28, 1853 (1973).
15. D.V. von Rosenberg, R.P. Chambers, G.A. Swan, Ind. Eng. Chem. Fundam. 16, 154 (1977).
16. E. Glueckauf, Proc. Roy. Soc., Ser A 186, 35 (1946).
17. D.O. Cooney and E.N. Lightfoot, Ind. Eng. Chem. Process Design Develop. 5, 25 (1966).
18. D.O. Cooney and F.P. Strusi, Ind. Eng. Chem. Fundam. 11, 123 (1972).
19. W.J. Weber, Jr., and J.C. Crittenden, J. Water Poll. Control Fed. 47, 924 (1975).
20. J.S.C. Hsieh, R.M. Turian, and Chi Tien, A.I.Ch.E. Journal 23, 263 (1977).
21. J.S. Shen and J.M. Smith, Ind. Eng. Chem. Fundam. 7, 106 (1968).
22. R. Madey and P.J. Photinos, Carbon 17, 93 (1979).
23. P.J. Photinos, A. Nordstrom and R. Madey, Carbon 17, 505 (1979)

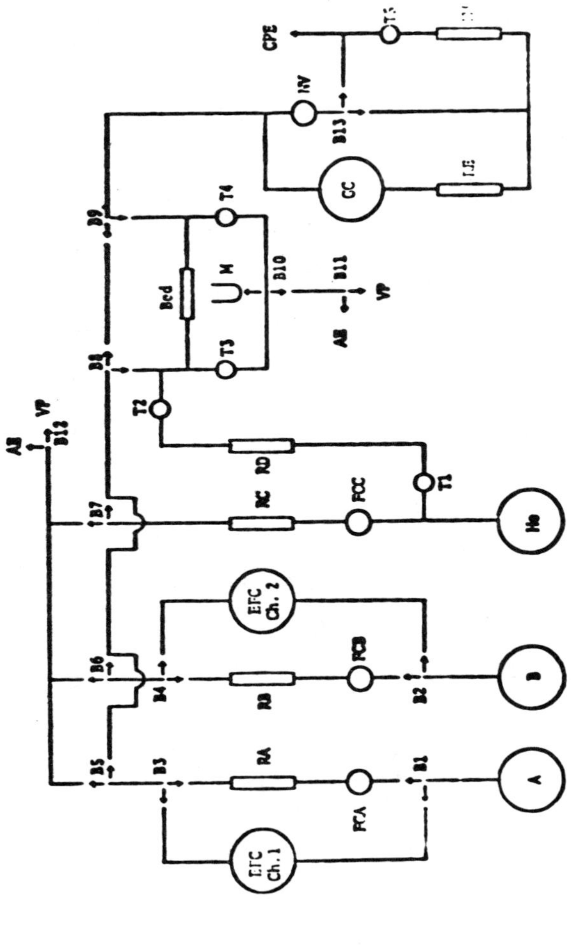

CPE
T5
INV
B13
GC
B9
T4
Bed
M
B10
B11
VP
AB
T3
B8
T2
AB
VP
B12
RD
T1
B7
RC
FCC
He
EFC
Ch. 2
B6
B4
RB
FCB
B2
B
B5
B3
RA
B1
A
FCA
EFC
Ch. 1

Fig. 1

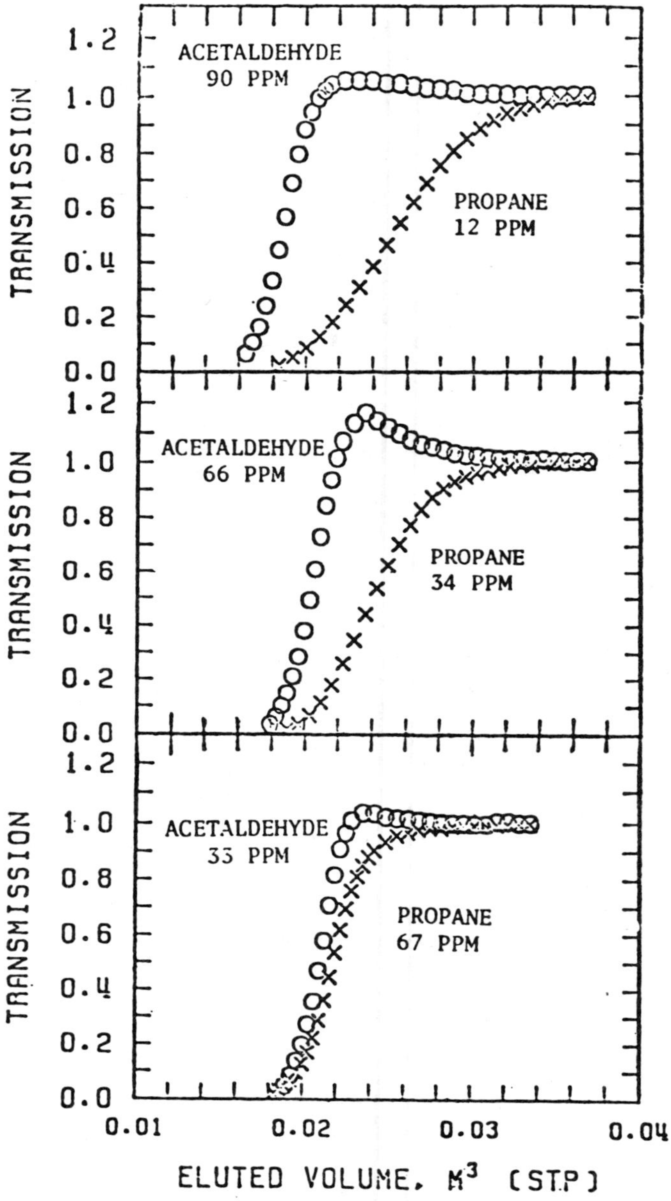

ACETALDEHYDE
90 PPM
PROPANE
12 PPM
ACETALDEHYDE
66 PPM
PROPANE
34 PPM
ACETALDEHYDE
33 PPM
PROPANE
67 PPM
TRANSMISSION
1.2
1.0
0.8
0.6
0.4
0.2
0.0
0.01
0.02
0.03
0.04
ELUTED VOLUME, M3 (STP)

Fig. 2

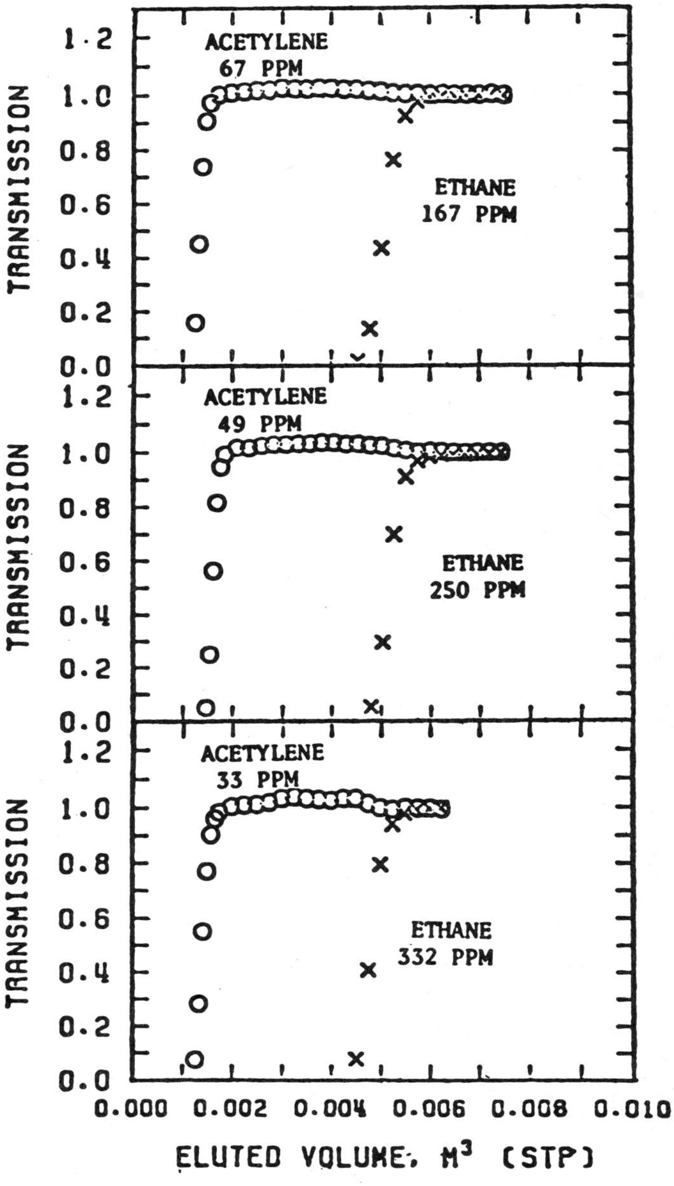

TRANSMISSION
ACETYLENE 67 PPM
ETHANE 167 PPM
ACETYLENE 49 PPM
ETHANE 250 PPM
ACETYLENE 33 PPM
ETHANE 332 PPM
0.0
0.2
0.4
0.6
0.8
1.0
1.2
0.000
0.002
0.004
0.006
0.008
0.010
ELUTED VOLUME, M3 (STP)

Fig. 3

AIP Conference Proceedings

		L.C. Number	ISBN
No.1	Feedback and Dynamic Control of Plasmas	70-141596	0-88318-100-2
No.2	Particles and Fields - 1971 (Rochester)	71-184662	0-88318-101-0
No.3	Thermal Expansion - 1971 (Corning)	72-76970	0-88318-102-9
No.4	Superconductivity in d-and f-Band Metals (Rochester, 1971)	74-18879	0-88318-103-7
No.5	Magnetism and Magnetic Materials - 1971 (2 parts) (Chicago)	59-2468	0-88318-104-5
No.6	Particle Physics (Irvine, 1971)	72-81239	0-88318-105-3
No.7	Exploring the History of Nuclear Physics	72-81883	0-88318-106-1
No.8	Experimental Meson Spectroscopy - 1972	72-88226	0-88318-107-X
No.9	Cyclotrons - 1972 (Vancouver)	72-92798	0-88318-108-8
No.10	Magnetism and Magnetic Materials - 1972	72-623469	0-88318-109-6
No.11	Transport Phenomena - 1973 (Brown University Conference)	73-80682	0-88318-110-X
No.12	Experiments on High Energy Particle Collisions - 1973 (Vanderbilt Conference)	73-81705	0-88318-111-8
No.13	π-π Scattering - 1973 (Tallahassee Conference)	73-81704	0-88318-112-6
No.14	Particles and Fields - 1973 (APS/DPF Berkeley)	73-91923	0-88318-113-4
No.15	High Energy Collisions - 1973 (Stony Brook)	73-92324	0-88318-114-2
No.16	Causality and Physical Theories (Wayne State University, 1973)	73-93420	0-88318-115-0
No.17	Thermal Expansion - 1973 (lake of the Ozarks)	73-94415	0-88318-116-9
No.18	Magnetism and Magnetic Materials - 1973 (2 parts) (Boston)	59-2468	0-88318-117-7
No.19	Physics and the Energy Problem - 1974 (APS Chicago)	73-94416	0-88318-118-5
No.20	Tetrahedrally Bonded Amorphous Semiconductors (Yorktown Heights, 1974)	74-80145	0-88318-119-3
No.21	Experimental Meson Spectroscopy - 1974 (Boston)	74-82628	0-88318-120-7
No.22	Neutrinos - 1974 (Philadelphia)	74-82413	0-88318-121-5
No.23	Particles and Fields - 1974 (APS/DPF Williamsburg)	74-27575	0-88318-122-3

No.24	Magnetism and Magnetic Materials - 1974 (20th Annual Conference, San Francisco)	75-2647	0-88318-123-1
No.25	Efficient Use of Energy (The APS Studies on the Technical Aspects of the More Efficient Use of Energy)	75-18227	0-88318-124-X
No.26	High-Energy Physics and Nuclear Structure - 1975 (Santa Fe and Los Alamos)	75-26411	0-88318-125-8
No.27	Topics in Statistical Mechanics and Biophysics: A Memorial to Julius L. Jackson (Wayne State University, 1975)	75-36309	0-88318-126-6
No.28	Physics and Our World: A Symposium in Honor of Victor F. Weisskopf (M.I.T., 1974)	76-7207	0-88318-127-4
No.29	Magnetism and Magnetic Materials - 1975 (21st Annual Conference, Philadelphia)	76-10931	0-88318-128-2
No.30	Particle Searches and Discoveries - 1976 (Vanderbilt Conference)	76-19949	0-88318-129-0
No.31	Structure and Excitations of Amorphous Solids (Williamsburg, VA., 1976)	76-22279	0-88318-130-4
No.32	Materials Technology - 1975 (APS New York Meeting)	76-27967	0-88318-131-2
No.33	Meson-Nuclear Physics - 1976 (Carnegie-Mellon Conference)	76-26811	0-88318-132-0
No.34	Magnetism and Magnetic Materials - 1976 (Joint MMM-Intermag Conference, Pittsburgh)	76-47106	0-88318-133-9
No.35	High Energy Physics with Polarized Beams and Targets (Argonne, 1976)	76-50181	0-88318-134-7
No.36	Momentum Wave Functions - 1976 (Indiana University)	77-82145	0-88318-135-5
No.37	Weak Interaction Physics - 1977 (Indiana University)	77-83344	0-88318-136-3
No.38	Workshop on New Directions in Mossbauer Spectroscopy (Argonne, 1977)	77-90635	0-88318-137-1
No.39	Physics Careers, Employment and Education (Penn State, 1977)	77-94053	0-88318-138-X
No.40	Electrical Transport and Optical Properties of Inhomogeneous Media (Ohio State University, 1977)	78-54319	0-88318-139-8
No.41	Nucleon-Nucleon Interactions - 1977 (Vancouver)	78-54249	0-88318-140-1
No.42	Higher Energy Polarized Proton Beams (Ann Arbor, 1977)	78-55682	0-88318-141-X
No.43	Particles and Fields - 1977 (APS/DPF, Argonne)	78-55683	0-88318-142-8
No.44	Future Trends in Superconductive Electronics (Charlottesville, 1978)	77-9240	0-88318-143-6

No.45	New Results in High Energy Physics - 1978 (Vanderbilt Conference)	78-67196	0-88318-144-4
No.46	Topics in Nonlinear Dynamics (La Jolla Institute)	78-057870	0-88318-145-2
No.47	Clustering Aspects of Nuclear Structure and Nuclear Reactions (Winnepeg, 1978)	78-64942	0-88318-146-0
No.48	Current Trends in the Theory of Fields (Tallahassee, 1978)	78-72948	0-88318-147-9
No.49	Cosmic Rays and Particle Physics - 1978 (Bartol Conference)	79-50489	0-88318-148-7
No.50	Laser-Solid Interactions and Laser Processing - 1978 (Boston)	79-51564	0-88318-149-5
No.51	High Energy Physics with Polarized Beams and Polarized Targets (Argonne, 1978)	79-64565	0-88318-150-9
No.52	Long-Distance Neutrino Detection - 1978 (C.L. Cowan Memorial Symposium)	79-52078	0-88318-151-7
No.53	Modulated Structures - 1979 (Kailua Kona, Hawaii)	79-53846	0-88318-152-5
No.54	Meson-Nuclear Physics - 1979 (Houston)	79-53978	0-88318-153-3
No.55	Quantum Chromodynamics (La Jolla, 1978)	79-54969	0-88318-154-1
No.56	Particle Acceleration Mechanisms in Astrophysics (La Jolla, 1979)	79-55844	0-88318-155-X
No. 57	Nonlinear Dynamics and the Beam-Beam Interaction (Brookhaven, 1979)	79-57341	0-88318-156-8
No. 58	Inhomogeneous Superconductors - 1979 (Berkeley Springs, W.V.)	79-57620	0-88318-157-6
No. 59	Particles and Fields - 1979 (APS/DPF Montreal)	80-66631	0-88318-158-4
No. 60	History of the ZGS (Argonne, 1979)	80-67694	0-88318-159-2
No. 61	Aspects of the Kinetics and Dynamics of Surface Reactions (La Jolla Institute, 1979)	80-68004	0-88318-160-6